Applied Bayesian Modeling and Causal Inference from Incomplete-Data Perspectives

Applied Bayesian Modeling and Causal Inference from Incomplete-Data Perspectives

An essential journey with Donald Rubin's statistical family

Edited by

Andrew Gelman
Department of Statistics, Columbia University, USA

Xiao-Li Meng
Department of Statistics, Harvard University, USA

John Wiley & Sons, Ltd

Other Wiley Editorial Offices

John Wiley & Sons Inc., 111 River Street, Hoboken, NJ 07030, USA

Jossey-Bass, 989 Market Street, San Francisco, CA 94103-1741, USA

Wiley-VCH Verlag GmbH, Boschstr. 12, D-69469 Weinheim, Germany

John Wiley & Sons Australia Ltd, 33 Park Road, Milton, Queensland 4064, Australia

John Wiley & Sons (Asia) Pte Ltd, 2 Clementi Loop #02-01, Jin Xing Distripark, Singapore 129809

John Wiley & Sons Canada Ltd, 22 Worcester Road, Etobicoke, Ontario, Canada M9W 1L1

Wiley also publishes its books in a variety of electronic formats. Some content that appears
in print may not be available in electronic books.

British Library Cataloguing in Publication Data

A catalogue record for this book is available from the British Library

ISBN 978-0-470-09043-5

Produced from LaTeX files supplied by the authors and processed by Laserwords Private Limited,
Chennai, India

This book is printed on acid-free paper responsibly manufactured from sustainable forestry
in which at least two trees are planted for each one used for paper production.

Contents

Preface   xiii

I   Casual inference and observational studies   1

1 An overview of methods for causal inference from observational studies, by Sander Greenland   3
   1.1   Introduction 3
   1.2   Approaches based on causal models 3
   1.3   Canonical inference 9
   1.4   Methodologic modeling 10
   1.5   Conclusion 13

2 Matching in observational studies, by Paul R. Rosenbaum   15
   2.1   The role of matching in observational studies 15
   2.2   Why match? 16
   2.3   Two key issues: balance and structure 17
   2.4   Additional issues 21

3 Estimating causal effects in nonexperimental studies, by Rajeev Dehejia   25
   3.1   Introduction 25
   3.2   Identifying and estimating the average treatment effect 27
   3.3   The NSW data 29
   3.4   Propensity score estimates 31
   3.5   Conclusions 35

4 Medication cost sharing and drug spending in Medicare, by Alyce S. Adams   37
   4.1   Methods 38
   4.2   Results 40
   4.3   Study limitations 45
   4.4   Conclusions and policy implications 46

5 A comparison of experimental and observational data analyses, by Jennifer L. Hill, Jerome P. Reiter, and Elaine L. Zanutto   49
   5.1   Experimental sample 50

    5.2 Constructed observational study 51
    5.3 Concluding remarks 60

**6 Fixing broken experiments using the propensity score,
by Bruce Sacerdote**   **61**
    6.1 Introduction 61
    6.2 The lottery data 62
    6.3 Estimating the propensity scores 63
    6.4 Results 65
    6.5 Concluding remarks 71

**7 The propensity score with continuous treatments,
by Keisuke Hirano and Guido W. Imbens**   **73**
    7.1 Introduction 73
    7.2 The basic framework 74
    7.3 Bias removal using the GPS 76
    7.4 Estimation and inference 78
    7.5 Application: the Imbens–Rubin–Sacerdote lottery sample 79
    7.6 Conclusion 83

8 Causal inference with instrumental variables, by Junni L. Zhang   **85**
    8.1 Introduction 85
    8.2 Key assumptions for the LATE interpretation of the IV estimand . 87
    8.3 Estimating causal effects with IV 90
    8.4 Some recent applications 95
    8.5 Discussion 95

9 Principal stratification, by Constantine E. Frangakis   **97**
    9.1 Introduction: partially controlled studies 97
    9.2 Examples of partially controlled studies 97
    9.3 Principal stratification 101
    9.4 Estimands 102
    9.5 Assumptions 104
    9.6 Designs and polydesigns 107

II Missing data modeling   **109**

**10 Nonresponse adjustment in government statistical agencies:
constraints, inferential goals, and robustness issues,
by John L. Eltinge**   **111**
    10.1 Introduction: a wide spectrum of nonresponse adjustment efforts in
        government statistical agencies 111
    10.2 Constraints 112
    10.3 Complex estimand structures, inferential goals, and utility functions 112

10.4 Robustness 113
10.5 Closing remarks 113

11 Bridging across changes in classification systems, by Nathaniel Schenker   **117**
11.1 Introduction 117
11.2 Multiple imputation to achieve comparability of industry and occupation codes 118
11.3 Bridging the transition from single-race reporting to multiple-race reporting 123
11.4 Conclusion 128

12 Representing the Census undercount by multiple imputation of households, by Alan M. Zaslavsky   **129**
12.1 Introduction 129
12.2 Models 131
12.3 Inference 134
12.4 Simulation evaluations 138
12.5 Conclusion 140

13 Statistical disclosure techniques based on multiple imputation, by Roderick J. A. Little, Fang Liu, and Trivellore E. Raghunathan   **141**
13.1 Introduction 141
13.2 Full synthesis 143
13.3 SMIKe and MIKe 144
13.4 Analysis of synthetic samples 147
13.5 An application 149
13.6 Conclusions 152

14 Designs producing balanced missing data: examples from the National Assessment of Educational Progress, by Neal Thomas   **153**
14.1 Introduction 153
14.2 Statistical methods in NAEP 155
14.3 Split and balanced designs for estimating population parameters . 157
14.4 Maximum likelihood estimation 159
14.5 The role of secondary covariates 160
14.6 Conclusions 162

15 Propensity score estimation with missing data, by Ralph B. D'Agostino Jr.   **163**
15.1 Introduction 163
15.2 Notation 165
15.3 Applied example: March of Dimes data 168
15.4 Conclusion and future directions 174

**16 Sensitivity to nonignorability in frequentist inference,
by Guoguang Ma and Daniel F. Heitjan**   **175**
16.1 Missing data in clinical trials 175
16.2 Ignorability and bias 175
16.3 A nonignorable selection model 176
16.4 Sensitivity of the mean and variance 177
16.5 Sensitivity of the power 178
16.6 Sensitivity of the coverage probability 180
16.7 An example 184
16.8 Discussion 185

III Statistical modeling and computation   **187**

17 Statistical modeling and computation, by D. Michael Titterington **189**
17.1 Regression models 190
17.2 Latent-variable problems 191
17.3 Computation: non-Bayesian 191
17.4 Computation: Bayesian 192
17.5 Prospects for the future 193

18 Treatment effects in before-after data, by Andrew Gelman   **195**
18.1 Default statistical models of treatment effects 195
18.2 Before-after correlation is typically larger for controls than for
treated units 196
18.3 A class of models for varying treatment effects 200
18.4 Discussion 201

19 Multimodality in mixture models and factor models, by Eric Loken **203**
19.1 Multimodality in mixture models 204
19.2 Multimodal posterior distributions in continuous latent variable
models 209
19.3 Summary 212

**20 Modeling the covariance and correlation matrix of repeated measures,
by W. John Boscardin and Xiao Zhang**   **215**
20.1 Introduction 215
20.2 Modeling the covariance matrix 216
20.3 Modeling the correlation matrix 218
20.4 Modeling a mixed covariance-correlation matrix 220
20.5 Nonzero means and unbalanced data 220
20.6 Multivariate probit model 221
20.7 Example: covariance modeling 222
20.8 Example: mixed data 225

21 Robit regression: a simple robust alternative to logistic and probit regression, by Chuanhai Liu   **227**
    21.1 Introduction 227
    21.2 The robit model 228
    21.3 Robustness of likelihood-based inference using logistic, probit, and robit regression models 230
    21.4 Complete data for simple maximum likelihood estimation 231
    21.5 Maximum likelihood estimation using EM-type algorithms 233
    21.6 A numerical example 235
    21.7 Conclusion 238

22 Using EM and data augmentation for the competing risks model, by Radu V. Craiu and Thierry Duchesne   **239**
    22.1 Introduction 239
    22.2 The model 240
    22.3 EM-based analysis 243
    22.4 Bayesian analysis 244
    22.5 Example 248
    22.6 Discussion and further work 250

23 Mixed effects models and the EM algorithm, by Florin Vaida, Xiao-Li Meng, and Ronghui Xu   **253**
    23.1 Introduction 253
    23.2 Binary regression with random effects 254
    23.3 Proportional hazards mixed-effects models 259

24 The sampling/importance resampling algorithm, by Kim-Hung Li   **265**
    24.1 Introduction 265
    24.2 SIR algorithm 266
    24.3 Selection of the pool size 267
    24.4 Selection criterion of the importance sampling distribution 271
    24.5 The resampling algorithms 272
    24.6 Discussion 276

IV   Applied Bayesian inference   **277**

25 Whither applied Bayesian inference?, by Bradley P. Carlin   **279**
    25.1 Where we've been 279
    25.2 Where we are 281
    25.3 Where we're going 282

26 Efficient EM-type algorithms for fitting spectral lines in high-energy astrophysics, by David A. van Dyk and Taeyoung Park   **285**
    26.1 Application-specific statistical methods 285

26.2 The Chandra X-ray observatory 287
26.3 Fitting narrow emission lines 289
26.4 Model checking and model selection 294

27 Improved predictions of lynx trappings using a biological model,
by Cavan Reilly and Angelique Zeringue   **297**
27.1 Introduction 297
27.2 The current best model 298
27.3 Biological models for predator prey systems 299
27.4 Some statistical models based on the Lotka-Volterra system 300
27.5 Computational aspects of posterior inference 302
27.6 Posterior predictive checks and model expansion 304
27.7 Prediction with the posterior mode 307
27.8 Discussion 308

28 Record linkage using finite mixture models, by Michael D. Larsen **309**
28.1 Introduction to record linkage 309
28.2 Record linkage 310
28.3 Mixture models 311
28.4 Application 314
28.5 Analysis of linked files 316
28.6 Bayesian hierarchical record linkage 317
28.7 Summary 318

29 Identifying likely duplicates by record linkage in a survey
of prostitutes, by Thomas R. Belin, Hemant Ishwaran, Naihua Duan,
Sandra H. Berry, and David E. Kanouse   **319**
29.1 Concern about duplicates in an anonymous survey 319
29.2 General frameworks for record linkage 321
29.3 Estimating probabilities of duplication in the Los Angeles Women's
      Health Risk Study 322
29.4 Discussion 328

30 Applying structural equation models with incomplete data,
by Hal S. Stern and Yoonsook Jeon   **331**
30.1 Structural equation models 332
30.2 Bayesian inference for structural equation models 334
30.3 Iowa Youth and Families Project example 339
30.4 Summary and discussion 342

31 Perceptual scaling, by Ying Nian Wu, Cheng-En Guo,
and Song Chun Zhu   **343**
31.1 Introduction 343
31.2 Sparsity and minimax entropy 347
31.3 Complexity scaling law 353

31.4 Perceptibility scaling law 356
31.5 Texture = imperceptible structures 358
31.6 Perceptibility and sparsity 359

References   **361**

Index   **401**

Preface

This volume came into existence because of our long-held desire to produce a "showcase" book on the ways in which complex statistical theories and methods are actually applied in the real world. By "showcase," we do not imply in any way that this volume presents the best possible analyses or applications—any such claim would only demonstrate grotesque lack of understanding of the complexity and artistic nature of statistical analysis. The world's top five statisticians, however selected, could never produce identical "solutions" to any real-life statistical problem. Putting it differently, if they were all to arrive at the same answer, in the usual mathematical sense, then the problem must be of a toy nature.

Just as objects displayed in a museum showcase are often collectibles from various sources attracting different degrees of appreciation by different viewers, readers of this volume may walk away with different degrees of intellectual stimulation and satisfaction. Nevertheless, we have tried to provide something for almost everyone. To put it another way, it would be difficult to find an individual, statistician or otherwise, who could successfully deal with a real-life statistical problem without having the frustration of dealing with missing data, or the need for some sophistication in modeling and computation, or the urge, possibly subconscious, to learn about underlying causal questions. The substantive areas touched upon by the chapters in this volume are also wide-ranging, including astrophysics, biology, economics, education, medicine, neuroscience, political science, psychology, public policy, sociology, visual learning, and so forth. The Summary of Contents below provides a more detailed account.

Like any showcase display, there is a general theme underlying the chapters in this volume. Almost all the methods discussed in this volume benefited from the incomplete-data perspective. This is certainly true for the counterfactual model for causal inference, for multiple imputation, for the EM algorithm and more generally for data augmentation methods, for mixture modeling, for latent variables, for Bayes hierarchical models, and so forth. Most of the chapters also share a common feature in that out of the total of 31 chapters, 24 are authored or coauthored by Donald Rubin's students and grandstudents. Their names are indicated in the "family tree" on page xix. Three of the remaining seven chapters are coauthored by Don's long-time collaborators: Guido Imbens, Rod Little, and Hal Stern. The remaining four chapters are written by specially invited distinguished experts who are not part of the "Rubin statistical family": Sander Greenland, John Eltinge, Mike Titterington, and Brad Carlin. Each of these "outsiders" provides an overview article to lead the four parts of the volume. No matter how large any statistical family is, it is obvious that readers will benefit from both within and between-family variability, sometimes dominantly so by the latter.

The immediate motivation for this volume is to celebrate Don Rubin's 60th birthday, and it is scheduled to appear just in time for the 2004 Joint Statistical Meetings in Toronto, during which Don will deliver the Fisher lecture. As his students, we obviously wish to dedicate this volume to Don, whose enormous contribution to statistics and impact on general quantitative scientific studies are more than evident from the chapters presented in this volume. We checked the Science Citation Index and found that his papers have been cited over 8,000 times— but Don claims that what he really likes is that his ideas such as ignorability and multiple imputation are so accepted that people use them without even citing him.[1] (A quick look through Parts 3 and 4 of this volume, along with the reference list, reveals that Bayes, and Metropolis are similarly honored but not cited by our contributors.)

Indeed, Don's work is so wide-ranging that it was not an easy task to come up with an accurate but attractive title for this volume. Titles we considered include "Blue-label Statistics (60 years): Sipping with Donald Rubin," "Defenders of Tobacco Companies: From R. A. Fisher to D. B. Rubin," and so forth. We finally settled on the current title, not as amusing as some of us would have liked, but conveying the serious objective of this volume: to showcase a range of applications and topics in applied statistics related to inference and missing data and to take the reader to the frontiers of research in these areas.

Summary of Contents

Part 1: Causal inference and observational studies

Part 1 contains nine chapters, leading with Sander Greenland's overview of three common approaches to causal inference from observational studies. Greenland's chapter is followed with a chapter on the role of matching in observational studies, and a chapter reviewing the basics of the most popular method of performing matching, based on propensity scores, with illustrations using data from the National Supported Work Demonstration and the Current Population Survey. Propensity score matching is in some ways as fundamental to observational studies as randomization is to experimental studies, for it provides "the next to the best thing"—a remarkably simple and effective method for reducing or even eliminating confounding factors when randomization is not possible or not used in the design stage.

The next three chapters apply the propensity score method to three studies in public health and economics: a Medicare cost-sharing and drug-spending study, an infant health development study, and a Massachusetts lottery study. Along the way,

---

[1] For the record, it's Rubin (1976) and Rubin (1978).

these chapters also demonstrate how to use propensity score matching to construct observational studies and to fix "broken experiments." The seventh chapter of Part 1 shows how propensity scores can be extended to continuous treatments, and the methods are applied to the aforementioned lottery study.

The eighth chapter provides an introduction to another popular method in causal inference, the method of instrumental variables. The last chapter of Part 1 investigates the use of instrumental variables for dealing with "partially controlled" studies, a rather difficult class of problems where extra caution is needed in order to arrive at meaningful estimates for treatment effects. The fundamental concept of "principal stratification" is introduced and illustrated with a study on the effectiveness of a needle exchange program in reducing HIV transmission.

Part 2: Missing data modeling

The second part of the book begins with a review by John Eltinge of methods used to adjust for nonresponse in government surveys. The next three chapters provide three accounts of applications of multiple imputation, one of the most popular methods for dealing with missing data, especially in the context of producing public-use data files. The first of the three applications concerns the use of multiple imputation for the purposes of bridging across changes in classification systems, from the earliest application of multiple imputation for achieving comparability between 1970 and 1980 industry and occupation codes, to one of the latest applications involving bridging the transition from single-race reporting to multiple-race reporting in the 2000 Census.

The second of the three applications concerns the use of multiple imputation for representing Census undercount, an extremely contentious issue due to the use of census data in allocation of congressional seats and federal funding. The third application touches on data confidentiality—another controversial issue that has received much attention among the public and in government. Multiple imputation provides a flexible framework for dealing with the conflict between confidentiality and the informativeness of released data by replacing parts or all of the data with synthetic imputations.

The remaining three chapters of Part 2 address design and estimation issues in the presence of missing data. The first of the three investigates the "missing by design" issue with the National Assessment of Educational Progress, an ongoing collection of surveys of students (and teachers) in the U.S. that uses a matrix sampling design to reduce the burden on each student—namely, different students are administered different small subsets of a large collection of test items. The next chapter presents models and computation methods for dealing with the problem of missing data in estimating propensity scores, with application to the March of Dimes observational study examining various effects of post-term birth versus term birth on preteen development. The last chapter of Part 2 describes a convenient method for diagnosing the sensitivity of inferences to nonignorability in missing data modeling, a thorny but essential issue in almost any real-life missing-data problem.

Part 3: Statistical modeling and computation

The third part of the book begins with an overview by Mike Titterington of modeling and computation, which between them cover much of applied statistics nowadays. As Titterington notes, although the more cerebral activity of modeling is rather different from the nuts-and-bolts issues of computation, the two lines of research are in practice closely interwoven. General ideas of modeling are immediately worthwhile only if they are computationally feasible. On the other hand, the need for fitting realistic models in complex settings (which essentially is the case for most applied activities when we take them seriously) has been the strongest stimulus for more advanced and statistical computational methods, the availability of which in turn promote investigators to consider models beyond what are traditionally available (to a point that there is a growing concern that we fit more complex models simply because we can).

The remaining chapters in Part 3 clearly demonstrate this interweaving, both in methodological research and in practice. The second chapter proposes a class of variance-component models for dealing with interactions between treatment and pretreatment covariates, a problem motivated by several examples including observational studies of the effects of redistricting and incumbency in electoral systems in the United States. The next chapter investigates a novel "preclassifying" method for dealing with the tricky computational issue of label-switching with mixtures and other models with latent categories. The investigation involves both the EM algorithm and Markov chain simulation, and the method is illustrated with a well-known factor analysis in educational testing. The following chapter deals with the complicated problem of modeling covariance and correlation matrices, giving specific steps for a Markov chain simulation to fit the proposed models in a variety of settings, and providing a detailed application to a repeated measurement problem arising from a study of long-term neuropsychological impacts after head trauma, using data from UCLA Brain Injury Research Center.

Part 3 continues with a new class of regression models for analyzing binary outcomes, the "robit regression" model, which replaces the normal model underlying the probit regressions by the more robust (hence the term "robit") t models. The models are then fitted by the EM algorithm and several of its recent extensions. The next two chapters detail how to use both the EM algorithm and Markov chain simulation methods for fitting competing risk models and mixed-effect models, including generalized mixed-effect models and the so-called frailty models for estimating hazard rates in survival analysis. Again, all methods are illustrated in detail using simulated and real data sets. The concluding chapter of Part 3 provides a comprehensive overview and investigation of the sampling/importance resampling algorithm for Bayesian and general computation.

Part 4: Applied Bayesian inference

The final part of the book begins with an entertaining survey by Brad Carlin on the past, present, and future of applied Bayesian inference, followed by six chapters on how Bayesian methods are applied to address substantive questions in natural and social sciences. The first study is an inference on emission lines in high-energy astrophysics, based on photon counts collected by the Chandra X-ray observatory. Carefully constructed and problem-specific hierarchical models are developed to handle the complex nature of the sources and the collection process of the data. Complications include, but are not limited to, the mixing of continuum and line emission sources, background contamination, restrictions due to "effective area" and instruments, and absorption by interstellar or intergalactic media. The next chapter demonstrates how and why statistical modeling should be integrated with scientific modeling in addressing substantive questions. In studying a famous example from time series analysis—the dynamic of the Canadian lynx population—a simple biological "prey-predator" type of model combined with empirical time-series models provides a more realistic depiction of the lynx population (with forecasts substantially outperforming previously proposed models), even without the availability of actual data on its prey, the snowshoe hare population!

The next two chapters apply Bayesian methods for record linkage—the problem of matching subjects from different data files or even within the same data file. The methods were originally developed for the purposes of linking various governmental files, such as for estimating undercount by identifying individuals who were counted in both the decennial Census and a Post-Enumeration Survey and those who were only counted in one of the canvasses. However, as Chapter 29 demonstrates, the methods are also useful in identifying duplicates in anonymous surveys. A case in point is the Los Angeles Women's Health Risk Study, where it was found that about 10% of surveyed prostitutes appeared more than once in the data source because of the monetary incentive given for drawing blood, which was necessary in order to estimate the prevalence of HIV and other diseases in this population.

The next chapter discusses Bayesian inference for structural equation models with incomplete data, as applied to a longitudinal study of rural families using data from the Iowa Youth and Families Project. The final chapter of the book provides a fascinating framework, inspired by the Bayesian philosophy, for studying a profound problem in visual learning, namely, how to model humans' perceptual transition over scale. For an image of, say, trees, at near distance, we perceive individual leaves, including their edges and shapes. For the same image at far distance, however, we only perceive a collective foliage impression, even though the natural scene itself is the same. The proposed entropy-based framework provides an elegant theoretical explanation of this common-sense perception change. More importantly, it leads to statistical methods for creating synthesized images by effectively separating sparse structures from collective textures in natural images. The pictures on the cover of this volume, supplied by Zijian Xu and Yingnian Wu, show an illustrative comparison of sketch images at two different levels of resolution.

Acknowledgments

Our foremost thanks, of course, go to all the contributors. We are particularly grateful to the four "outsiders" for their willingness to write capsule overviews of huge chunks of statistics, especially given the stringent time constraints. Dale Rinkel's editorial assistance cannot be over-thanked, for without it, we simply could not have been able to make the deadline for publication. We also thank our editors, Elizabeth Johnston, Rob Calver, and Kathryn Sharples, and the others at Wiley who have helped make this book happen, along with the National Science Foundation for partial support.

 And of course we thank Don for turning both of us from students of textbooks into editors of this volume.

— Andrew Gelman and Xiao-Li Meng, April, 2004

THE RUBIN STATISTICAL FAMILY TREE

William G. Cochran, Advisor†

Donald B. Rubin

*Paul Rosenbaum ('80)	*Michael Larsen ('96)
David Lax ('81)	Heng Li ('96)
*Nathaniel Schenker ('85)	Sadek Wahba ('96)
*Kim-Hung Li ('85)	*Ying Nian Wu ('97)
*Daniel Heitjan ('85)	*Rajeev Dehejia ('97)
**Guoguang Ma	*Bruce Sacerdote ('97)
*T.E. Raghunathan ('87)	Jiangtao Du ('98)
Leisa Weld ('87)	Leonard Roseman ('98)
*Neal Thomas ('88)	*Keisuke Hirano ('98)
*Alan Zaslavsky ('89)	*Elaine Zanutto ('98)
*Andrew Gelman ('90)	*Constantine Frangakis ('99)
**John Boscardin	*Jerome Reiter ('99)
**Cavan Reilly	Jaap P.L. Brand ('99)
*Xiao-Li Meng ('90)	*Alyce Adams ('99)
**Radu Craiu	*Jennifer Hill ('00)
**Florin Vaida	C.J. Zijin Shen ('00)
**David van Dyk	*Eric Loken ('01)
*Thomas Belin ('91)	*Junni Zhang('02)
Joseph Schafer ('92)	Samantha Cook ('04)
*Ralph D'Agostino, Jr. ('94)	Elizabeth Stuart ('04)
*Chuanhai Liu ('94)	

*   Rubin's advisees (with year of Ph.D. graduation) contributing chapters to this book.
** Rubin Statistical Family "grandchildren" contributing chapters. (Other "grandchild-
     dren" are not shown.)
†   William G. Cochran's statistical mentors were Ronald A. Fisher, John Wishart, and
     Frank Yates.

Part I

Casual inference and observational studies

1

An overview of methods for causal inference from observational studies

Sander Greenland[1]

1.1 Introduction

This chapter provides a brief overview of causal-inference methods found in the health sciences. It is convenient to divide these methods into a few broad classes: Those based on formal models of causation, especially potential outcomes; those based on canonical considerations, in which causality is a property of an association to be diagnosed by symptoms and signs; and those based on methodologic modeling. These are by no means mutually exclusive approaches; for example, one may (though need not) base a methodologic model on potential outcomes, and a canonical approach may use modeling methods to address specific considerations. Rather, the categories reflect historical traditions that until recently had only limited intersection.

1.2 Approaches based on causal models

Background: potential outcomes

Most statistical methods, from orthodox Neyman–Pearsonian testing to radical subjective Bayesianism, have been labeled by their proponents as solutions to problems

[1] Departments of Epidemiology and Statistics, University of California, Los Angeles. The author is grateful to Katherine Hoggatt, Andrew Gelman, James Robins, Marshall Joffe and Donald Rubin for helpful comments.

Applied Bayesian Modeling and Causal Inference from Incomplete-Data Perspectives.
Edited by A. Gelman and X-L. Meng © 2004 John Wiley & Sons, Ltd ISBN: 0-470-09043-X

of inductive inference (Greenland, 1998), and causal inference may be classified as a prominent (if not the major) problem of induction. It would then seem that causal-inference methods ought to figure prominently in statistical theory and training. That this has not been so has been remarked on by other reviewers (Pearl, 2000). In fact, despite the long history of statistics up to that point, it was not until the 1920s that a formal statistical model for causal inference was proposed (Neyman, 1923), the first example of a *potential-outcome* model.

Skeptical that induction in general and causal inference in particular could be given a sound logical basis, David Hume nonetheless captured the foundation of potential-outcome models when he wrote:

> "We may define a cause to be an object, followed by another, ...where, if the first object had not been, the second had never existed." (Hume, 1748, p. 115)

A key aspect of this view of causation is its *counterfactual* element: It refers to how a certain outcome event (the "second object," or effect) would not have occurred if, *contrary to fact*, an earlier event (the "first object," or cause) had not occurred. In this regard, it is no different from standard frequentist statistics (which refer to sample realizations that might have occurred, but did not) and some forms of competing-risk models (those involving a latent outcome that would have occurred, but for the competing risk). This counterfactual view of causation was adopted by numerous philosophers and scientists after Hume (e.g., Mill, 1843; Fisher, 1918; Cox, 1958; Simon and Rescher, 1966; MacMahon and Pugh, 1967; Lewis, 1973).

The development of this view into a statistical theory for causal inference is recounted by Rubin (1990), Greenland, Robin, and Pearl (1999), Greenland (2000), and Pearl (2000). To describe that theory, suppose we wish to study the effect of an intervention variable X with potential values (range) $x_1, \ldots, x_J$ on a subsequent outcome variable Y defined on an observational unit or a population. The theory then supposes that there is a vector of *potential outcomes* $\mathbf{y} = (y(x_1), \ldots, y(x_J))$, such that if $X = x_j$ then $Y = y(x_j)$; this vector is simply a mapping from the X range to the Y range for the unit. To say that intervention x_i causally affects Y relative to intervention x_j then means that $y(x_i) \neq y(x_j)$; and the effect of intervention x_i relative to x_j on Y is measured by $y(x_i) - y(x_j)$ or (if Y is strictly positive) by $y(x_i)/y(x_j)$. Under this theory, assignment of a unit to a treatment level x_i is simply a choice of which coordinate of y to attempt to observe; regardless of assignment, the remaining coordinates are treated as existing pretreatment covariates on which data are missing (Rubin, 1978a). Formally, if we define the vector of potential treatments $\mathbf{x} = (x_1, \ldots, x_J)$, with treatment indicators $r_i = 1$ if the unit is given treatment x_i, 0 otherwise, and $\mathbf{r} = (r_1, \ldots, r_J)$, then the actual treatment given is $x_a = \mathbf{r}'\mathbf{x}$ and the actual outcome is $y_a = y(x_a) = \mathbf{r}'\mathbf{y}$. Viewing $\mathbf{r}$ as the item-response vector for the items in $\mathbf{y}$, causal inference under potential outcomes can be seen as a special case of inference under item nonresponse in which $\Sigma_i r_i = 0$ or 1, that is, at most one item in $\mathbf{y}$ is observed per unit (Rubin, 1991).

The theory extends to stochastic outcomes by replacing the $y(x_i)$ by probability mass functions $p_i(y)$ (Greenland, 1987; Robins, 1988; Greenland, Robin, and Pearl, 1999), so the mapping is from X to the space of probability measures on Y. This extension is embodied in the "set" or "do" calculus for causal actions (Pearl, 1995, 2000) described briefly below. The theory also extends to continuous X by allowing the potential-outcome vector to be infinite-dimensional with coordinates indexed by X, and components $y(x)$ or $p_x(y)$. Finally, the theory extends to complex longitudinal data structures by allowing the treatments to be different event histories or processes (Robins, 1987, 1997).

Limitations of potential-outcome models

The power and controversy of this formalization derives in part from defining cause and effect in simple terms of interventions and potential outcomes, rather than leaving them informal or obscure. Judged on the basis of the number and breadth of applications, the potential-outcome approach is an unqualified success, as contributions to the present volume attest. Nonetheless, because only one of the treatments x_i can be administered to a unit, for each unit at most one potential outcome $y(x_i)$ will become an observable quantity; the rest will remain counterfactual, and hence in some views less than scientific (Dawid, 2000). More specifically, the approach has been criticized for including structural elements that are in principle unidentifiable by randomized experiments alone. An example is the correlation among potential outcomes: Because no two potential outcomes $y(x_i)$ and $y(x_j)$ from distinct interventions $x_i \neq x_j$ can be observed on one unit, nothing about the correlation of $y(x_i)$ and $y(x_j)$ across units can be inferred from observing interventions and outcomes alone; the correlation becomes unobservable and hence by some usage "metaphysical."

This sort of problem has been presented as if it is a fatal flaw of potential outcomes models (Dawid, 2000). Most commentators, however, regard such problems as indicating inherent limits of inference on the basis of unrepeatable "blackbox" observation: For some questions, one must go beyond observations of unit responses, to unit-specific investigation of the mechanisms of action (e.g., dissection and physiology). This need is familiar in industrial statistics in the context of destructive testing, although controversy does not arise there because the mechanisms of action are usually well understood. The potential-outcomes approach simply highlights the limits of what statistical analyses can show without background theory about causal mechanisms, even if treatment is randomized: standard statistical analyses address only the magnitude of associations and the average causal effects they represent, not the mechanisms underlying those effects.

Translating potential outcomes into statistical methodology

Among the earliest applications of potential outcomes were the randomization tests for causal effects. These applications illustrate the transparency potential outcomes can bring to standard methods, and show their utility in revealing the assumptions needed to give causal interpretations to standard statistical procedures.

Suppose we have N units indexed by n and we wish to test the strong (sharp) null hypothesis that treatment X has no effect on Y for any unit, that is, for all $i, j, n, y_n(x_i) = y_n(x_j)$. Under this null, the observed distribution of Y among the N units would not differ from its observed value, regardless of how treatment is allocated among the units. Consequently, given the treatment-allocation probabilities (propensity scores), we may compute the exact null distribution of any measure of differences among treatment groups. In doing so, we can and should keep the marginal distribution of Y at its observed value, for with no treatment effect on Y, changes in treatment allocation cannot alter the marginal distribution of Y.

The classic examples of this reasoning are permutation tests based on uniform allocation probabilities across units (simple randomization), such as Fisher's exact test (Cox and Hinkley, 1974, sec. 6.4). For these tests, the fixed Y-margin is often viewed as a mysterious assumption by students, but can be easily deduced from the potential-outcome formulation, with no need to appeal to obscure and controversial conditionality principles (Greenland, 1991). Potential-outcome models can also be used to derive classical confidence intervals (which involve nonnull hypotheses and varying margins), superpopulation inferences (in which the N units are viewed as a random sample from the actual population of interest), and posterior distributions for causal effects of a randomized treatment (Robins, 1988; Rubin, 1978). The models further reveal hidden assumptions and limitations of common procedures for instrumental-variable estimation (Angrist, Imbens, and Rubin, 1996), for intent-to-treat analyses (Goetghebeur and van Houwelingen, 1998), for separating direct and indirect effects (Robins and Greenland, 1992, 1994; Frangakis and Rubin, 2002), for confounding identification (Greenland, Robins, and Pearl, 1999), for estimating causation probabilities (Greenland and Robins, 2000), for handling time-varying covariates (Robins, 1987, 1998; Robins et al., 1992), and for handling time-varying outcomes (Robins, Greenland, and Hu, 1999a).

A case study: g-estimation

Potential-outcome models have contributed much more than conceptual clarification. As documented elsewhere in this volume, they have been used extensively by Rubin, his students, and his collaborators to develop novel statistical procedures for estimating causal effects. Indeed, one defense of the approach is that it stimulates insights which lead not only to the recognition of shortcomings of previous methods but also to development of new and more generally valid methods (Wasserman, 2000).

Methods for modeling effects of time-varying treatment regimes (generalized treatments, or "g-treatments") provide a case study in which the potential-outcome approach led to a very novel way of attacking an exceptionally difficult problem. The difficulty arises because a time-varying regime may not only be influenced by antecedent causes of the outcome (which leads to familiar issues of confounding) but may also influence later causes, which in turn may influence the regime. Robins (1987) identified a recursive "g-computation" formula as central to modeling treatment effects under these feedback conditions and derived nonparametric tests on the basis of this formula (a special case of which was first described by Morrison, 1985). These tests proved impractical beyond simple null-testing contexts, which led to the development of semiparametric modeling procedures for inferences about time-varying treatment effects (Robins, 1998).

The earliest of these procedures were based on the structural-nested failure-time model (SNFTM) for survival time Y (Robins, Blevins et al., 1992; Robins and Greenland, 1994; Robins, 1998), a generalization of the strong accelerated-life model (Cox and Oakes, 1984). Suppressing the unit subscript n, suppose a unit is actually given fixed treatment $X = x_a$ and fails at time $Y_a = y(x_a)$, the potential outcome of the unit under $X = x_a$. The basic accelerated-life model assumes the survival time of the unit when given $X = 0$ instead would have been $Y_0 = e^{x_a \beta} Y_a$, where Y_0 is the potential outcome of the unit under $X = 0$, and the factor $e^{x_a \beta}$ is the amount by which setting $X = 0$ would have expanded (if $x_a \beta > 0$) or contracted (if $x_a \beta < 0$) survival time relative to setting $X = x_a$.

Suppose now X could vary and the actual survival interval $S = (0, Y_a)$ is partitioned into K successive intervals of length $\Delta t_1, \ldots, \Delta t_K$, such that $X = x_k$ in interval k, with a vector of covariates $Z = z_k$ in the interval. A basic SNFTM for the survival time of the unit had X been held at zero over time is then $Y_0 = \Sigma_k \exp(x_k \beta) \Delta t_k$; the extension to a continuous treatment history $x(t)$ is $Y_0 = \int_S e^{x(t) \beta} dt$. The model is semiparametric insofar as the distribution of Y_0 across units is unspecified or incompletely specified, although this distribution may be modeled as a function of covariates, for example, by a proportional-hazards model for Y_0.

Likelihood-based inference on β is unwieldy, but testing and estimation can be easily done with a clever two-step procedure called g-estimation (Robins et al., 1992; Robins and Greenland, 1994; Robins, 1998). To illustrate the basic idea, assume no censoring of Y, no measurement error, and let X_k and Z_k be the treatment and covariate random variables for interval k. Then, under the model, a hypothesized value β_h for β produces for each unit a computable value $Y_0(\beta_h) = \Sigma_k \exp(x_k \beta_h) \Delta t_k$ for Y_0. Next, suppose that for all k, Y_0 and X_k are independent given past treatment history $X_1, \ldots, X_{k-1}$ and covariate history $Z_1, \ldots, Z_k$ (as would obtain if treatment were sequentially randomized given these histories). If $\beta = \beta_h$, then $Y_0(\beta_h) = Y_0$ and so must be independent of X_k given the histories. One may test this conditional independence of $Y_0(\beta_h)$ and the X_k with any standard method. For example, one could use a permutation test or some approximation to one (such as the usual logrank test) stratified on histories; subject to further modeling assumptions, one could instead use a test that the coefficient of $Y_0(\beta_h)$ is zero in a model for the regression of X_k on $Y_0(\beta_h)$ and the histories. In either case, α-level rejection of conditional independence of X_k and $Y_0(\beta_h)$ implies α-level rejection of $\beta = \beta_h$, and the set of all β_h not so rejected form a $1 - \alpha$

confidence set for β. Furthermore, the random variable corresponding to the value b for β that makes $Y_0(b)$ and the X_k conditionally independent is a consistent, asymptotically normal estimator of β (Robins, 1998).

Of course, in observational studies, g-estimation shares all the usual limitations of standard methods. The assignment mechanism is not known, so inferences are only conditional on an uncertain assumption of "no sequential confounding"; more precisely, that Y_0 and the X_k are independent given the treatment and covariate histories used for stratification or modeling of Y_0 and the X_k. If this independence is not assumed, then rejection of β_h only entails that either $\beta \neq \beta_h$ or that Y_0 and the X_k are dependent given the histories (i.e., there is residual confounding). Also, inferences are conditional on the form of the model being correct, which is not likely to be exactly true, even if fit appears good. Nonetheless, as in many standard testing contexts (such as the classical t-test), under broad conditions the asymptotic size of the stratified test of the no-effect hypothesis $\beta = 0$ will not exceed α if Y_0 and the X_k are indeed independent given the histories (i.e., absent residual confounding), even if the chosen SNFTM for Y_0 is incorrect, although the power of the test may be severely impaired by the model misspecification (Robins, 1998). In light of this "null-robustness" property, g-null testing can be viewed as a natural extension of classical null testing to time-varying treatment comparisons.

If (as usual) censoring is present, g-estimation becomes more complex (Robins, 1998). As a simpler though more restrictive approach to censored longitudinal data with time-varying treatments, one may fit a marginal structural model (MSM) for the potential outcomes using a generalization of Horvitz–Thompson inverse-probability-of-selection weighting (Robins, 1999; Hernan, Brumback, and Robins, 2001). Unlike standard time-dependent Cox models, both SNFTM and MSM fitting require special attention to the censoring process, but make weaker assumptions about that process. Thus their greater complexity is the price one must pay for the generality of the procedures, for both can yield unconfounded effect estimates in situations in which standard models appear to fit well but yield very biased results (Robins et al., 1992; Robins and Greenland, 1994; Robins, Greenland, and Hu, 1999a; Hernan, Brumback, and Robins, 2001).

Other formal models of causation

Most statistical approaches to causal modeling incorporate elements formally equivalent to potential outcomes (Pearl, 2000). For example, the sufficient-component cause model found in epidemiology (Rothman and Greenland, 1998, Chapter 2) is a potential-outcome model. In structural-equation models (SEMs), the component equations can be interpreted as models for potential outcomes (Pearl, 1995, 2000), as in the SNFTM example. The identification calculus based on graphical models of causation (causal diagrams) has a direct mapping into the potential-outcomes framework, and yields the g-computation algorithm as a by-product (Pearl, 1995).

These and other connections are reviewed by Pearl (2000), and Greenland and Brumback (2002).

It appears that causal models lacking a direct correspondence to potential outcomes have yet to yield generally accepted statistical methodologies for causal inference, at least within the health sciences. This may represent an inevitable state of affairs arising from a counterfactual element at the core of all commonsense or practical views of causation (Lewis, 1973; Pearl, 2000). Consider the problem of predictive causality: We can recast causal inferences about future events as predictions conditional on specific intervention or treatment-choice events. The choice of x for X is denoted "set $X = x$" in Pearl (1995) and "do $X = x$" in Pearl (2000); the resulting collection of predictive probabilities $P\{Y = y | \text{set}(X = x_i)\}$ or $P\{Y = y | \text{do}(X = x_i)\}$ is isomorphic to the set of stochastic potential outcomes $p_i(y)$. As Hume (1748) and Lewis (1973) noted, for causal inferences about past events, we are typically interested in questions of the form "what would have happened if X had equaled x_c rather than x_a," where the alternative choice x_c does not equal the actual choice x_a and so must be counterfactual; thus, consideration of potential outcomes seems inescapable when confronting historical causal questions, a point conceded by thoughtful critics of counterfactuals (Dawid, 2000).

1.3   Canonical inference

Some approaches to causal inference bypass definitional controversy by not basing their methods on a formal causal model. The oldest of these approaches is traceable to John Stuart Mill in his to attempt to lay out a system of inductive logic on the basis of canons or rules, which causal associations were presumed to obey (Mill, 1843). Perhaps the most widely cited of such lists today are the Austin Bradford Hill considerations (misnamed "criteria" by later writers) (Hill, 1965), which are discussed critically in numerous sources (e.g., Koepsell and Weiss, 2003; Phillips and Goodman, 2003; Rothman and Greenland, 1998, Chapter 2), and which will be the focus here.

The canonical approach usually leaves terms like "cause" and "effect" as primitives (formally undefined concepts) around which the self-evident canons are built, much like axioms are built around the primitives of "set" and "is an element of" in mathematics, although the terms may be defined in terms of potential outcomes. Only the canon of proper temporal sequence (cause must precede effect) is a necessary condition for causation. The remaining canons or considerations are not necessary conditions; instead, they are like diagnostic symptoms or signs of causation—that is, properties an association is assumed more likely to exhibit if it is causal than if it is not (Hill, 1965; Susser, 1988, 1991). Thus, the canonical approach makes causal inference appear more akin to clinical judgment than experimental science, although experimental evidence is among the considerations (Hill, 1965; Rothman and Greenland, 1998, Chapter 2; Susser, 1991). Some of the considerations (such as temporal sequence, association, dose–response or predicted gradient, and specificity) are empirical signs and thus subject to conventional statistical analysis; others (such as plausibility) refer to prior belief and thus (as with disease symptoms) require elicitation, and could be used to construct priors for Bayesian analysis.

The canonical approach is widely accepted in the health sciences, subject to many variations in detail. Nonetheless, it has been criticized for its incompleteness and informality, and the consequent poor fit it affords to the deductive or mathematical approaches familiar to classic science and statistics (Rothman and Greenland, 1998, Chapter 2). Although there have been some interesting attempts to reinforce or reinterpret certain canons as empirical predictions of causal hypotheses (e.g., Susser, 1988; Weed, 1986; Weiss, 1981, 2002; Rosenbaum, 2002b), there is no generally accepted mapping of the entire canonical approach into a coherent statistical methodology; one simply uses standard statistical techniques to test whether empirical canons are violated. For example, if the causal hypothesis linking X to Y predicts a strictly increasing trend in Y with X, a test of this statistical prediction may serve as a statistical criterion for determining whether the hypothesis fails the dose–response canon. Such usage falls squarely in the falsificationist/frequentist tradition of twentieth century statistics, but leaves unanswered most of the policy questions that drive causal research (Phillips and Goodman, 2003).

1.4 Methodologic modeling

In the second half of the twentieth century, a third approach emerged from battles over the policy implications of observational data, such as those concerning the epidemiology of cigarette smoking and lung cancer. One begins with the idea that, conditional on some set of concomitants or covariates Z, there is a population association or relation between X and Y that is the target of inference, usually because it is presumed to accurately reflect the effect of X on Y in that population (as in the canonical approach, "cause" and "effect" may be left undefined or defined in other terms such as potential outcomes). Observational and analytic shortcomings then distort or bias estimates of this effect: Units may be selected for observation in a nonrandom fashion; conditioning on additional unmeasured covariates U may be essential for the X–Y association to approximate a causal effect; inappropriate covariates may be entered into the analysis; components of X or Y or Z may not be adequately measured; and so on.

One can parametrically model these methodologic shortcomings and derive effect estimates on the basis of the models. If (as is usual) the data under analysis cannot provide estimates of the methodologic parameters, one can fix the parameters at specific values, estimate effects based on those values, and see how effect estimates change as these values are varied (sensitivity analysis). One can also assign the parameters prior to distributions on the basis of background information, and summarize the effect estimates over these distributions (e.g., with the resulting posterior distribution). These ideas are well established in engineering and policy research and are covered in many books, albeit in a wide variety of forms and specialized applications. Little and Rubin (2002) focus on missing-data problems; Eddy, Hasselblad, and Schachter (1992) focus on medical and health-risk assessment; and Vose (2000) covers general risk assessment. Nonetheless, general methodologic or bias modeling has only recently begun to appear in epidemiologic research (Robins, Rotnitzky, and Scharfstein, 1999b; Graham, 2000; Gustafson, 2003; Lash and Fink, 2003; Phillips, 2003; Greenland, 2003), although more basic sensitivity analyses have been employed sporadically since the 1950s (see Rothman and Greenland, 1998, Chapter 19, for citations and an overview).

Consider again the problem of estimating the effect of X on Y, given a vector of antecedent covariates Z. Standard approaches are based on estimating $E(Y|x, z)$ and taking the fitted (partial) regression of Y on X given Z as the effect of X on Y. Usually a parametric model $r(x, z; \beta)$ for $E(Y|x, z)$ is fit and the coefficient for X is taken as the effect (this approach is reflected in common terminology that refers to such coefficients as "main effects"). The fitting is almost always done as if (1) within levels of X and Z, the data are a simple random sample and any missingness is completely at random, (2) the causal effect of X on Y is accurately reflected by the association of X and Y given Z (i.e., there is no residual confounding—as might be reasonable to assume if X were randomized within levels of Z), and (3) X, Y, and Z are measured without error. But, in reality, (1) sampling and missing-data probabilities may jointly depend on X, Y, and Z in an unknown fashion, (2) conditioning on certain unmeasured (and possibly unknown) covariates U might be essential for the association of X and Y to correspond to a causal effect of X on Y, and (3) X, Y, and Z components may be mismeasured.

Let $V = (X, Y, Z)$. One approach to sampling (selection) biases is to posit a model $s(v; \sigma)$ for the probability of selection given v, then use this model in the analysis along with $r(x, z; \beta)$, for example, by incorporating $s(v; \sigma)$ into the likelihood function (Eddy, Hasselblad, and Schachter, 1992; Little and Rubin, 2002; Gelman, Carlin, Stern, and Rubin, 2003) or by using $s(v; \sigma)^{-1}$ as a weighting factor (Robins, Rotnitzky, and Zhao, 1994; Robins, Rotnitzky, and Scharfstein, 1999b). The joint parameter (β, σ) is usually not fully identified from the data under analysis, so one must either posit various fixed values for σ and estimate β for each chosen σ (sensitivity analysis), or else give (β, σ) a prior density and conduct a Bayesian analysis. A third approach, Monte-Carlo risk analysis or Monte-Carlo sensitivity analysis (MCSA), repeatedly samples σ from its marginal prior, resamples (bootstraps) the data, and reestimates β using the sampled σ and data; it then gives the distribution of results obtained from this repeated sampling-estimation cycle. MCSA can closely approximate Bayesian results under certain (though not all) conditions (Greenland, 2001, 2004), most notably that β and σ are *a priori* independent and the prior for β is vague. The basic selection-modeling methods can be generalized (with many technical considerations) to handle missing data

(Little and Rubin, 2002; Robins, Rotnitzky, and Zhao, 1994; Robins, Rotnitzky, and Scharfstein, 1999b).

One approach to problem (2) is to model the joint distribution of U, V with a parametric model $p(u, v|\beta, \gamma) = p(y|u, x, z, \beta)p(u, x, z|\gamma)$. Again, one can estimate β by likelihood-based or by weighting methods, but because U is unmeasured (latent), the parameter (β, γ) will not be fully identified from the data and so some sort of sensitivity analysis or prior distribution will be needed (e.g., Yanagawa, 1984; Robins, Rotnitzky, and Scharfstein, 1999b; Greenland, 2003, 2004). Results will depend heavily on the prior specification given U. For example, U may be a specific unmeasured covariate (e.g., smoking status) with well studied relations to X, Y, and Z, which affords straightforward Bayesian and MCSA analyses (Steenland and Greenland, 2004). On the other hand, U may represent an unspecified aggregation of latent confounders, in which case the priors and hence inferences are more uncertain (Greenland, 2003).

Next, suppose that the "true" variable vector $V = (X, Y, Z)$ has the corresponding measurement or surrogate W (a vector with subvectors corresponding to measurements of components of X, Y, and Z). The measurement-error problem (problem 3) can then be expressed as follows: For some or all units, at least one of the V components is missing, but the measurement (subvector of W) corresponding to that missing V component is present. If enough units are observed with both V and W complete, the problem can be handled by standard missing-data methods. For example, given a model for the distribution of (V, W) one can use likelihood-based methods (Little and Rubin, 2002), or impute V components where absent and then fit the model $r(x, z; \beta)$ for $E(Y|x, z)$ to the completed data (Cole, Chu, and Greenland, 2004), or fit the model to the complete records using weights derived from all records using a model for missing-data patterns (Robins, Rotnitzky, and Zhao, 1994; Robins, Rotnitzky, and Scharfstein, 1999b). Alternatively, there are many measurement-error correction procedures that directly modify β estimates obtained by fitting the regression using W as if it were V; this is usually accomplished with a model relating V to W fitted to the complete records (Ruppert, Stefanski, and Carroll, 1995).

If a component of V is never observed on any unit (or, more practically, if there are too few complete records to support large-sample missing-data or measurement-error procedures), one may turn to latent-variable methods (Berkane, 1997). For example, one could model the distribution of (V, W) or a sufficient factor from that distribution by a parametric model; the unobserved components of V are the latent variables in the model. The parameters will not be fully identified, however, and sensitivity analysis or prior distributions will again be needed. In practice, a realistic specification can become quite complex, with subsequent inferences displaying extreme sensitivity to parameter constraints or prior distribution choices (e.g., Greenland, 2004). Nonetheless, display of this sensitivity can help provide an honest accounting for the large uncertainty that can be generated by apparently modest and realistic error distributions.

1.5 Conclusion

The three approaches described above represent separate historical streams rather than distinct methodologies, and can be blended in various ways. For example, methodologic models for confounding or randomization failure are often based on potential outcomes; the result of any modeling exercise is simply one more input to larger, informal judgments about causal relations; and those judgments may be guided by canonical considerations. Insights and innovations in any approach can thus benefit the entire process of causal inference, especially when that process is seen as part of a larger context. Finally, other traditions or approaches (some perhaps yet to be imagined) may contribute to the process. Hence, I would advise against regarding any one approach or blending as a complete solution or algorithm for problems of causal inference; the area remains one rich with open problems and opportunities for innovation.

2

Matching in observational studies

Paul R. Rosenbaum[1]

2.1 The role of matching in observational studies

In their review of methods for controlling bias in observational studies, Cochran and Rubin (1973, p. 417–8) described the role of matching as follows:

> An observational study differs from an experiment in that the random assignment of treatments (i.e., agents, programs, procedures) to units is absent. As has been pointed out by many writers since Fisher (1925), this randomization is a powerful tool in that many systematic sources of bias are made random. If randomization is absent, it is virtually impossible in many practical circumstances to be convinced that the estimates of the effects of treatments are in fact unbiased. This follows because other variables that affect the dependent variable besides the treatment may be differently distributed across treatment groups, and thus any estimate of the treatment is confounded by these extraneous **x**-variables. ... In dealing with the presence of confounding variables, a basic step in planning an observational study is to list the major confounding variables, design the study to record them, and find some method of removing or reducing the biases that they may cause. In addition, it is useful to speculate about the size and direction of any

[1] Department of Statistics, University of Pennsylvania, Philadelphia, Pa. This work was supported by a grant from the U.S. National Science Foundation.

Applied Bayesian Modeling and Causal Inference from Incomplete-Data Perspectives.
Edited by A. Gelman and X-L. Meng © 2004 John Wiley & Sons, Ltd ISBN: 0-470-09043-X

remaining biases. . . There are two principal strategies for reducing bias in observational studies. In matching or matched sampling, the samples are drawn from the populations in such a way that the distributions of the confounding variables are similar in some respects in the samples. Alternatively, random samples may be drawn, the estimates of the treatment being adjusted by means of a model relating the dependent variable y to the confounding variable $\mathbf{x}$. . . .A third strategy is to control bias due to the $\mathbf{x}$-variables by both matched sampling and statistical adjustment.

2.2   Why match?

Matching is used to accomplish several objectives.

Matched sampling. In matched sampling, there is a treated group of moderate size and a large reservoir of potential controls, and some or all of the information about the covariates $\mathbf{x}$ is available. However, additional information must be collected at significant cost for subjects included in the study, perhaps their responses y, or perhaps additional covariate information. In this case, cost considerations may require some form of sampling of the large reservoir of potential controls. In matched sampling, potential controls are drawn from the reservoir to be similar to the treated group in terms of available covariates, so the sampling process both reduces cost and begins to remove bias due to $\mathbf{x}$.

Matching increases robustness of model-based adjustments. Matching can be used as a method for sampling controls or alternatively as an analytical method that retains many or all available controls. In either case, model-based adjustment of matched treated and control groups is more robust to inaccuracies of the model than is model-based adjustment of unmatched groups (Rubin, 1973b, 1979; Table 2 versus Table 3).

A persuasive method of adjustment. Matching attempts to compare the outcomes y of treated and control subjects who were comparable in terms of the observed covariates $\mathbf{x}$ before treatment. The success of matching is often indicated in published reports by a table showing that the distribution of $\mathbf{x}$ is similar for matched treated and control groups (e.g., Rosenbaum and Rubin, 1985a; Table 2). Nontechnical audience quickly grasp the importance of comparing comparable subjects, and they also understand simple comparisons demonstrating the groups are comparable, at least in terms of the observed covariates $\mathbf{x}$. In contrast, model-based adjustments that intend to accomplish the same goal indirectly without matching may be far less persuasive to nontechnical audiences, because those adjustments require more technical knowledge to understand and appraise. Moreover, it is straightforward to present both (i) matched analyses and (ii) model-based adjustment of matched analyses, and because the conclusions typically do not differ greatly, (i) may aid in making (ii) palatable and plausible to nontechnical audiences.

Avoiding inappropriate comparisons. Even when data for potential controls are available without cost, so that all controls may be used without sampling, it may be wise to set aside the data for some potential controls. If a treatment is typically given only to people who have some need of it, then the support of the distribution of **x** for the treated group may be only a portion of the support of the distribution of **x** in the general population of potential controls. For instance, if the treated group consisted of children in the Head Start program in a particular city, then all of these children would come from low-income homes, while the untreated children in the same city would come from homes with a wide range of incomes. Any attempt to estimate the effect of Head Start on children from upper-income homes is pure extrapolation, pure speculation. Moreover, when the reservoir of potential controls is many times larger than the treated group, there is only a slight increase in the standard error of an estimated effect owing to setting aside some controls (Rosenbaum and Rubin, 1985a, p. 33). Two nice case studies illustrating this point are Smith (1997) and Dehejia and Wahba (1999).

Aiding thick description. A thick description is a narrative account that would attempt to make the reader feel familiar with the situation and acquainted with individuals involved. Thick descriptions are common in the social sciences (e.g., Athens, 1997; Bosk, 1981; Estroff, 1985; Katz, 1999) and are familiar in medicine from the "Case Reports from the Massachusetts General Hospital" published in the *New England Journal of Medicine*. The techniques of qualitative research, such as thick description, are not easily combined with model-based adjustments, because the parameters of the model do not typically refer to intact human beings and the situations in which they live. In contrast, it is straightforward to coordinate quantitative and narrative accounts using matching. Moreover, such narrative investigations can improve matching through deeper insight into the relationship between the measured **x** and the underlying reality. See Rosenbaum and Silber (2001) for discussion of this aspect of matching and a case study.

2.3   Two key issues: balance and structure

Matching is sometimes conceived as forming pairs of subjects, one treated, one control, with nearly the same values of the observed covariates **x**; however, this conception, or misconception, is a substantial handicap when using matching to control bias. There are two key issues: covariate balance and the structure of matched sets. The simulation study by Gu and Rosenbaum (1993) considered a wide variety of issues and proposals, and concluded that balance and structure had both the largest and most consistent effect on the quality of matched samples.

Covariate balance using propensity scores

When $\mathbf{x}$ is of moderate to high dimension k—that is, when there are many covariates—it will be difficult if not impossible to match most treated subjects to controls with the same value of $\mathbf{x}$. It is easy to see why. Imagine dividing each covariate into just two levels, say above or below its median, and trying to match exactly for the level of each of the k covariates. There will be 2^k patterns of levels with a k-dimensional $\mathbf{x}$, or about a million patterns with $k = 20$ covariates. Even if one had thousands or tens of thousands of potential controls available for matching, with $k = 20$ covariates, it would be difficult to find a control to match the levels for many treated subjects—there are too many patterns and too few controls. Moreover, with just two levels for each covariate, even an exact match for the two levels formed from the covariates would not be adequate to control bias from continuous covariates. Specifically, Cochran (1968) found that four or five levels for each continuous covariate would be needed, making matching for level much more difficult, with $4^{20} \doteq 10^{12}$ or $5^{20} = 9.5 \times 10^{13}$ patterns for $k = 20$ covariates.

The alternative to closely matching individuals for $\mathbf{x}$ is to balance $\mathbf{x}$, that is, to form matched treated and control groups with similar distributions of $\mathbf{x}$. More precisely, write $Z = 1$ for a treated subject, $Z = 0$ for a control. Covariate balance means that the observed covariates $\mathbf{x}$ and the treatment Z are conditionally independent within matched sets. If there is covariate balance, then a treated subject, $Z = 1$, may be matched to a control, $Z = 0$, with a different value of $\mathbf{x}$, but their two values of $\mathbf{x}$ will not help to identify the treated subject.

Covariate balance may be achieved by matching on a scalar, the propensity score, as proposed by Rosenbaum and Rubin (1983a). The propensity score is the conditional probability of exposure to treatment given the observed covariates, that is, $e(\mathbf{x}) = \Pr(Z = 1 \,|\, \mathbf{x})$. They showed that

$$\mathbf{x} \;\|\; Z \;|\; e(\mathbf{x}), \tag{2.1}$$

where $A \;\|\; B \;|\; C$ is Dawid's (1979) notation for A is conditionally independent of B given C. In other words, the covariates $\mathbf{x}$ may strongly predict who will receive treatment, $Z = 1$, and who will receive control, $Z = 0$, but (2.1) asserts that among subjects with the same value of the propensity score, $e(\mathbf{x})$, the covariates $\mathbf{x}$ no longer predict treatment assignment Z. Moreover, Rosenbaum and Rubin (1983) also showed that if adjustments for the k-dimensional $\mathbf{x}$ suffices to permit estimation of treatment effects, then adjustments for scalar propensity score $e(\mathbf{x})$ also suffices. Typically, the propensity score $e(\mathbf{x})$ is estimated from a model, such as a logit model, $\log\left[e(\mathbf{x}) / \{1 - e(\mathbf{x})\}\right] = \alpha + \mathbf{x}^T \beta$, and then one matches on the estimated propensity score or on $\mathbf{x}^T \widehat{\beta}$. Two case studies of matching or stratifying on the propensity score are given by Rosenbaum and Rubin (1984, 1985a), one balancing 20 covariates, the other balancing 74 covariates, using the scalar estimated propensity score. Somewhat surprisingly, these case studies and some theory shows that estimated propensity scores are slightly better than true propensity scores at balancing covariates, apparently because estimated propensity scores cannot distinguish systematic imbalances from chance imbalances and the esti-mated scores work to remove both. In a simulation, Gu and Rosenbaum (1993) found that matching on the propensity score was much better than other multi-variate matching methods, such as Mahalanobis metric matching, when there were $k = 20$ covariates.

If the treatment, Z, has more than two versions, say several doses, but the con-ditional distribution Pr$(Z \,|\mathbf{x})$ depends on $\mathbf{x}$ only through a function $e(\mathbf{x})$, then $e(\mathbf{x})$ acts as a generalized propensity score, with parallel properties (Joffe and Rosen-baum, 1999a, 1999b). For instance, if Z takes on ordinal values $j = 0, 1, \ldots, J$, and Pr$(Z \,|\mathbf{x})$ follows McCullagh's (1980) ordinal logit model,

$$\log \left[\frac{\Pr(Z \geq j \,|\mathbf{x})}{1 - \Pr(Z \geq j \,|\mathbf{x})} \right] = \alpha_j + \mathbf{x}^T \beta, \quad j = 0, 1, \ldots, J,$$

then $\mathbf{x}^T \beta$ has the key properties of the propensity score for the two treatment levels described above. See Lu, Zanutto, Hornik, and Rosenbaum (2001) for an application of this approach, and see Imbens (2000) for an alternative approach with separate propensity scores for each level of Z.

For many treatments, the decision is to treat now or to wait and see, possibly treating later. For instance, this is common with many surgical procedures, in which the decision not to operate today does not foreclose the possibility of operating at a future date. In this case, the treatment/control indicator $Z(t)$ varies with time $t \geq 0$ for each person, starting with $Z(0) = 0$, and possibly stepping up to $Z(t) = 1$ at some later $t > 0$. The covariates $\mathbf{x}(t)$, too, may vary with time up to the moment of treatment, and changes in the covariates may increase the chance of being switched to treatment. In this case, one might match a subject newly treated at time t to a similar subject still untreated at t. This is known as *risk-set matching*, where the risk set resembles that in Cox's proportional hazards model. If the hazard of treatment varies as a function $h\{\mathbf{x}(t)\}$ of the observed covariates, $\mathbf{x}(t)$, then the hazard $h\{\mathbf{x}(t)\}$ has properties similar to the propensity score. See Li, Propert, and Rosenbaum (2001, §4) for detailed discussion.

Propensity scores can also be used in other ways. For instance, propensity scores can be used in exact or approximate permutation inference, alone or in con-junction with covariate adjustment (Rosenbaum, 1984a, 2002a). The reciprocal of the propensity score may be used as a form of weighting adjustment (Rosenbaum, 1987a; Robins, Rotnitzky, and Zhao, 1995; Imbens, 2000).

Structure of matched sets

There are limits to what can be accomplished with pair matching, in which one treated subject is matched to one control. Rubin (1973a, 1976b) provides formal upper bounds on bias reduction with matched pairs using expected order statistics of the covariates, but the intuition behind these bounds may be described briefly. For simplicity, imagine just $k = 1$ covariate x, and two superimposed histograms: a red histogram describing its distribution among treated subjects, and a blue histogram describing its distribution in the reservoir of potential controls, where the height of the histogram equals the *number* of subjects. If the reservoir of potential controls is larger than the treated group, then the blue histogram will be higher than the red histogram for at least some values of x. However, an exact pair matching for x will exist only if the blue histogram is higher than the red for *all* values of x. If the x's for treated subjects tend to be larger than those for potential controls, if the red histogram is higher than the blue one for large x, it may not be possible to construct matched pairs that balance x even in the sense of equating their means. Again, Rubin (1973a, 1976b) gives numerical values of these upper bounds on bias reduction with pair matching. The problem does not disappear as the sample size increases if the treated group and the reservoir of potential controls grow at the same rate. When there are several covariates, $k > 1$, the same considerations apply with histograms describing the distributions of scalar propensity scores. Generally, it is unwise to overcome this problem by discarding treated subjects who are difficult to match; see Rosenbaum and Rubin (1985b) who show that this "solution" creates substantial problems of its own.

A second limitation of pair matching concerns efficiency. If data on controls are inexpensive or free, it may be possible to closely match some treated subjects to more than one control, thereby reducing sampling variability.

More flexible matching structures overcome these limitations. The form of an optimal stratification is a *full matching* in which a matched set may contain one treated subject and one or more controls, or one control and one or more treated subjects (Rosenbaum, 1991). Where the blue histogram (for controls) is above the red histogram (for treated subjects), treated subjects are matched to more than one control, whereas where the red histogram is above the blue histogram, control subjects are matched to more than one treated subject. Provided the two distributions have the same support, as the sample size increases, full matching can remove all of the bias due to x, and it can use as many controls as desired. The analysis of data from a full matching is only slightly more complex than the analysis from a pair matching; it must take account of the varied sample sizes in distinct matched sets. In a simulation, Gu and Rosenbaum (1993) found that full matching was much better than pair matching or matching with a fixed number of controls.

When matching with multiple controls, Ming and Rosenbaum (2000) calculated upper bounds on bias reductions similar to Rubin's (1973a, 1976b) for pair matching. They found substantially greater bias reduction when the number of controls is allowed to vary from set to set than when that number is fixed, the same for all sets. Tables 2.1–2.2 present an illustration adapted from the case–control study of mortality after surgery by Silber et al. (2001). Deaths following surgery were optimally matched to survivors using estimated probability of death based on baseline covariates measured upon admission to the hospital. The estimated probabilities of death were from a model predicting mortality from baseline covariates and fitted to separate data from previous years. Table 2.1 shows the estimated mortality risks for the first five matched sets, matching three survivors to each death. Some of the

Set	Death	Survivors
1	.034	.034, .034, .034
2	.274	.141, .114, .104
3	.227	.194, .158, .150
4	.024	.024, .024, .024
5	.485	.439, .198, .155

Table 2.1 Optimal matching with three controls: risk scores for the first five matched sets. *Source*: Ming and Rosenbaum (2000), *Biometrics*.

Set	Death	Survivors
1	.034	.034, .034, .034, .034
2	.274	.216
3	.227	.218
4	.024	.024, .024, .024, .024
5	.485	.439

Table 2.2 Optimal matching with one to four controls: risk scores for the first five matched sets. *Source*: Ming and Rosenbaum (2000), *Biometrics*.

matches are very poor: some of the people who died were at much higher risk than their matched controls. In contrast, Table 2.2 shows the same first five matched sets, optimally matched, but with variable set sizes, and the same total number of controls in the complete matched sample. Here, the matching is much closer. Ming and Rosenbaum (2000) also examine increases in variance owing to unequal rather than equal set sizes, finding the gains in bias reduction are often large, while the increases in variance are often small.

2.4 Additional issues

Matching algorithms

There is competition among treated subjects for the best control matches. The control who is the best match for the first treated subject may also be the best match for the second treated subject. Should that control be assigned to the first treated subject or to the second or to someone else? Table 2.3 illustrates the problem, with three treated subjects, three controls, and a table of covariate distances between them. Control "a" is closest to all three treated subjects, α, β, and γ. A bad strategy is a greedy algorithm: it picks the best match, sets that match aside, picks the best from what is left, and so on. Here, greedy pairs (α, a) at a cost of 0, sets α and a aside, then picks (β, b) at a cost of 5, and is forced to accept (γ, c) at a cost

		a	b	c
	α	0	1	1
Treated	β	1	5	6
	γ	1	6	100

Controls

Table 2.3 A distance matrix between three treated subjects and three controls.

of 100, for an average cost per comparison of $35 = (0 + 5 + 100)/3$. An optimal match has (α, c), (β, b), (γ, a), for an average cost of $2.33 = (1 + 5 + 1)/3$. An optimal full matching does better still: it has (α, b, c) at a cost of $2 = 1 + 1$ and (β, γ, a) at a cost of $2 = 1 + 1$, for a total cost of $1 = (1 + 1 + 1 + 1)/4$.

Fast algorithms exist for optimal matching problems. See Rosenbaum (1989, 2002b, §11) for discussion of optimal matching in observational studies. See Bertsekas (1991) for good general algorithms, Fortran and C code. See Bergstralh, Kosanke, and Jacobsen (1996) and Ming and Rosenbaum (2001) for implementations in SAS. Matching with doses, as in Lu, Zanutto, Hornik, and Rosenbaum (2001), requires "nonbipartite matching" for which Derigs (1988) presents an algorithm and Fortran code. An alternative general algorithm is available in C; see Galil (1986).

Covariance adjustment of matched data

Rubin (1973b, 1979) found using simulations that covariance adjustment of matched pairs was more efficient than matching alone and more robust to model misspecification than covariance adjustment alone. In particular, covariance adjustment of matched pair differences consistently reduced bias, even when the covariance adjustment model was wrong, but covariance adjustment alone sometimes increased the bias compared to no adjustment when the model was wrong. Using propensity scores in an appropriate way, it is possible to draw valid inferences even though the covariance model is incorrect (Rosenbaum, 2002a).

Overmatching

Matched sampling builds adjustments for covariates into the research design. Unlike analytical adjustments for covariates, adjustments that are built into the design are difficult to undo at a later stage.

A covariate is a variable measured prior to treatment and hence unaffected by the treatment. An outcome is measured after treatment and may be affected by the treatment. If one adjusts for an outcome as if it were a covariate, the adjustment itself may introduce a bias where none existed previously (Rosenbaum, 1984b; Wainer, 1989). Matching on outcomes as if they were covariates is sometimes called "overmatching," but that term is not ideal because it does not emphasize the important distinction between covariates and outcomes, and vaguely hints that the number of covariates is important. In certain special situations, adjustments for outcomes may confer benefits, reducing biases from important unobserved covariates (see Rosenbaum, 1984b, §1.2 and §1.3 for examples and §3.6 for theory).

Although analyses that adjust one outcome for another are sometimes useful for specific purposes, it is often best to maintain flexibility, to avoid building those adjustments irrevocably into the research design, and to avoid matching on outcomes. If needed, by applying analytical adjustments to matched pairs or sets, some analyses can go on to adjust for certain outcomes, while other analyses refrain from such adjustments.

Matching with two control groups

Matching removes or reduces visible bias from observed covariates, but those biases are the tip of the iceberg. There may also be biases that cannot be seen in the data from covariates that were not measured. Much of the effort in the design and analysis of an observational study is devoted to addressing concerns about possible hidden biases.

Perhaps the simplest and most common tactic is the use of two control groups selected to systematically vary certain unobserved covariates (Campbell, 1969; Rosenbaum, 1987b, 2002b, §8). That is, although a certain covariate u is not measured, u is known to be substantially higher in one control group than in the other. If the two control groups have similar outcomes, which are very different from the outcomes in the treated group, then this is consistent with a treatment effect as opposed to bias from u, whereas substantial differences in outcomes between the two control groups cannot be explained as a treatment effect and are consistent with bias from u.

An interesting example is found in Card and Krueger's (1994) study of the effects of the minimum wage on employment in the fast food industry. On April 1, 1992, New Jersey raised its minimum wage by about 20% from $4.25 per hour to $5.05 per hour. Economic theory is generally understood to predict a decline in consumption when prices are forced up, here a decline in employment among minimum wage earners when the minimum wage is increased. Card and Krueger looked at changes in employment at fast-food restaurants, such as Burger King or Wendy's, from the year before the wage increase to the year after, comparing New Jersey to eastern Pennsylvania where the minimum wage remained at $4.25 per hour. They found no sign of a decline in employment following the increase in the minimum wage. In certain analyses, they used two control groups, that is two groups of restaurants not required by law to materially increase wages. One control group consisted of restaurants in Pennsylvania, the other consisted of restaurants in New Jersey whose lowest wage was already at least $5.00 per hour. Although Burger Kings are much the same throughout New Jersey and Pennsylvania they are not identical, and one could raise concerns about either control group. For example, taxes and regulations differ somewhat in New Jersey and Pennsylvania. Also, Burger Kings in New Jersey paying the minimum wage are likely to be in different labor markets than Burger Kings in New Jersey paying substantially more than the minimum wage—one thinks of the wealthy suburb of Princeton and the poor city of Camden. However, Card and Krueger found similar results for both control groups.

When two control groups are used to systematically vary an unobserved covariate u, it is still important to control for observed covariates $\mathbf{x}$. In Card and Krueger's study, one would like to compare Burger Kings to Burger Kings, Wendy's to Wendy's, company owned restaurants to other company owned restaurants, restaurants open long hours to other restaurants open long hours, and so on. Matching may be used to form an incomplete block design to compare the treated group and the two control groups. Lu and Rosenbaum (2004) use Derigs' (1988) Fortran algorithm to construct optimal matched designs with two control groups, using data from Card and Krueger to illustrate. For an algorithm in C, see Galil (1986).

Other tactics for addressing unobserved covariates, such as sensitivity analyses, known effects, and coherence are discussed in Rosenbaum (2002b, §4-§9, 2003, 2004).

3

Estimating causal effects in nonexperimental studies

Rajeev Dehejia[1]

3.1 Introduction

This chapter discusses the use of propensity score methods to estimate causal effects in nonexperimental studies. Statisticians and social scientists have studied this question since at least the early part of the twentieth century, and it remains a central question even today. Though a randomized experiment is the gold standard in estimating treatment effects, there are many settings in which an experiment cannot be performed, owing to cost limitations or an obligation to provide treatment or because evaluation is undertaken after the treatment program has already been offered. In such settings, evaluation of the treatment effect is undertaken using a nonexperimental comparison group, in addition to individuals who have received the treatment.

The fundamental difficulty in estimating the treatment effect in an observational study, as noted in Rubin (1974), is controlling for those pretreatment variables that determine assignment to treatment and that affect the outcome of interest. These variables can be of two types: those that are observed by the researcher and those that are not observed. The methods discussed in this chapter deal exclusively with controlling for the former. Methods that deal with unobservable differences between the treatment and the comparison group are discussed in Imbens and Angrist (1992) and Angrist, Imbens, and Rubin (1996).

[1]Department of Economics and SIPA, Columbia University, New York. I am grateful to Don Rubin for his support, suggestions, and encouragement over the last 10 years. His energy, ideas, and creativity have been inspirational.

Applied Bayesian Modeling and Causal Inference from Incomplete-Data Perspectives.
Edited by A. Gelman and X-L. Meng   © 2004 John Wiley & Sons, Ltd   ISBN: 0-470-09043-X

The key insight for estimating treatment effects in nonexperimental settings, when assignment to treatment is based on observed variables, is identified in Rubin (1978a): conditional on the pretreatment covariates that determine assignment to treatment, assignment to treatment is essentially random. When there are only a few relevant variables, this provides a simple means of estimating the treatment effect: by matching or grouping observations on the basis of pretreatment covariates, estimating the treatment effect within each group, and then averaging over these treatment effects to obtain the overall treatment effect.

The difficulty with implementing this strategy in practice is that in many settings of interest there are a large number of variables that determine assignment to treatment. Controlling for a high-dimensioned set of pretreatment covariates poses several difficulties: (i) matching or grouping requires a metric or rule to order the covariates, (ii) matching has been shown to be inconsistent when there are four or more continuous covariates (Abadie and Imbens, 2002), (iii) standard nonparametric methods encounter the curse of dimensionality, and (iv) standard regression-based methods linearly extrapolate between the treatment and comparison groups and typically (though not necessarily) assume a constant treatment effect.

The central contribution of Rosenbaum and Rubin (1983a) is to provide a means of eschewing the higher-dimensional issues discussed above by focusing on a uni-dimensional summary of the pretreatment covariates, namely the propensity score. By estimating the propensity score, one can (i) check the balance between the treatment and comparison groups in terms of pretreatment covariates, (ii) create a comparison group that is well balanced in the same sense, and (iii) provided that assignment to treatment is based on the observed covariates, obtain an unbiased estimate of the treatment effect.

In this chapter, I illustrate the use of propensity score methods by applying them to a widely studied data set constructed by Lalonde (1986) (see also Heckman and Hotz, 1989; Dehejia and Wahba 1999, 2002). Lalonde combined data from the National Supported Work (NSW) Demonstration, a randomized trial described below, with data from two nonexperimental comparison groups. The randomized trial provides a benchmark estimate of the treatment effect. By combining the experimental treatment group with the nonexperimental comparison groups, Lalonde was able to evaluate the efficacy of a range of nonexperimental estimators. His finding was that in general nonexperimental methods do not succeed in robustly replicating the benchmark experimental estimate. Lalonde's result was one among many within economics, the social sciences, and statistics that was influential in building a consensus for randomized trials as the key method to reliably evaluate treatment effects. In this chapter, I reevaluate this finding by applying propensity score methods to Lalonde's data set.

Section 3.2 provides a concise, self-contained overview of propensity score methods. Section 3.3 outlines a few salient features of the NSW Data. Section 3.4 presents estimates of the treatment effect. Section 3.5 concludes.

3.2   Identifying and estimating the average treatment effect

Identification

Let Y_{1i} represent the value of the outcome when unit i is subject to regime 1 (called treatment), and Y_{0i} the value of the outcome when unit i is exposed to regime 0 (called control). Only one of Y_{0i} or Y_{1i} can be observed for any unit, since we cannot observe the same unit under both treatment and control. Let T_i be a treatment indicator ($= 1$ if exposed to treatment, $= 0$ otherwise). Then the observed outcome for unit i is $Y_i = T_i Y_{1i} + (1 - T_i) Y_{0i}$. The treatment effect for unit i is $\tau_i = Y_{1i} - Y_{0i}$. The average treatment effect for this population is: $\tau = E(Y_{1i}) - E(Y_{0i})$, where the expectation is over the population of treated and control individuals.

In an experimental setting, where assignment to treatment is randomized, the treatment and control groups are randomly drawn from the same population. This implies that $\{Y_{1i}, Y_{0i} \perp\!\!\!\perp T_i\}$ (using Dawid's (1979) notation, $\perp\!\!\!\perp$ represents independence), so that, for $j = 0, 1$:

$$E\left(Y_{ji} \mid T_i = 1\right) = E\left(Y_{ji} \mid T_i = 0\right) = E\left(Y_i \mid T_i = j\right),$$

and $\tau = E(Y_{1i}) - E(Y_{0i}) = E(Y_i | T_i = 1) - E(Y_i | T_i = 0)$, which is readily estimated.

In a nonexperimental setting, this expression cannot be estimated directly since Y_{0i} is not observed for treated units. Assuming that assignment to treatment is based on observable covariates, X_i, namely that $\{Y_{1i}, Y_{0i} \perp\!\!\!\perp T_i\} | X_i$ (Rubin, 1974, 1977), we obtain:

$$E\left(Y_{ji} \mid X_i, T_i = 1\right) = E\left(Y_{ji} \mid X_i, T_i = 0\right) = E\left(Y_i \mid X_i, T_i = j\right),$$

for $j = 0, 1$. Conditional on the observables, X_i, there is no systematic pretreatment difference between the groups assigned to treatment and control. This allows us to identify the treatment effect:

$$\tau = E\{E\left(Y_i | X_i, T_i = 1\right) - E\left(Y_i | X_i, T_i = 0\right)\}, \tag{3.1}$$

where the outer expectation is over the distribution of X, namely the distribution of preintervention variables.

One method for estimating the treatment effect that stems from (3.1) is estimating $E(Y_i | X_i, T_i = 1)$ and $E(Y_i | X_i, T_i = 0)$ as two nonparametric equations. This estimation strategy becomes difficult, however, if the covariates, X_i, are high dimensional. The propensity score theorem provides an intermediate step:

Proposition 1 (Rosenbaum and Rubin, 1983a) *Let $p(X_i)$ be the probability of unit i having been assigned to treatment, defined as $p(X_i) \equiv p(T_i = 1 | X_i) =$*

$E(T_i|X_i)$. Assume $0 < p(X_i) < 1$, $\forall X_i$, and $Pr(T_1, T_2, \ldots T_n|X_1, X_2, \ldots X_n) = \prod_{i=1,\ldots,N} p(X_i)^{T_i}(1 - p(X_i))^{(1-T_i)}$ for the N units in the sample. Then:

$$\{(Y_{1i}, Y_{0i}) \perp\!\!\!\perp T_i\} \mid X_i \Rightarrow \{(Y_{1i}, Y_{0i}) \perp\!\!\!\perp T_i\} \mid p(X_i).$$

If $\{(Y_{1i}, Y_{0i}) \perp\!\!\!\perp T_i\}|X_i$ and the assumptions of Proposition 1 hold, then:

$$\tau = E\{E\,(Y_i|T_i = 1, p\,(X_i)) - E\,(Y_i|T_i = 0, p\,(X_i))\}. \qquad (3.2)$$

The outer expectation is over the distribution of $p(X_i)$, namely the propensity score in the treated and control groups.

One intuition for the propensity score is that, whereas in equation (3.1) we are trying to condition on X (intuitively, to find observations with similar covariates), in equation (3.2) we are trying to condition just on the propensity score, because the proposition implies that observations with the same propensity score have the same distribution of the full vector of covariates X.

In an observational study, we are often interested in the treatment effect for the treated group, rather than the overall average treatment effect. In our application, the treatment group is selected from the population of interest, namely welfare recipients eligible for the program. The (nonexperimental) comparison group is drawn from a different population (in our application both the Current Population Survey [CPS] and Panel Survey of Income Dynamics [PSID] are more representative of the general US population). Thus, the treatment effect we are trying to identify is the average treatment effect for the treated population:

$$\tau \mid_{T=1} = E(Y_{1i}|T_i = 1) - E(Y_{0i}|T_i = 1). \qquad (3.3)$$

The arguments above are readily extended to identifying the treatment effect on the treated:

$$\tau \mid_{T=1} = E\,\{E\,(Y_i|T_i = 1, p(X_i)) - E\,(Y_i|T_i = 0, p(X_i))\,|T_i = 1\}, \qquad (3.4)$$

where the outer expectation is over the distribution of $p(X_i)|T_i = 1$, namely the distribution of the propensity score in the treated group.

The estimation strategy

Estimation is in two steps. First, we estimate the propensity score, separately for each nonexperimental sample consisting of the experimental treatment units and the specified set of comparison units (PSID or CPS). We use a logistic probability model, but other standard models yield similar results. One issue is what functional form of the preintervention variables to include in the logit. We rely on the following proposition:

Proposition 2 (Rosenbaum and Rubin, 1983) *If $p(X_i)$ is the propensity score, then:*

$$X_i \perp\!\!\!\perp T_i \mid p(X_i).$$

Proposition 2 asserts that, conditional on the propensity score, the covariates are independent of assignment to treatment, so that, for observations with the same propensity score, the distribution of covariates should be the same across the treatment and comparison groups. Conditioning on the propensity score, each individual has the same probability of assignment to treatment, as in a randomized experiment.

We use this proposition to assess estimates of the propensity score. For any given specification (we start by introducing the covariates linearly), we group observations into strata defined on the estimated propensity score and check whether we succeed in balancing the covariates within each stratum. We use tests for the statistical significance of differences in the distribution of covariates, focusing on first and second moments (see Rosenbaum and Rubin, 1984). If there are no significant differences between the two groups within each stratum, then we accept the specification. If there are significant differences, we add higher-order terms and interactions of the covariates until this condition is satisfied. Failure to satisfy this condition under all specifications would lead to the conclusion that the treatment and control groups do not overlap along all dimensions.

In the second step, given the estimated propensity score, we need to estimate a univariate nonparametric regression $E(Y_i|T_i = j, p(X_i))$, for $j = 0, 1$. We focus on two simple methods for obtaining a flexible functional form, stratification, and matching, but in principle one could use any of the standard array of nonparametric techniques (e.g., see Härdle and Linton, 1994; Heckman, Ichimura, and Todd, 1997) or weighting (see Hirano, Imbens, and Ridder, 2002).

With stratification, observations are sorted from lowest to highest estimated propensity score. We discard the comparison units with an estimated propensity score less than the minimum (or greater than the maximum) estimated propensity score for treated units. The strata, defined on the estimated propensity score, are chosen so that the covariates within each stratum are balanced across the treatment and comparison units (we know such strata exist from step one). On the basis of equations (3.2) and (3.4), within each stratum we take a difference in means of the outcome between the treatment and comparison groups, and weight these by the number of (treated) observations in each stratum. We also consider matching on the propensity score. Each treatment unit is matched with replacement to the comparison unit with the closest propensity score; the unmatched comparison units are discarded (see Dehejia and Wahba, 2002 for more details; also Rubin, 1979).

3.3 The NSW data

The NSW was a US federally and privately funded program that aimed to provide work experience for individuals who had faced economic and social problems

Sample	No. of Obs	Age	Education	Black	Hispanic	No Degree	Married	RE74 US$	RE75 US$
NSW:									
Treated	185	25.81	10.35	0.84	0.059	0.71	0.19	2,096	1,532
		(0.35)	(0.10)	(0.02)	(0.01)	(0.02)	(0.02)	(237)	(156)
Control	260	25.05	10.09	0.83	0.1	0.83	0.15	2,107	1,267
		(0.34)	(0.08)	(0.02)	(0.02)	(0.02)	(0.02)	(276)	(151)
Comparison groups:[a]									
PSID	2,490	34.85	12.11	0.25	0.032	0.31	0.87	19,429	19,063
		[0.78]	[0.23]	[0.03]	[0.01]	[0.04]	[0.03]	[991]	[1,002]
CPS	15,992	33.22	12.02	0.07	0.07	0.29	0.71	14,016	13,650
		[0.81]	[0.21]	[0.02]	[0.02]	[0.03]	[0.03]	[705]	[682]

Age = age in years; Education = number of years of schooling;
Black = 1 if black, 0 otherwise; Hispanic = 1 if Hispanic, 0 otherwise; Nodegree = 1
if no high school degree, 0 otherwise; Married = 1 if married, 0 otherwise;
REx = earnings in calendar year 19x.
[a]Definition of Comparison Groups (Lalonde, 1986):
PSID: All male household heads less than 55 years old who did not classify
themselves as retired in 1975.
CPS: All CPS males less than 55 years of age.

Table 3.1 Sample means of characteristics for National Support Work Demonstration and comparison samples. (Standard errors in parentheses.) [Standard error on difference in means with NSW subset/Treated in brackets.]

prior to enrollment in the program (see Hollister, Kemper, and Maynard, 1984; Manpower Demonstration Research Corporation, 1983).[2] Candidates for the experiment were selected on the basis of eligibility criteria, and then were either randomly assigned to, or excluded from, the training program. Table 3.1 provides the characteristics of the sample we use, Lalonde's male sample (185 treated and 260 control observations).[3] The table highlights the role of randomization: the distribution of

---

[2]Four groups were targeted: women on Aid to Families with Dependent Children (AFDC), former addicts, former offenders, and young school dropouts. Several reports extensively document the NSW program. For a general summary of the findings, see Manpower Demonstration Research Corporation (1983).

[3]The data we use are a subsample of the data used in Lalonde (1986). The analysis in Lalonde (1986) is based on one year of pretreatment earnings. But as Ashenfelter (1978), and Ashenfelter and Card (1985) suggest, the use of more than one year of pretreatment earnings is key in accurately estimating the treatment effect, because many people who volunteer for training programs experience a drop ("Ashenfelter's dip") in their earnings just prior to entering the training program. Using the Lalonde sample of 297 treated and 425 control units, we exclude the observations for which earnings in 1974 could not be obtained, thus arriving at a reduced sample of 185 treated observations and 260 control observations. Because we obtain this subset by looking at pretreatment covariates, we do not

the covariates for the treatment and control groups is not significantly different. We use the two nonexperimental comparison groups constructed by Lalonde (1986), drawn from the CPS and PSID.[4]

Table 3.1 presents the sample characteristics of the two comparison groups and the treatment group. The differences are striking: the PSID and CPS sample units are eight to nine years older than those in the NSW group; their ethnic and racial composition is different; they have on average completed high school degrees, while NSW participants were by and large high school dropouts; and, most dramatically, pretreatment earnings are much higher, by more than $10,000$, for the comparison units than for the treated units, by more than $10,000.

3.4   Propensity score estimates

Comparing the treatment and comparison samples

One of the simplest, and most powerful, uses of the propensity score is as a diagnostic on the quality of a nonexperimental comparison group. Whereas Table 3.1 convincingly established significant overall differences between the treatment and comparison groups, the propensity score allows us to focus directly on the comparison units that are not well matched with the treatment group. Using the method outlined in Section 3.2, we estimate the propensity score for the two composite samples (NSW-CPS and NSW-PSID), incorporating the covariates linearly and with higher-order terms.[5]

Figures 3.1 and 3.2 provide a simple diagnostic on the data examined, plotting the histograms of the estimated propensity scores for the NSW-CPS and NSW-PSID samples. The histograms do not include the comparison units (11,168 units for the CPS and 1,254 units for the PSID) whose estimated propensity score is less than the minimum estimated propensity score for the treated units. Also, the first bins of both diagrams contain most of the remaining comparison units (4,398 for the CPS and 1,007 for the PSID). Hence, it is clear that very few of the comparison units are similar to the treated units. In fact, one of the strengths of the propensity score method is that it dramatically highlights this fact. On comparing the other bins, we note that the number of comparison units in each bin is greater than or (approximately) equal to the number of treated units in the NSW-CPS sample, but in the NSW-PSID sample many of the upper bins have far more treated units than comparison units.

---

disturb the balance in observed and unobserved characteristics between the experimental treated and control groups. See Dehejia and Wahba (1999) for a comparison of the two samples.

[4]These are the CPS-1 and PSID-1 comparison groups from Lalonde's paper.

[5]We use the following specifications for the propensity score. For the PSID, $\text{Prob}(T_i = 1) = F(\text{age, age}^2, \text{education, education}^2, \text{married, no degree, black, Hispanic, RE74, RE75, RE74}^2, \text{RE75}^2, \text{u74} \times \text{black})$. For the CPS, $\text{Prob}(T_i = 1) = F(\text{age, age}^2, \text{education, education}^2, \text{no degree, married, black, Hispanic, RE74, RE75, u74, u75, educ} \times \text{RE74, age}^3)$.

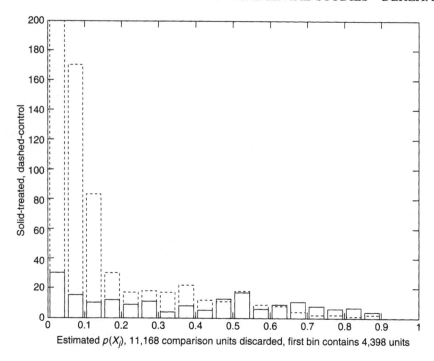

Figure 3.1   Histogram of estimated propensity score, National Support Work Demonstration and Current Population Survey.

Estimating the treatment effect

We use stratification and matching on the propensity score to group the treatment units with the comparison units whose estimated propensity scores are greater than the minimum—or less than the maximum—propensity score for treatment units. We estimate the treatment effect by summing the within-stratum difference in means between the treatment and comparison observations (of earnings in 1978), where the sum is weighted by the number of treated observations within each stratum (Table 3.2, column (4)). An alternative is a within-stratum regression, again taking a weighted sum over the strata (Table 3.2, column (5)). When the covariates are well balanced, such a regression should have little effect, but it can help to eliminate the remaining within-stratum differences. Likewise for matching, we can estimate a difference in means between the treatment and matched comparison group for earnings in 1978 (column (7)), and also perform a regression of 1978 earnings on covariates (column (8)).

For comparison, in columns (1) and (2), we estimate the treatment effect using a difference in means and a regression over the full sample.[6] We also estimate a least

---

[6]We estimate a regression of the form,

$$Y_i = \alpha + \beta x_i + \delta T_i + \epsilon_i,$$

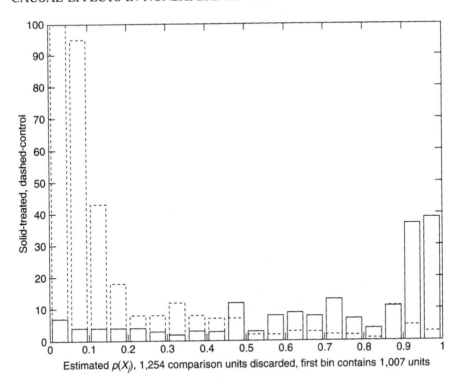

Figure 3.2   Histogram of estimated propensity score, National Support Work Demonstration and Panel Survey of Income Dynamics.

squares regression of earnings in 1978 on a quadratic of the estimated propensity score and a treatment indicator, for those observations used in stratification and matching.

Table 3.2 presents the results. For the PSID sample, the stratification estimate is $1,608 and the matching estimate is $1,691, compared to the benchmark randomized-experiment estimate of $1,794. The estimates from a difference in means or regression on the full sample are −$15,205 and $731. In columns (5) and (8), controlling for covariates has little impact on the stratification and matching estimates. Likewise for the CPS, the propensity-score-based estimates from the CPS—$1,713 and $1,582—are much closer to the experimental benchmark than estimates from the full comparison sample: −$8,498 and $972.

Column (3) in Table 3.2 illustrates the value of allowing both for a heterogeneous treatment effect and for a nonlinear functional form in the propensity score. The estimators in columns (4) to (8) have both these characteristics, whereas in

where δ is the treatment effect and we include age, age^2, education, no degree, black, Hispanic, RE74, and RE75 as controls. We use the same covariates for within-stratum regressions and the post-matching weighted regression.

| | NSW Earnings Less Comparison Group Earnings | | NSW Treatment Earnings Less Comparison Group Earnings, Propensity Score Estimates | | | | | |
| | | | Quadratic in the Estimated p-score | Stratifying on the Estimated p-score | | | Matching on the Estimated p-score | |
	(1) Difference in Means	(2) Regression Adjusted	(3)	(4) Difference in Means	(5) Regression Adjusted	(6) [Observations][a]	(7) Difference in Means	(8) Regression Adjusted[b]
NSW	1794 (633)	1672 (638)						
PSID	−15205 (1154)	731 (886)	294 (1389)	1608 (1571)	1494 (1581)	[1255]	1691 (2209)	1473 (809)
CPS	−8498 (712)	972 (550)	1117 (747)	1713 (1115)	1774 (1152)	[4117]	1582 (1069)	1616 (751)

[a] Number of observations refers to the actual number of comparison and treatment units used for (3) to (5), namely all treatment units and those comparison units whose estimated propensity score is greater than the minimum, and less than the maximum, estimated propensity score for the treatment group.

[b] Weighted Least Squares: treatment observations weighted as 1, and control observations weighted by the number of times they are matched to a treatment observation.

Table 3.2 Estimated training effects for the National Support Work Demonstration male participants using comparison groups from Panel Survey of Income Dynamics and Current Population Survey.

column (3) we regress 1978 earnings on a less nonlinear function (quadratic as opposed to the step function, that is, within-stratum constant, in columns (4) and (5)) of the estimated propensity score and a treatment indicator. The estimates are comparable to those in column (2), where we regress the outcome on all preintervention characteristics, and are further from the experimental benchmark than the estimates in columns (4) to (8). This demonstrates the ability of the propensity score to summarize all preintervention variables, but underlines the importance both of using the propensity score in a sufficiently nonlinear functional form and of allowing for a heterogeneous treatment effect.

Finally, it must be noted that even though the propensity score estimates presented in columns (3) to (8) are closer to the experimental benchmark than those presented in columns (1) and (2), with the exception of the adjusted matching estimator, their standard errors are higher: in Table 3.2, column (5), the standard errors are 1,152 and 1,581 for the CPS and PSID, compared with 550 and 886 in Table 3.2, column (2). This is because the propensity score estimators use fewer observations. When stratifying on the propensity score, we discard irrelevant controls, so that the strata may contain as few as seven treated observations. The standard errors for the adjusted matching estimator (751 and 809) are similar to those in column (2). However, in this application, given the extent of the bias in the regression estimates, the bias-variance trade-off is probably not of paramount concern.

3.5 Conclusions

This chapter illustrates the use of propensity score methods on the Lalonde data. The results demonstrate that propensity score methods are able to identify the subset of units from the comparison groups that are comparable to the treatment group in terms of pretreatment covariates and to accurately estimate the treatment effect. It is important to bear in mind that the first of these is a general property of propensity score methods, whereas the second is a feature of the data we examine. By comparing units on the basis of the propensity score, one will always be able to assess the quality of a comparison group and to extract from the comparison group the subset that is most comparable to the treatment group in terms of pretreatment covariates. However, whether this is sufficient to obtain an accurate estimate of the treatment effect depends on the particular application and on whether one observes the covariates that determine assignment to treatment. As such, the conclusion from this analysis is not that propensity score methods can always estimate the treatment effect, but that these methods are (i) a useful tool for judging the quality of, and creating a well-matched, comparison group; and (ii) in applications in which the assignment to treatment is based on observable covariates, a simple and accurate means of estimating the treatment effect.

4

Medication cost sharing and drug spending in Medicare

Alyce S. Adams[1]

In 2003, Congress passed the Medicare Prescription Drug, Improvement, and Modernization Act (Public Law 108-173), which provides an outpatient prescription benefit under Medicare, effective January 1, 2006, to be delivered by private health plans as a stand-alone drug benefit. Beneficiaries would pay out of pocket for 100% of prescription drug costs up to the $250 deductible and 25% thereafter, up to the $2,500 coverage limit. For beneficiaries whose out-of-pocket costs exceed $3,600, cost sharing of 5% will be imposed. For a detailed summary of the law, see Kaiser Family Foundation (2004).

Unfortunately, little is known about the impact of various levels of cost sharing on prescription drug use in the Medicare population. Findings from previous studies (Blustein, 2000; Adams, Soumerai, and Ross-Degnan, 2001; Federman et al., 2001; Stuart and Zacker, 1999; Artz, 2002) indicate that having any drug coverage and specific types of coverage (i.e., Medicaid, employer-sponsored) are associated with higher levels of overall and essential medication use. However, these studies did not explicitly control for possible selection bias resulting from self-selection into

[1]Department of Ambulatory Care and Prevention, Harvard Medical School, Boston, Mass. I would like to thank Drs Soumerai, Ross-Degnan, and Federman for their contributions to the design and conduct of our previous studies of coverage effects using the MCBS. These earlier papers provided an ample foundation for the conduct of this study. I am also indebted to Professor Donald Rubin and Drs Jennifer Hill and Ken Kleinman for their assistance in the application of the propensity score to this context.

coverage categories (Ettner, 1996). Specifically, individuals choose to purchase or are eligible for drug coverage based on factors that are likely related to medication spending (e.g., health status, income, anticipated use of medications), irrespective of coverage status.

The purpose of this chapter is to demonstrate the application of propensity score methods to the estimation of the association between prescription drug cost sharing and medication spending in the Medicare population. To accomplish this, I first use logistic regression to estimate the propensity to have any drug coverage and specific levels of cost sharing (e.g., 20–40% of total drug expenditures; 40–60%). I then estimate the relationships between coverage, cost sharing, and total and antihypertensive drug expenditures for community dwelling beneficiaries, stratifying by quintiles of the propensity score. Rosenbaum and Rubin (1983a, 1984) have demonstrated the effectiveness of subclassification on the propensity score in reducing bias in observational studies. I then compare these results to those obtained from models in which the propensity score is not used and when it is included as a covariate to assess the impact of the method of propensity score adjustment on the estimates.

4.1   Methods

Data source

The data source for the analysis was the Medicare Current Beneficiary Survey (MCBS) Cost and Use Files for 1995. The MCBS is a national longitudinal survey of approximately 12,000 Medicare beneficiaries. Respondents were interviewed four times during the course of the year regarding their health care status, utilization, and access. The sample was stratified according to geographic region and post stratified by age, gender, region, metropolitan residence, and year of entry into the sample. Additional information on sample selection for the MCBS can be found in the article by Adler (1994).

Study sample

For the purposes of this study, I included those beneficiaries most likely to voluntarily enroll in the new Medicare prescription drug plans. They included all beneficiaries with fee for service (i.e., no drug coverage and no private insurance) or private insurance (i.e., employer sponsored, self-purchased), with or without drug coverage. Excluded were beneficiaries with VA, state drug coverage plans, local drug coverage plans, which are typically more generous than the new benefit. Dual enrollees, those with both Medicaid and Medicare, will be moved from Medicaid drug coverage to Medicare drug coverage. However, they are not comparable to the other groups included in this analysis due to high rates of disability and health services use and have therefore been excluded from the analysis.

Drug coverage was determined from self-reports, administrative records, and self-reported payments for medication expenditures as described by Davis et al.

(1999). Individuals were classified as having employer or Medigap drug coverage if they reported having drug coverage through a private health plan in the insurance record or if any of their expenditures were paid by these entities. Drug coverage through the state, the VA, and other public sources was defined using similar methods. Medicaid enrollment was determined by beneficiary self-report, Medicare administrative records, and payment source for medication expenditures.

Outcome variables

The outcomes of interest in this study were total and antihypertensive drug spending. Drug expenditures per person were estimated from the prescription drug file of the MCBS. Owing to the skewed distribution of expenditures, a natural log transformation was employed. Antihypertensive drugs were identified using the U.S. Pharmacopeia Drug Information (USP DI) (1999). Medicines classified as antihypertensive were diuretics (including loop diuretics), calcium channel blockers, angiotensen converting enzyme (ACE) inhibitors, and beta-blockers. For more on the selection of antihypertensive medications, please see Adams, Soumerai, and Ross-Degnan (2001).

Predictor variables

The primary covariates of interest were drug coverage (1/0) and two indicators of beneficiary cost sharing (i.e., = 1 if cost sharing = 20–40% vs >40% paid out of pocket; = 1 if cost sharing = 40–60% vs >60% paid out of pocket). Drug coverage was defined as described above. Level of cost sharing was defined as the proportion of medication expenditures reportedly paid out of pocket by the beneficiary. Individuals without drug coverage were coded as having cost sharing of 100%. Other control variables included age, race, gender, marital status, education, region of residence, income, health status, functional status (limitations in the following activities of daily living (ADLs): bathing/showering, dressing, eating, getting in or out of a chair, and using the toilet), the number of other chronic conditions, and the number of medical provider visits, which captures medical provider visits, separately billing physicians, separately billing labs, and other medical services. For the analysis of antihypertensive drug expenditures, we also included an indicator for whether the individual self-reported having one of five conditions that were likely to increase their exposure to antihypertensive drugs (i.e., high blood pressure, previous myocardial infarction, coronary heart disease, other heart related conditions, and diabetes). All covariates were represented by dichotomous and categorical variables.

Statistical analysis

The analysis was conducted using STATA statistical software (Version 7.0). In the first stage of the analysis, propensity scores were estimated using logistic regression for the likelihood of having any drug coverage and of having specific levels of cost sharing (20–40% and 40–60%, respectively). Covariates were included in the models if Wald tests produced p-values <0.25. Inclusion of the region variables was determined using a joint test of significance. Once the basic models were constructed, the statistical significance ($p < .25$) of interaction terms was assessed. The predicted value from each of the models was saved and labeled as a propensity score (i.e., predicted probability of each of the outcomes). I then compared the groups of interest (i.e., coverage vs no coverage; copay of 20–40% vs copay > 40%; copay of 40–60% vs copay > 60%) on their covariates controlling only for propensity score quintile. The propensity modeling process was repeated until there were no significant ($p < .10$) differences between the groups of interest.

Three models were constructed to estimate the impact of coverage on total drug expenditures for individuals with positive drug expenditures using weighted ordinary least squares. Covariates for inclusion in the drug expenditure models were identified using univariate statistics. The key covariate of interest in the first model was a dichotomous indicator of the individual's coverage status (1/0). The key covariate of interest in the second model was a dichotomous indicator of whether the individual paid between 20 and 40% (vs greater than 40%) of their drug expenditures out of pocket. The third model estimated the impact of paying between 40 to 60% out of pocket (vs greater than 60%) on total drug expenditures.

The significance of covariates included in the final models was determined at the 0.05 level for both the full and reduced model to assure the robustness of the findings to the modeling procedure. Each model was stratified by the corresponding propensity score quintile. These results were then compared to estimates obtained when the propensity score quintile was included in the final regression models as a covariate and when the propensity score was not controlled for in the analysis. For comparative purposes, I calculated the weighted average from the stratified models by weighting the estimate for each subclass by the proportion of the study population represented in that subclass. Standard estimates for the weighted average were calculated assuming zero covariance between the variances of the estimates. The entire procedure was repeated for antihypertensive drug expenditures. These and other applications of the propensity score in observational studies can be found in D'Agostino (1998).

4.2   Results

Drug spending by coverage and cost sharing

Drug spending varied considerably by both enrollee characteristics and drug coverage status (Table 4.1). Those without coverage had considerably lower spending patterns compared to those with coverage, regardless of health status. Spending also varied by level of cost sharing, with those with lower cost sharing spending more on prescription drugs. Enrollee characteristics varied considerably by coverage status and level of cost sharing. Controlling for propensity score quintile resulted in statistically significant reductions in these differences (Table 4.2).

Characteristics	No Drug Coverage (N = 2,504)	Some Drug Coverage (N = 2,433)	Cost Sharing: 20–40% (N = 674)	Cost Sharing: 40–60% (N = 372)
Age < 65	$700 (12%)	$1600 (7%)	$2000 (7%)	$1200 (8%)
65–74	$528 (32%)	$770 (42%)	$720 (40%)	$700 (43%)
75–84	$551 (38%)	$780 (38%)	$910 (39%)	$630 (37%)
85+	$527 (18%)	$760 (13%)	$870 (14%)	$560 (12%)
Male	$540 (41%)	$820 (44%)	$850 (44%)	$660 (46%)
Female	$570 (59%)	$830 (56%)	$940 (56%)	$730 (54%)
Non-white	$480 (10%)	$770 (6%)	$890 (6%)	$710 (10%)
White	$560 (90%)	$830 (94%)	$910 (94%)	$700 (90%)
Not married	$550 (52%)	$820 (40%)	$930 (40%)	$710 (38%)
Married	$560 (48%)	$830 (60%)	$890 (60%)	$690 (62%)
<12 yrs educ.	$570 (46%)	$800 (32%)	$910 (35%)	$670 (30%)
12 or more	$540 (54%)	$840 (68%)	$900 (65%)	$710 (70%)
Income < 10,000	$560 (32%)	$740 (11%)	$980 (10%)	$970 (13%)
$10 k–20 k	$550 (38%)	$850 (36%)	$980 (37%)	$610 (32%)
>$20,000	$560 (30%)	$830 (53%)	$840 (53%)	$690 (54%)
No ADL limits	$460 (65%)	$690 (70%)	$740 (67%)	$580 (67%)
1	$670 (14%)	$980 (13%)	$1100 (16%)	$690 (14%)
2	$620 (7%)	$1100 (7%)	$1100 (6%)	$970 (7%)
>2	$830 (14%)	$1400 (10%)	$1400 (12%)	$1200 (12%)
Poorer health	$770 (30%)	$1300 (24%)	$1300 (26%)	$990 (26%)
Good or better	$460 (70%)	$690 (76%)	$750 (74%)	$600 (74%)
0 Chronic cond.	$440 (12%)	$470 (11%)	$540 (9%)	$410 (11%)
1	$430 (25%)	$610 (26%)	$710 (24%)	$410 (26%)
>1	$640 (63%)	$970 (63%)	$1000 (66%)	$870 (63%)

Table 4.1 Distribution of individual characteristics and drug expenditures by drug coverage status using data from the 1995 Medicare Current Beneficiary Survey. The percentages represent the proportion of the population matching that specific characteristic.

Characteristics	Drug Coverage Before	Drug Coverage After	Cost Sharing: 20–40% Before	Cost Sharing: 20–40% After	Cost Sharing: 40–60% Before	Cost Sharing: 40–60% After
<65 yrs of age	39***	0.88	4.6*	0.06	2.9	0.23
>84 yrs of age	25***	0.04	5*	0.02	6.5**	0.06
Male gender	5.1*	0.88	1.3	0.92	3.7	0.02
White race	28***	1.4	5*	0.94	0.49	0.25
Married	73***	0.92	19*	0.08	22***	0.59
Education > 12 yrs	104***	0.98	9.9**	0.23	24***	0.003
Income	137***	1.8	15***	0.05	14***	0.58
Poorer health	26***	0.13	0.79	0.23	1.3	0.06
# Chronic conditions	0.56	3.5	5.4*	0.56	0.22	0.53
ADLs	20***	0.98	0.72	0.50	0.24	0.01
Region 1	10**	1.1	0.003	0.004	0.67	0.0004
Region 2	34***	0.01	16***	0.04	7.1**	0.35
Region 3	3.8*	0.01	2.1	0.46	0.03	0.41
Region 4	11**	1.4	14***	0.71	7.7**	0.33
Region 5	11**	0.0004	0.0001	0.40	2.5	0.37
Region 6	1.9	0.06	3.1	0.08	0.25	0.006
Region 7	15*	0.04	1.6	0.14	0.02	0.15
Region 8	0.56	0.08	0.02	0.08	0.76	0.25
Region 9	5.3*	2.2	0.11	0.07	2.1	0.21
Region 10	13**	0.42	1.8	0.04	0.27	0.12

Table 4.2 F statistics for differences in individual characteristics by drug coverage status before and after controlling for propensity score quintile. *$0.05 > p > 0.01$; **$0.01 > p > 0.001$; ***$0.001 > $ p-value.

Effect of coverage on total drug spending within propensity score subclass

Having drug coverage of any kind was positively associated with drug spending in all five subclasses. However, the magnitude of the effect varied considerably by drug class and the relationship between the effect and the propensity score quintiles was not linear. The smallest effect ($\beta = 0.17$; se $= 0.081$) was in the second subclass, while the largest effect was in the third subclass ($\beta = 0.58$; se $= 0.070$).

The level of cost sharing was also significantly associated with drug spending. Copayments between 20 to 40% were associated with dramatically higher rates of drug spending and the magnitude of the association varied by propensity score quintile. Estimates ranged from $\beta = 0.36$ (0.11) to $\beta = 0.71$ (0.13). Cost sharing between 40 to 60% was also significantly associated with higher drug spending relative to those with cost sharing greater than 60%, though the magnitude of the effect was considerably lower than that for individuals with lower cost sharing (Range: 0.13–0.55). Other factors positively associated with drug spending included age less than 65 (i.e., disability), white race, poorer health status, the number of chronic conditions, and the number of medical provider visits. Region of residence was also associated with the level of drug expenditures.

Effect of coverage on antihypertensive drug spending within propensity score subclass

The results of the antihypertensive spending models are presented in the bottom half of Table 4.3. Any drug coverage was significantly associated with higher antihypertensive drug spending in all subclasses. However, the magnitude of the effect varied by subclass and the relationship was again nonlinear. The values ranged between 0.13 in the first and second subclass and 0.73 in the fifth subclass. Copayments between 20 to 40% were also associated with higher antihypertensive drug expenditures ($\beta : 0.18$(5th subclass) $- 0.57$(2nd subclass)). However, the association between antihypertensive spending and cost sharing between 40 to 60% was not statistically significant. In fact, for three of the five subclasses, the relationship was negative. Other factors positively associated with antihypertensive drug spending included functional disability, the number of chronic conditions, and having at least one of the five conditions for which antihypertensives are generally prescribed.

Comparison of estimates with and without propensity score adjustment

Table 4.4 compares the coefficients obtained from the multivariate model with no adjustment for the likelihood of being in a given coverage group (βols), the

Propensity Score Quintiles	Any Drug Coverage (se)	20–40% Cost Sharing (se)	40–60% Cost Sharing (se)
Total drug spending			
Q1	0.33 (0.09)	0.63 (0.17)	0.55 (0.21)
Q2	0.17 (0.08)	0.71 (0.13)	0.23 (0.16)
Q3	0.58 (0.07)	0.51 (0.11)	0.33 (0.15)
Q4	0.45 (0.08)	0.36 (0.11)	0.14 (0.15)
Q5	0.035 (0.088)	0.36 (0.09)	0.13 (0.13)
Antihypertensive spending			
Q1	0.13 (0.11)	0.38 (0.21)	0.36 (0.23)
Q2	0.13 (0.10)	0.57 (0.15)	−0.12 (0.22)
Q3	0.33 (0.09)	0.30 (0.13)	0.08 (0.17)
Q4	0.26 (0.11)	0.21 (0.14)	−0.05 (0.18)
Q5	0.73 (0.10)	0.18 (0.11)	−0.27 (0.16)

Table 4.3   Estimated effect of coverage obtained from the expenditure models stratified by propensity score quintile. Other covariates included in the multivariate models included age, gender, race, income, functional status, health status, number of chronic conditions, and the number of medical provider visits.

Covariates	βols	βstrat	βcov
Total drug spending			
Any coverage	0.41 (0.04)	0.39 (0.04)	0.39 (0.04)
Copay:20–40%	0.48 (0.05)	0.51 (0.06)	0.48 (0.05)
Copay:40–60%	0.22 (0.07)	0.28 (0.07)	0.22 (0.07)
Antihypertensive drug spending			
Any coverage	0.20 (0.04)	0.19 (0.05)	0.19 (0.05)
Copay:20–40%	0.31 (0.06)	0.33 (0.07)	0.30 (0.06)
Copay:40–60%	−0.04 (0.08)	0.002 (0.09)	−0.03 (0.08)

Table 4.4   Estimated effect of coverage status and cost sharing without propensity score adjustment (βols) to those obtained after subclassification (βstrat) and covariance adjustment (βcov) using the propensity score. The coefficient from the analysis using subclassification on the propensity score is the weighted average of the coefficients from each of the models, where the weights equal the proportion of the population in each propensity score quintile.

weighted average of the estimates from the multivariate model stratified by propensity score quintile (βstrat), and the coefficients from the multivariate model that included the propensity score quintile as a covariate (βcov).

The analysis of the effect of having any drug coverage on drug expenditures were consistent across all and antihypertensive expenditures. Specifically, the stratified analysis and the covariance adjustment approaches produced the same coefficient, which was smaller than that produced by the linear regression models. However, in the analysis of the impact of cost sharing, the coefficient obtained from the unadjusted linear regression and the covariance adjusted models were more similar and consistently smaller than the estimates from the stratified model.

4.3   Study limitations

Self-reported spending and coverage status

Reliance on self-reported data in this study may have led to underestimation of drug use. I may have also overestimated out-of-pocket payments for beneficiaries who were awaiting reimbursement at the time the survey was conducted. To the extent that individuals with coverage are more likely to underestimate use (i.e., because they are not responsible for paying the bill), these results would underestimate the impact of coverage on use. Likewise, if individuals with coverage anticipated reimbursement of expenditures or considered premium costs in their spending decisions, these findings may underestimate the impact of total cost sharing on use.

It is possible that some beneficiaries with coverage were classified as having no coverage. Further, no distinctions were made between individuals with part year and full year coverage. Overestimation of actual coverage rates is likely to have led to underestimation of the impact of coverage on expenditures.

Expenditures as a measure of medication use

Expenditures represent not only use but also prescription price. Therefore, it is possible that I overestimated the impact of coverage on use owing to variations in cost by level of cost sharing. In a previous analysis, we (Adams, Soumerai, and Ross-Degnan, 2001) found that expenditures were a good indicator of actual use. However, I may have underestimated use for individuals who typically receive medication at lower cost. In this analysis, our main focus was on spending rather than overall use. Additional research is needed to estimate the impact of the new drug benefit on the price of medications to beneficiaries.

Antihypertensive drugs were chosen to represent essential drug spending in this analysis owing to high rates of nonadherence and considerable evidence of their efficacy in reducing morbidity and mortality in this population (Sanson-Fisher and Clover, 1995; Morrell, Park, Kidder, and Martin, 1997). However, the price elasticity of other essential medications may differ from that of hypertension. Therefore, evidence of a negative relationship between cost sharing and drug spending for other essential medications would lend additional support to these findings.

Selection bias and use of the propensity score

It is likely that individuals with coverage differ from those without coverage in ways that are related to their medication spending, but are unrelated to their coverage status. As a result, this study may contain some degree of selection bias. I attempted to reduce the effect of selection bias by including propensity scores in the estimation models. Stratification on the propensity score resulted in considerable improvement in the comparability of the groups of interest with respect to the observed covariates. However, to the extent that the selection bias is due to unmeasured factors, selection bias still possesses a threat to the validity of the above findings.

To estimate the impact of the method of propensity score correction, I compared the results from the stratified models to those from models employing covariance adjustment using the propensity score. While the weighted estimates from the stratified analysis were quite similar to those from the covariance adjusted models for the analysis of any drug coverage, they were consistently higher than the covariance adjusted model estimates for the subanalysis of cost sharing.

Lack of control for sampling design

Ordinary least squares were used to assess the impact of cost sharing on drug spending due to the appearance of strata with only one psu (primary sampling unit) in reduced samples. One option would have been to delete observations included in strata with only one psu, a limitation of the software program. Rather than delete observations, I chose to not control for the sampling design in the analysis. As a result, the findings from the antihypertensive drug analysis may not be generalizable to the entire Medicare population.

4.4   Conclusions and policy implications

As in previous research, this study found that having any drug coverage was significantly associated with greater use of all and essential medications. Analysis of cost sharing rather than source of drug coverage allowed for estimation of the impact of specific levels of cost sharing on drug use. While cost sharing levels of 20 to 40% and 40 to 60% were both positively associated with total drug spending, the magnitude of the effect was greatest for those with cost sharing between 20 and 40%. Further, while lower cost sharing was associated with greater use of essential medications, higher cost sharing rates were not. These results indicate that higher rates of cost sharing may not only slow the growth of total drug spending but may also reduce the use of life saving medications. These results are consistent with our previous findings showing that patients with coverage characterized by higher than average cost sharing, spending less on essential medications.

An interesting methodological finding from this study is that relative to the estimates using subclassification on the propensity score, ordinary least squares estimates consistently overestimated the impact of drug coverage on spending and underestimated the impact of the level of cost sharing on overall and antihypertensive drug spending. These results provide additional evidence that propensity scores are a useful and accessible method for reducing selection bias in studies of the impact of insurance coverage on health care spending. Further, disparities between the results obtained from subclassification versus covariance adjustment with the propensity score for the cost sharing analysis suggest that subclassification may be a more economical method for reducing selection bias. At the very least, researchers should use both approaches to confirm that their results are not dependent upon the method employed.

5

A comparison of experimental and observational data analyses

Jennifer L. Hill, Jerome P. Reiter, and Elaine L. Zanutto[1]

For obtaining inferences about causality, randomized experiments are the gold standard. Random assignment of treatments ensures, in large samples, that the background characteristics in the treatment groups are similar, so that comparisons of the groups' outcome variables estimate differences in the effects of the treatment assignments. For some causal questions, however, it is not possible to assign treatments to units at random, perhaps for ethical or practical reasons. Typically, such observational studies involve collecting and comparing units from existing databases that have nonrandom treatment assignments. Unlike in randomized experiments, there is no assurance that background characteristics are similar across treatment groups, and simple comparisons of the outcome variables can be confounded by such differences.

Researchers use a variety of methods to deal with confounding in observational studies. One approach is to fit linear regressions that include causally-relevant background characteristics as covariates. Typically, such models include indicator

[1] School of International and Public Affairs, Columbia University, New York, Institute of Statistics and Decision Sciences Duke University, Durham, N.C., and Department of Statistics, University of Pennsylvania, Philadelphia, Pa.

Applied Bayesian Modeling and Causal Inference from Incomplete-Data Perspectives.
Edited by A. Gelman and X-L. Meng   © 2004 John Wiley & Sons, Ltd   ISBN: 0-470-09043-X

variables for the treatments. Another approach, developed by Rosenbaum and Rubin (1983b, 1984) specifically to deal with the problem of confounding in observational studies, is to use propensity scores. Here, the goal is to create two groups of units closely balanced on causally-relevant background characteristics. Importantly, both approaches can mitigate only confounding from observed background variables; the groups may still differ on variables not controlled for in the models.

In this chapter, we illustrate the potential efficacy of these types of analyses. The causal question we address concerns the effects on intelligence test scores of a particular intervention that provided very high quality childcare for children with low birth weights. We have data from the randomized experiment performed to evaluate the causal effect of this intervention, as well as observational data from the National Longitudinal Survey of Youth on children not exposed to the intervention. Using these two datasets, we compare several estimates of the treatment effect from the observational data to the estimate of the treatment effect from the experiment, which we treat as the gold standard. This general strategy of evaluating the efficacy of competing nonexperimental techniques by creating a "constructed" observational study using a randomized experiment was first used by Lalonde (1986). Other studies using the same or similar strategies include Lalonde and Maynard (1987), Fraker and Maynard (1987), Friedlander and Robins (1995), Heckman, Ichimura, and Todd (1997), and Dehejia and Wahba (1999). We also demonstrate the use of propensity scores with data that has been multiply imputed to handle pretreatment and posttreatment missingness. To our knowledge, these other constructed observational studies performed analyses using only units with fully observed data.

In the end, for these data, we find that the propensity score approaches yield estimated treatment effects consistent with the effects in the experiment, whereas the regression approach does not. The analyses also illustrate the importance of matching on geographic characteristics, something that can be easily overlooked when using propensity score approaches.

5.1   Experimental sample

Low birth weight infants have elevated risks of cognitive impairment and academic failures later in life (Klebanov, Brooks-Gunn, and McCormick, 1994a, 1994b). One approach to reduce these risks is to provide extraordinary support for the families of low birth weight infants, for example, intensive early childhood education for the infants and access to trained specialists for the parents.

To assess the effectiveness of such interventions, in 1985 researchers designed the Infant Health Development Program (IHDP). The IHDP involved randomizing 985 low birth weight infants to one of two groups: (1) a treated group assigned to receive weekly visits from specialists and to attend daily childcare at child development centers and (2) a control group that did not have access to the weekly visits or child development centers. There were 377 infants assigned to the treated group and 608 assigned to the control group. The IHDP provided transportation to the childcare centers to reduce the risk of noncompliance. More details on the design of the experiment can be found in IHDP (1990), Brooks-Gunn, Liaw, and Klebanov (1992), and Hill, Brooks-Gunn, and Waldfogel (2003).

The outcome variable is the infant's score on the Peabody Picture Vocabulary Test Revised (PPVT-R) administered at age three or four. Other outcome variables were analyzed in the experiment, but this is the only outcome measured at the same time point in the IHDP and NLSY. The PPVT-R scores are available for all but 173 infants (17.6%).

There are many background variables associated with PPVT-R scores. We limit the variables in our analyses to those measured in both datasets, but a rich set of variables remains. These include characteristics of the infant's mother measured at the time of birth of her child: age, marital status, race (Hispanic, black or other), educational attainment (less than high school, high school, some college, completed college), whether she worked during her pregnancy, and whether she received prenatal care. They also include characteristics of the child: sex, whether the child was first born, the birth weight, age of the child in 1990, the number of weeks the child was born preterm, and the number of days the child had to stay in the hospital after birth. In addition to these sociodemographic variables, we have geographic data: county level unemployment rates and state indicators. In the experimental data, these variables are all fully observed, except for whether or not the mother worked during pregnancy, which is missing for 50 infants (5.1%).

All missing data are handled using multiple imputation (Rubin, 1987b). Imputation methods are described in detail in Section 5.2.

As expected, randomization balances the distributions of the background variables in the treated and control groups. This is evident in the first panel of Table 5.1, which displays the covariates' means and standard deviations across the five imputed datasets for both the treated and control groups. The second panel of Table 5.1 displays similar summaries for the observational study, which is discussed in Section 5.2.

The experimental estimate of the intention-to-treat effect for the intervention relative to the control is 6.39 with a standard error of 1.17. This suggests that the combination of intensive childcare and home visits had a significant positive average effect on children's test scores.

5.2 Constructed observational study

We now construct an observational study to assess the same question that the experiment addressed: what is the impact of the IHDP treatment? We use the treated infants from the IHDP as the treatment group, and a sample of infants from the National Longitudinal Survey of Youth (NLSY) as the comparison group. This "constructed" observational study reflects the type of data researchers might have access to in the absence of a randomized experiment.

The NLSY is a panel survey that began in 1979 with a sample of approximately 12,000 teenagers who, appropriately weighted, were nationally representative at

| | Experimental Sample | | Observational Sample | |
| | Control | Treated | Full NLSY | Treated |
	Mean (SD)	Mean (SD)	Mean (SD)	Mean (SD)
Mother				
Age (yrs.)	24.74 (6.11)	24.39 (5.93)	23.76 (3.15)	24.59 (5.93)
Hispanic	0.12 (0.33)	0.10 (0.30)	0.21 (0.41)	0.10 (0.30)
Black	0.54 (0.50)	0.55 (0.50)	0.29 (0.45)	0.53 (0.50)
White	0.34 (0.47)	0.34 (0.48)	0.50 (0.50)	0.37 (0.48)
Married	0.46 (0.50)	0.41 (0.49)	0.69 (0.46)	0.42 (0.49)
No HS degree	0.40 (0.49)	0.45 (0.50)	0.30 (0.46)	0.43 (0.50)
HS degree	0.27 (0.44)	0.28 (0.45)	0.42 (0.49)	0.28 (0.45)
Some college	0.21 (0.41)	0.16 (0.37)	0.19 (0.39)	0.17 (0.37)
College degree	0.12 (0.33)	0.11 (0.31)	0.08 (0.27)	0.13 (0.33)
Working	0.57 (0.50)	0.57 (0.50)	0.62 (0.49)	0.59 (0.49)
Prenatal care	0.95 (0.21)	0.95 (0.22)	0.99 (0.11)	0.95 (0.22)
Child				
Birth weight	1769 (473)	1819 (436)	3314 (604)	1819 (439)
Days in hospital	26.6 (24.7)	23.4 (22.3)	4.47 (7.63)	23.7 (22.6)
Age 1990 (mos.)	56.8 (2.13)	56.8 (2.04)	56.3 (29.1)	56.8 (2.03)
Weeks preterm	7.04 (2.77)	6.91 (2.52)	1.24 (2.18)	6.96 (2.52)
Sex (1=female)	0.51 (0.50)	0.50 (0.50)	0.50 (0.50)	0.50 (0.50)
First born	0.43 (0.50)	0.47 (0.50)	0.42 (0.49)	0.47 (0.50)
Geography				
Unemployment	0.08 (0.05)	0.08 (0.06)	0.09 (0.04)	0.08 (0.05)
Lives in state 1	0.14 (0.35)	0.13 (0.34)	0.01 (0.11)	0.13 (0.33)
Lives in state 2	0.11 (0.32)	0.12 (0.33)	0.02 (0.14)	0.12 (0.33)
Lives in state 3	0.10 (0.30)	0.12 (0.33)	0.05 (0.21)	0.12 (0.32)
Lives in state 4	0.14 (0.35)	0.12 (0.32)	0.02 (0.12)	0.12 (0.32)
Lives in state 5	0.16 (0.37)	0.13 (0.34)	0.06 (0.23)	0.12 (0.33)
Lives in state 6	0.09 (0.28)	0.12 (0.32)	0.04 (0.19)	0.13 (0.33)
Lives in state 7	0.16 (0.37)	0.14 (0.35)	0.09 (0.28)	0.13 (0.34)
Lives in state 8	0.10 (0.30)	0.12 (0.30)	0.01 (0.12)	0.14 (0.34)

Table 5.1 Means and standard deviations for both the experimental and observational studies. Dichotomous variables equal one for "yes" answers and equal zero for "no" answers. Differences in the experimental and observational samples for the IHDP treateds reflect differences due to independent imputation of missing data.

that time. These participants were interviewed every year thereafter until 1994 and biannually after that. Children of women in the NLSY have also been followed since 1986. Given that the IHDP began in 1985, we restrict our NLSY sample to the 4,511 children born from 1981 to 1989. The IHDP treatment was very intensive and extraordinary, so that the NLSY controls are unlikely to have received similar treatments.

As in the experimental data, the observational data contain missing values. The outcome variable, PPVT-R scores, is missing for 870 infants (19.3%). Twelve of the covariates have missing data, ranging from a minimum of 4 infants (.1%) for mother's education to a maximum of 212 infants (4.7%) for child's birth weight. Most covariates have missing data rates around 4%. Missing data were handled using multiple imputation; see below.

Panel 2 of Table 5.1 displays the means and standard deviations of the potentially confounding covariates for the treatment group and full NLSY comparison group (we reserve the term "control" for experimental control group). The treated children and the NLSY comparison group look quite different on a number of the covariates measured.

Analyses

There are several ways to control for differences in the groups' sociodemographic background variables. One approach is to fit a multiple regression of PPVT-R scores on the background variables, including an indicator variable for treatment. With this "Regression" approach, when the model describes relationships in the data well, the resulting estimated coefficient of the treatment indicator is a reasonable estimate of the average causal effect of the treatment. However, the estimate can be badly biased when the model fits the data poorly. When the data in the treated and comparison groups have different characteristics, the fitted regression involves extrapolations over much of the multidimensional covariate space (Rubin, 1997). Such violations of model assumptions can be difficult to detect.

A second approach is to match units on the basis of estimated propensity scores to attempt to construct groups balanced on the confounding covariates. Treatment effects can be estimated by differencing the sample averages of the treated and matched comparison groups; we call this the "P-score Direct" approach. Or, they can be estimated by using the treated and matched groups in a multiple regression of the outcome on the confounding covariates and an indicator for treatment; we call this the "P-score Regression" approach. Alternative propensity score approaches, not considered here, include subclassification on propensity scores (Rosenbaum and Rubin, 1984; D'Agostino, 1998; Dehejia and Wahba, 1999) and propensity score weighted estimation (Rosenbaum, 1987a, 1987b; Schneider, Cleary, Zaslavsky, and Epstein, 2001; Hirano, Imbens, and Ridder, 2003).

The P-score Direct approach avoids the specification of regression models for the relationship between the outcome and the covariates. Although models must be fit to estimate propensity scores, estimates of treatment effects are generally less sensitive to misspecification of the propensity score model than the Regression approach is to misspecification of the regression model (Drake, 1993; Rubin, 1997). With close matching on estimated propensity scores, the groups should be balanced on the observed background characteristics. Part of the model-fitting process is checking this balance so that the researcher can discern whether the groups are too different for resulting treatment effect estimates to be trustworthy. Assuming close balance, direct comparisons of the average for the treated group and the average for the matched comparison group should be mostly free of confounding due to the matched variables (Rosenbaum and Rubin, 1983, 1984).

The P-score Regression approach in a sense combines the other two approaches. It is less likely to be subject to extrapolations than the Regression approach, because the treated and matched comparison units are in similar regions of the covariate space. But, it adjusts for slight imbalances in the groups' background characteristics with a regression model, thereby potentially reducing bias and increasing precision (Rubin, 1973a, 1973b, 1979; Rubin and Thomas, 2000).

The Regression approach and the matched-sample approaches estimate different quantities. The Regression approach estimates the average treatment effect across the full sample, whereas the matched-sample approaches estimate the effect of the treatment on the treated (IHDP) group. These estimands can differ when the treatment effect is a nonconstant function of the covariates, in which case the estimated treatment effects can differ even if each method produces unbiased estimates. In this study, we seek to estimate the effect of the treatment on the IHDP-treated group.

Importantly, both the Regression and propensity score approaches work well only when we have controlled for all confounding covariates. When there are important confounding variables that have not been controlled for, either method can lead to biased estimates of treatment effects.

Missing data

Many social science researchers handle missing outcome data by restricting analyses to complete cases, sometimes in conjunction with other fixes such as dummy variables for missing data. This strategy leaves analyses open to bias because there may be systematic differences between the treated and control units with observed outcomes (Little and Rubin, 2002). This is even true in randomized experiments, unless the outcome data are missing completely at random (Frangakis and Rubin, 1999). For the constructed observational study, we therefore do not use experimental complete case estimates as benchmarks when comparing the regression and propensity score matching strategies.

Instead, we handle missing values using multiple imputation (Rubin, 1987). This retains the full sample for calculating intention-to-treat estimates and, under appropriate assumptions, should yield unbiased estimates of the intention-to-treat effect with the experimental data. The complete case estimate of the experimental estimate is 5.7, roughly half a standard error smaller than the multiple imputation estimate of 6.4. In the observational study, using only complete cases forces us to exclude large numbers of children when implementing the strategies (more than 3,000 children for the most comprehensive strategy). As a result, when using only complete cases, all the strategies perform poorly and without distinction.

For the experimental data, we assume the missing PPVT-R scores and mother's work status are missing at random (Rubin, 1976a). We believe that the number and breadth of the covariates measured makes this assumption plausible. We then generate multiple imputations from chained regression models (van Buuren, Boshuizen, and Knook, 1999; Raghunathan, Lepkowski, Van Hoewyk, and Solenberger, 2001). The models are fit with the MICE software (www.multiple-imputation.com) for S-Plus. The imputation model for PPVT-R scores is a linear regression, fit using main effects for all covariates and the treatment indicator, as well as interactions between the treatment variable and all covariates. The imputation model for mother's working status is a logistic regression, fit using all covariates, the treatment indicator, and the outcome variable as predictors. For both models, we include all the main effects and interactions to reduce the risk of generating imputations that are not consistent with the relationships in the data.[2] Five imputations are independently generated for each missing value.

For the observational data, we assume data are missing at random and use MICE to generate five imputations for each missing value, using chained linear, logistic, and polytomous logistic regression models as appropriate. For PPVT-R scores, the linear regression is fit using all covariates and the treatment indicator, as well as all interactions between covariates and the treatment indicator. For all other variables, predictors for the imputation models include all covariates, the treatment indicator, and the outcome variable. The missing at random assumption is more tenuous in the observational sample than in the experimental sample, because of the increase in the number of variables with missing data.

Propensity score analyses are performed in a two-step process. First, within each of the five completed datasets, we estimate propensity scores, find a matched control group, and calculate treatment effect estimates and their standard errors. Second, we combine these five estimates and their standard errors using Rubin's (1987) combining rules for multiple imputation. Other examples of propensity score analysis of multiply imputed data can be found in Hill, Waldfogel, and Brooks-Gunn (2002) and Hill, Brooks-Gunn, and Waldfogel (2003), and its underlying assumptions and potential efficacy are discussed in Hill (2004). Analyses for the Regression strategy are performed in the standard way using Rubin's combining rules.

Results of analyses

We consider several model specifications for the regression and propensity scores, controlling for different background variables. All regression models are of the

---

[2] Imputing missing data for the purpose of causal analyses is a bit more complicated than standard imputation but the discussion is too detailed for the confines of this chapter and will be reserved for future work.

form $Y \sim N(X\beta, \sigma^2)$, where X contains covariates. All propensity scores are estimated using the fitted values from the logistic regression of treatment on the same X included in the regressions. Matches for each treated child are determined by finding the NLSY child with the closest propensity score to that child. We use matching with replacement because evidence suggests that it can lead to smaller bias than matching without replacement (Dehejia and Wahba, 2002).

The first set of models, labeled DE, controls only for the sociodemographic variables. The second set of models, labeled DE + U, controls for the sociodemographic variables and the unemployment rate of the county the infant resides. Adding unemployment rate should help control for the economic conditions in which the child was raised. The third set of models, labeled DE + U + S, controls for the sociodemographic variables, the county unemployment rate, and the state the infant was born in. The state variable should help control for differences in the availability and quality of healthcare, childcare, and other services, as well as for differences in lifestyles across states. Ideally, we would control for county-level effects; however, there are not sufficient numbers of children in our study to do so. The fourth set of models, labeled DE + U + X, controls for the same variables as in DE + U + S but, additionally, performs exact matching on state. That is, each treated child is required to be matched with an NLSY child from the same state.

Many of the 4,511 children in the full NLSY sample reside in states other than the eight states from the IHDP. We exclude the children from these "other" states when fitting the logistic and linear regressions for the DE + U + S and DE + U + X analyses. This reduces the pool of potential matches to about 1,500 children, which could make close matching more difficult. Including children from non-IHDP states, however, forces a linear dependency with the group of treatment children in the logistic regressions if we try to include indicator variables for all states but one, making these models inestimable. An alternative to excluding the children from non-IHDP states is to combine data from two arbitrarily selected states into one category. In this case, the estimated propensity scores and the resulting treatment effect estimates depend critically on which states are selected for this combination, which is undesirable. We do not exclude children from the non-IHDP states for the corresponding Regression analyses because it seems unlikely that a researcher unaccustomed to matching would think of doing this. Excluding the children from non-IHDP states in the Regression analyses changes the estimates by roughly one-quarter of the standard error.

Table 5.2 displays summary statistics reflecting the balance in the covariates for the different logistic regression models. The entries are standardized differences between the treated and comparison group means, defined in the caption. Large absolute values indicate that the means are far apart, whereas absolute values near zero suggest close balance. This metric was used by Rosenbaum and Rubin (1984, 1985a) to display covariate balance.

When comparing the treated group to the full NLSY sample, without any matching, we see that the groups' means differ greatly, especially for birth weight and weeks preterm. Matching on sociodemographic variables through propensity scores

Variable	Full NLSY	DE	DE + U	DE + U + S	DE + U + X
Mother					
Age (yrs.)	0.17	0.19	0.23	0.14	0.25
Hispanic	−0.32	−0.07	−0.10	−0.39	−0.34
Black	0.52	0.13	0.04	0.31	0.40
White	−0.27	−0.08	0.04	0.01	−0.11
Married	−0.55	−0.19	−0.07	−0.23	−0.02
No HS degree	0.27	0.08	0.07	0.28	−0.19
HS degree	−0.32	−0.19	−0.20	−0.15	−0.06
Some college	−0.07	−0.03	0.00	−0.43	0.02
College degree	0.15	0.21	0.21	0.35	0.36
Working	−0.06	−0.04	0.01	0.27	0.10
Prenatal care	−0.22	−0.13	−0.17	−0.27	−0.25
Child					
Birth weight	−2.83	0.18	0.17	0.42	0.17
Days in hospital	1.14	0.01	−0.01	−0.69	−0.44
Age 1990 (mos.)	0.03	0.14	0.06	−0.06	−0.09
Weeks preterm	2.43	−0.09	−0.06	−0.90	−0.23
First born	0.10	0.15	0.07	0.03	0.20
Sex (1 = female)	0.00	−0.02	0.01	0.08	0.01
Geography					
Unemployment	−0.06	−0.08	−0.06	0.06	−0.08
Lives in state 1	0.47	0.50	0.48	0.33	0.00
Lives in state 2	0.40	0.42	0.39	0.06	0.00
Lives in state 3	0.26	0.16	0.21	−0.43	0.00
Lives in state 4	0.42	0.46	0.46	0.19	0.00
Lives in state 5	0.24	0.32	0.27	−0.24	0.00
Lives in state 6	0.34	0.30	0.31	0.12	0.00
Lives in state 7	0.14	0.23	0.22	−0.08	0.00
Lives in state 8	0.47	0.44	0.44	0.24	0.00
Method controls for:					
Demographics		X	X	X	X
Unemployment			X	X	X
States				X	X
Exact state match					X

Table 5.2 Summaries of covariate balance in treated and control groups. The entries equal $(\bar{x}_t - \bar{x}_c)/\sqrt{(s_t^2 + s_{0c}^2)/2}$, where $\bar{x}_t$ and $\bar{x}_c$ are the sample means of the treated and comparison groups' covariates, and s_t^2 and s_{0c}^2 are the sample variances of the 377 treated and 4,511 nontreated children's covariates. The common denominator facilitates comparisons of the balance in unmatched and matched comparison groups. DE + U + S includes state in the propensity score model, whereas DE + U + X forces state to be exactly balanced.

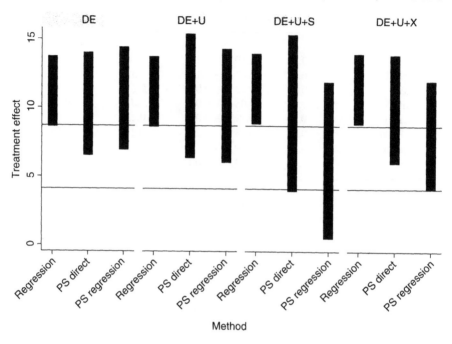

Figure 5.1   The bars represent approximate 95% confidence intervals for the average treatment effect using the various methods. The horizontal lines at 8.68 and 4.10 are the upper and lower limits, respectively, of the 95% confidence interval from the randomized experiment.

improves balance considerably, reducing most standardized differences. Matching additionally on unemployment rate does not substantially change balance. Exact matching on state results arguably in the best balance across the spectrum of variables. Exact matching on state gives better balance than simply including state indicators in the propensity score model, which is done by including indicator variables for state in the logistic regression used to estimate the propensity scores.

We now turn to the analysis of PPVT-R scores for each of these models. The point estimates and standard errors of the treatment effects are summarized in Figure 5.1 for each analysis. We calculate standard errors for P-score Direct estimates using $\sqrt{\mathrm{Var}(y_t)/n_t + \sum_i (w_i/n_c)^2 \mathrm{Var}(y_c)}$, where $\mathrm{Var}(y_t)$ is the variance of the treated units, $\mathrm{Var}(y_c)$ is the variance of the distinct matched control units, and w_i is the number of times matched control unit i is used. We calculate point estimates and standard errors for P-score Regression using weighted least squares, with weights equal to w_i. These variance estimates are somewhat *ad hoc*; however, there are no commonly accepted and statistically validated estimators of treatment effect variances when matching on propensity scores with replacement. This is a subject for future research. The corresponding approximate 95% confidence intervals are displayed in Figure 5.1.

We treat the result from the IHDP experiment as the target for comparison, since the estimated treatment effect is unbiased with relatively small standard error and the resulting confidence intervals are inferentially valid. The Regression approach, which always uses the full NLSY sample as the comparison group, consistently results in biased estimates of the treatment effect and little overlap with the confidence intervals from the randomized experiment. As we see in Panel 2 of Table 5.1, the treated group and full NLSY sample infants have very different covariate distributions, so that linear models fit using the full NLSY sample are especially prone to model misspecification caused by extrapolations. In contrast, once all sociodemographic and geographic variables are included in the matching, the P-score Direct and P-score Regression approaches result in estimates and intervals that more closely track those from the randomized experiment. The P-score Regression is better in this case than the P-score Direct, most likely because the regression model in the P-score Regression controls for residual imbalances in the covariates due to incomplete matching.

These analyses suggest that P-score Regression is the most effective for this study. However, generalizing this conclusion to say that propensity score matching is always the best approach, or always outperforms regression, would not be appropriate. We obtain reasonable estimates only after including the state variables in the propensity score models. If we had used the analyses based only on the sociodemographic variables, for example if the geographic variables were unavailable due to confidentiality restrictions, it would not have been easy for us to detect that those inferences are so strongly biased, since the socioeconomic variables are well balanced for the DE and DE + U propensity score analyses.

The analyses are sensitive to the specification of the model for the propensity scores, as illustrated by the similarities of the results in Figure 5.1 until state is included in the models. Additionally, when we restrict the sample to the infants at the higher end of the range of birth weights, who presumably are easier to find matches for than infants at the lower end of the range, we do not find an identical ordering in terms of which method comes closest to the experimental estimate for this subgroup of 7.4. For DE + U + S, the P-score Direct estimate is 11.2, and the P-score Regression estimate is 6.0. For DE + U + X, the P-score Direct estimate of 8.8 is slightly more reliable than the P-score Regression estimate of 5.0. This contrasts with the results in Figure 5.1, where the P-score Regression estimate did better across the board.

Since the propensity score analyses appear to outperform the unmatched Regression analyses for these data, one might wonder to what extent bias is reduced by limiting the sample to only those control observations most similar to the treated observations, and to what extent bias is reduced by the "reweighting" of the control sample that occurs when matching units. To explore this issue, we perform a Regression analysis on a sample that removes all control children who are from non-IHDP states or whose propensity scores are below the lowest propensity score among the treated units. The predictors include the demographic variables, unemployment rate, and the state indicators. The resulting estimate is 8.85 with a standard error of 1.89. This is closer to the experimental estimate than the regressions using the full sample, but not as close as the matched regression-adjusted results. Nonetheless, in these data, it appears that a large share of the bias reduction comes from reducing the sample space to observations that are similar to each other.

Finally, we illustrate the effect of including the children from "other' " states in the DE + U + S and DE + U + X models. If we arbitrarily combine the "other" states with state 8 (Washington) as the baseline for the dummy variables, the estimated treatment effects for the P-score Regression models are 10.1 for the DE + U + S and 5.9 for the DE + U + X. If we instead combine "other" states with the second to last state (Texas), the P-score Regression treatment effect estimates are 7.4 for the DE + U + S and 8.0 for the DE + U + X. The exact match effects change because the propensity score estimates change, even though afterwards we force exact matches on state. This artificial dependence on the specification of the dummy variables led us to exclude the children in "other" states for the DE + U + S and DE + U + X models.

5.3   Concluding remarks

By comparing the results of an experiment and observational study, we have shown the potential advantage of propensity score approaches over regressions fit using the full comparison sample. Our study also revealed an important finding: it is useful to control for geographic variables. Doing so resulted in estimates from the observational study that more closely matched those from the experiment. This reinforces the importance of controlling for as many variables as possible in a propensity score analysis (Rosenbaum and Rubin, 1983). The sensitivity of these estimates to model specification—all of which led to reasonable balance on the included covariates—suggests that a range of treatment effect estimates should be presented when performing propensity score analyses.

6

Fixing broken experiments using the propensity score

Bruce Sacerdote[1]

6.1 Introduction

"Let's not think of this as a problem with the data. This is an opportunity."
—Guido Imbens to disillusioned junior coauthor, 1996.

Suppose that we are interested in estimating a causal effect by comparing the outcomes for a treatment group and a control group. Rosenbaum and Rubin (1983b, 1985a) show that conditioning on the propensity score removes any bias in estimated treatment effects that may arise from observable pretreatment differences between the two groups.

Though much attention has been focused on the use of propensity score methods in nonexperimental settings, the method also works well in small experiments and imperfect natural experiments if the process of selection into treatment group is nonignorable, but known, and the researcher has the data needed to estimate the likelihood that an observation would be assigned to the treatment group (i.e., the propensity score). In our lottery data example, the treatment and control groups are not well matched, but the assignment process is known and the relevant covariates are present in the data set. This is precisely the case discussed in Rubin (1977).[2]

---

[1] Department of Economics, Dartmouth College, Hanover, N.H., and NBER.

[2] Estimation of the propensity score can be less straightforward if the treatment and control groups are drawn from purely observational data (e.g., the National Longitudinal Survey of Youth (NLSY) or the Panel Survey of Income Dynamics (PSID)) and selection into the treatment group is via an unknown and potentially complex process. See Dehejia and Wahba (1999) and Heckman, Ichimura, and Todd (1997) for some examples.

Applied Bayesian Modeling and Causal Inference from Incomplete-Data Perspectives.
Edited by A. Gelman and X.-L. Meng   © 2004 John Wiley & Sons, Ltd   ISBN: 0-470-09043-X

In the example below, we take the mismatched treatment and control groups and address the selection problem in two ways. First, we reweight the data using the estimated propensity score as in Hirano, Imbens, and Ridder (2000), Hahn (1998), and Dehejia and Wahba (1999). This balances the pretreatment observables in the data while still using all the available observations. We also show a second set of results in which we stratify (block) on the estimated propensity score. This amounts to dropping observations in the control group, which had a very low probability of being in the treatment group and dropping observations in the treatment group, which were unlikely to be control observations.

Estimation of the propensity score in this case is straightforward and delivers plausible estimated causal effects from winning the lottery on people's labor income, savings, housing consumption, probability of divorce and their expenditure on their children's education.

6.2   The lottery data

In 1996, Imbens, Rubin, and Sacerdote ran a survey of winners and players of Massachusetts Megabucks, which was at that time the most popular of the jackpot lottery games.[3] We limited the sample to people who had played Megabucks during 1984 to 1988 because we wanted to examine long run treatment effects, that is, 7 to 11 years after winning.

The survey asked subjects a variety of questions on their and their spouse's labor supply, educational attainment, savings, and consumption. We also asked subjects about their happiness, whether they were divorced, expenditures on children's education, and their children's educational attainment. Furthermore, we obtained Social Security Earnings Records for each subject by requesting that she sign a release form, which we then forwarded to the Social Security Administration.

One major issue with the study design was finding an appropriate control group to match the treatment group (the winners). The Massachusetts State Lottery does not automatically collect names and addresses of people who played Megabucks but did not win. However, we dealt with this problem in two ways. First, we surveyed people who had won small prizes of $100 to 5,000 and therefore had provided their names to the lottery. We also surveyed people whom we knew had played Megabucks in a given year because they had purchased a "season ticket", which is good every week that the lottery game is held.

Predictably, this led to a moderate amount of mismatch between the treatment and control groups and hence some probable biases in the uncorrected estimates of treatment effects. Our control group is older and has higher pretreatment education and income than the treatment group. This stems from the fact that younger and lower income players buy more tickets. The season ticket holders in the control group are more likely to be older, have higher incomes, and to buy fewer tickets per game.

---

[3]For details on the survey, see Imbens, Rubin, and Sacerdote (2001).

In Imbens, Rubin, and Sacerdote (2001), we addressed this problem in part by using a differences-in-differences estimator to remove pretreatment differences between the two groups. The second approach we used was to limit the sample to just the winners (the treatment group) and use the variation in the size of the prize won to identify the effect of unearned income on labor supply and other outcomes.

Here we address the selection problem using estimated propensity scores to adjust for the likelihood of an observation being in the treatment group. We report two separate treatment and control group comparisons. First, we compare the winners to the nonwinners. Second, following the earlier paper, we split the sample of winners into large and small winners. Even within the subset of winners, there are observable differences between the big winners and the small winners. In particular, men are relatively more likely to play (and buy more tickets) in response to a large advertised jackpot. Thus, men actually appear to win larger prizes in the lottery, conditional on winning. But this appearance of the lottery's bias against women is just an artifact of selection into the different weeks during which Megabucks is run.

We estimate separate propensity scores for the two separate experiments. One propensity score is for the likelihood that a player is a winner versus a nonwinner and the second is for big winners versus small winners. We define a big winner as someone who wins a nominal prize greater than $650,000, which is the sample median.[4]

6.3   Estimating the propensity scores

We argue above that selection into the treatment group is a function of the number of tickets bought, which itself depends on age, income, and gender. In the data set, we have each person's estimate of his average number of tickets bought per week around the time of winning. In theory, the propensity score could be estimated using just the number of tickets bought. But as Rubin and Thomas (1996) show, there is an efficiency gain from using all of the pretreatment variables in estimating the score. In column (1) of Table 6.1, we show the results from a probit regression of the indicator for treatment status (winning = 1) on a series of pretreatment covariates including age, gender, pretreatment income, and education, and the average number of tickets bought. Marginal coefficients are shown.

Not surprisingly, the self reported, average number of tickets purchased is strongly related to the probability of winning a jackpot prize (i.e., being in the treatment group). Pretreatment earnings tend to be negatively related to the propensity to be in the treatment group, with the coefficient on earnings in 1983 being large, negative, and statistically significant. Each additional $1,000 of earnings is associated with a 1.2% decrease in the likelihood of winning a jackpot prize.

Men are 6.4% less likely to be in the treatment group. However, in column (2), we see that conditional on winning a jackpot prize, men are 17.8% more likely to

---

[4]This nominal prize is paid out in equal installments over twenty years, meaning that the net present value is less than the stated amount. Many modern lottery games pay out the full amount immediately, or allow winners to choose between an annuity and a lump sum.

Variable	(1) Treated Group (Lottery Winner)	(2) Large Prize (Cond. on Winning)
Age	−0.01	
	(0.002)	
Male	−0.06	0.18
	(0.06)	(0.06)
Own years of high school	−0.04	
	(0.03)	
Own years of college	−0.11	
	(0.02)	
Number of tickets bought	0.04	
	(0.01)	
Working at time won	0.08	
	(0.07)	
Indicator for low income	−0.20	
	(0.11)	
Indicator for high income	−0.01	
	(0.13)	
Earnings 1978 × 1000	−0.004	
	(0.01)	
Earnings 1979 × 1000	−0.01	
	(0.01)	
Earnings 1980 × 1000	0.02	
	(0.01)	
Earnings 1981 × 1000	−0.01	
	(0.01)	
Earnings 1982 × 1000	0.006	
	(0.01)	
Earnings 1983 × 1000	−0.01	
	(0.006)	
Observations	593	291
Pseudo r-squared	0.34	0.02

Table 6.1 Estimation of the propensity scores. This table shows probit regressions used to estimate the propensity scores. The dependent variable in column (1) is an indicator for being in the original treatment versus control group. The dependent variable in column (2) is an indicator for winning a large versus a small jackpot, conditional on winning Megabucks. The large jackpots range from a nominal prize of $651,000–10.3 million, versus a nominal prize of $20,000–650,000 for the smaller jackpots. $\frac{\partial \hat{p}(x)}{\partial x}$ is shown rather than probit coefficients. Regression (1) also includes a set of dummies for the number of tickets bought. Standard errors are shown in parentheses.

win a prize greater than $650,000. Column (2) shows the probit used to estimate the propensity to win a large prize among all those who won a jackpot prize. We use the estimated propensity scores from this second regression in calculating the treatment effects from winning a large versus a small prize.

6.4   Results

Table 6.2 demonstrates the use of the propensity score to balance the pretreatment covariates across the treatment and control groups. In the raw data (columns 1–3), there are 302 control observations and 291 treatment observations. The two groups are statistically different on all of the pretreatment measures. For example, the control group is on average 67% male and 61 years old in 1996. In contrast, the treatment group is 55% male and has an average age of 55 years. The treatment group has earnings that are roughly $2,000 per year lower than the earnings of the control group, and the treatment group has on average fewer years of education.

Columns (4) to (6) show that most of these imbalances are removed by blocking on the estimated propensity score. Here we simply limit the observations in both

Variable	Raw Data			Blocking on the Score			Weighting by Score
	Mean Control	Mean Treat.	p-value for Diff.	Mean Control	Mean Treat.	p-value for Diff.	p-value for Diff.
Prize value (mil.)	0.0	1.1	0.00	0.0	1.0	0.00	0.00
Age of winner	61.2	54.7	0.00	57.0	55.9	0.47	0.10
Male	0.7	0.6	0.01	0.6	0.5	0.36	0.88
4 yrs hs?	0.9	0.8	0.01	0.9	0.9	0.61	0.19
4 yrs college?	0.5	0.2	0.00	0.3	0.3	0.11	0.12
Yrs of hs	3.8	3.6	0.00	3.8	3.7	0.81	0.13
Spouse's yrs of hs	3.9	3.7	0.01	3.8	3.7	0.66	0.19
Yrs of college	2.3	1.1	0.00	1.8	1.5	0.26	0.07
Spouse's yrs of coll.	2.0	1.4	0.00	1.8	1.8	0.89	0.23
#tix/wk, pretreat.	2.5	5.9	0.00	2.6	3.4	0.03	0.73
Work at time won?	0.7	0.8	0.03	0.8	0.8	0.22	0.40
Earnings (thous.)							
1978	8.3	6.6	0.01	7.5	6.9	0.49	0.35
1979	9.6	7.6	0.00	8.6	8.1	0.64	0.23
1980	10.7	8.6	0.01	9.8	9.6	0.81	0.20
1981	11.9	9.5	0.01	10.7	10.4	0.79	0.15
1982	12.8	10.6	0.03	11.9	11.9	0.99	0.10
1983	13.8	11.3	0.02	13.0	12.6	0.80	0.06
N	302	291		149	147		

Table 6.2   Balance in pretreatment covariates when blocking on and weighting by the score. This table shows pretreatment variables for two groups: (1) people who won prizes with nominal amounts of $22,000–10,000,000 in Mass Megabucks (the Treatment Group) and (2) people who played Megabucks but did not win a jackpot prize.

groups to those with an estimated treatment propensity greater than 0.20 and less than 0.80. After blocking on the score, none of the pretreatment differences between the treatment and the control groups remain statistically significant. For example, the average age is now 57 for the treatment group and 56 for the control group.

Column (7) shows the resulting p-values for the differences between groups when we weight by the estimated propensity score rather than blocking on the score. The chief advantage to weighting is that all the observations can then be used in estimating treatment effects, and with an appropriate weighting scheme, the estimator is efficient. (See Hirano, Imbens, and Ridder (2000)). We weight the treatment observations by $1/\hat{p}(x)$ and the control group observations by $1/(1 - \hat{p}(x))$, where $\hat{p}(x)$ is the estimated propensity score. In this example, weighting by the estimated score also achieves pretreatment balance between the treatment and control groups, although the balance is less perfect than with the blocking method.

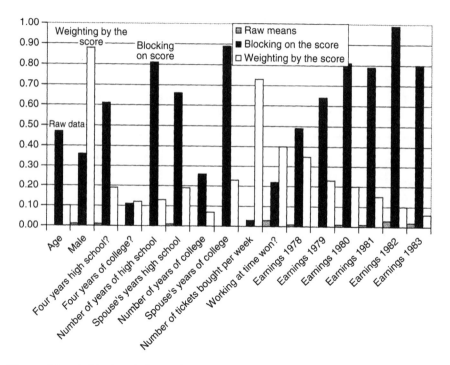

Figure 6.1   Achieving balance in pretreatment covariates using the propensity score: p-values for difference in means between treatment in control groups. The height of each bar represents the p-value for a test of the hypothesis that the difference in means between the treatment and control groups equals zero. The x-axis shows 16 different pretreatment variables. In the raw data (light grey), the treatment and control groups are statistically different on every pretreatment variable. In contrast, after blocking on the score, only one difference between the groups is statistically significant at the 10% level.

Figure 6.1 shows graphically that blocking on the propensity score or weighting by the propensity score balances the pretreatment covariates. The heights of the bars show p-values for differences in means between the treatment and control groups being equal to zero. The blue bars are the p-values for the raw (unadjusted) means. In the raw data, the treatment and control group are statistically different on all of the pretreatment measures. Once we adjust by weighting or blocking on the score, the p-values all rise dramatically, meaning that the large and statistically significant differences between the treatment and control groups are greatly reduced.

Either method of adjustment will remove the selection bias in the estimated treatment effects if we have unbiased estimates of the propensity score (Rosenbaum and Rubin (1983)). Propensity score adjustment is likely to reduce the selection bias in the lottery example because we know the assignment mechanism and we observe the covariates that determine the selection process.

Figure 6.2 shows the earnings of the treatment and control groups over time, weighting by the estimated propensity score. Pretreatment earnings (1978–1983) for the two groups are perfectly matched. The people in the treatment group win the lottery during 1984 to 1988 and their labor market earnings drop sharply below those of the control group.

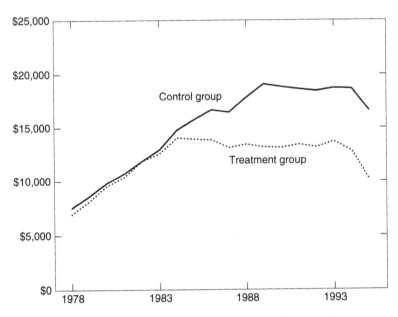

Figure 6.2   Effects of winning Mass Megabucks on labor supply treatment and control group social security earnings over time. Selection into treatment group is controlled for by weighting on the estimated propensity score. Subjects in the treatment group won Megabucks during 1984–1988.

Table 6.3 shows treatment effects from winning the lottery on earnings, consumption of cars and housings, savings, and divorce. Most of the treatment effects are statistically different from zero and do not change appreciably when we use weighting by the score or blocking (stratifying) on the score. Treatment subjects have labor income of about $6,000 less per year than the control subjects. Not surprisingly, the treatment subjects have about $20,000 less in their Individual Retirement Accounts, since they are working less.

The treated individuals drive cars worth $2,000 to $4,000 more, and own homes worth $6,000 more, relative to the control group. One of the strongest and most robust results is the effect on divorce. The treatment people are 9% more likely to get divorced relative to the control people. This may imply that divorce is a normal good (meaning that its consumption rises with income), or that the additional

| Variable | Weighting by the Score | | | | Blocking on the Score | | | |
	Mean Control	Mean Treated	Treat. Effect	(t-stat)	Mean Control	Mean Treated	Treat. Effect	(t-stat)
Earnings 1989 (thous.)	16.5	11.1	−5.4	(−3.8)	19.1	13.2	−5.9	(−2.7)
Earnings 1990 (thous.)	17.2	11.1	−6.1	(−4.2)	18.8	13.1	−5.7	(−2.6)
Earnings 1991 (thous.)	17.1	11.0	−6.2	(−4.1)	18.6	13.4	−5.2	(−2.3)
Earnings 1992 (thous.)	15.7	10.9	−4.8	(−3.2)	18.5	13.2	−5.3	(−2.4)
Earnings 1993 (thous.)	17.5	11.3	−6.3	(−4.0)	18.7	13.7	−5.0	(−2.1)
Earnings 1994 (thous.)	17.7	11.0	−6.7	(−4.2)	18.7	12.8	−5.8	(−2.4)
Earnings 1995 (thous.)	15.7	9.2	−6.5	(−4.2)	16.6	10.2	−6.5	(−2.8)
Working now?	0.7	0.4	−0.2	(−5.9)	0.6	0.5	−0.1	(−1.6)
Value of cars (thous.)	16.8	19.1	2.3	(1.8)	15.2	19.8	4.6	(2.6)
Sum all savings (thous.)	182.5	224.4	41.9	(2.0)	175.9	196.2	20.3	(0.7)
IRA value (thous.)	60.3	38.5	−21.9	(−2.9)	65.2	41.9	−23.3	(−2.5)
Savings acct. val. (thous.)	32.1	46.4	14.4	(2.3)	31.6	46.7	15.1	(1.3)
Value small bus. (thous.)	47.8	72.2	24.3	(2.0)	35.1	40.3	5.1	(0.3)
Value ins. policies (thous.)	80.7	93.6	12.9	(0.9)	67.6	85.5	17.9	(0.9)
Oth. major assets (thous.)	10.0	12.5	2.5	(0.5)	7.4	15.7	8.2	(1.6)
Home value (thous.)	161.4	167.3	5.9	(0.7)	160.4	159.7	−0.7	(−0.1)
Divorced since won?	0.0	0.1	0.1	(3.6)	0.0	0.1	0.1	(2.5)
Generally happy? (1 = yes)	0.9	1.0	0.1	(3.5)	1.0	1.0	0.0	(0.0)
N	302	291			149	147		

Table 6.3 Treatment effects from winning the lottery: blocking on and weighting by the propensity score. This table shows treatment effects (from winning the Massachusetts State Lottery) on income, consumption, savings, happiness, and the probability of divorce. In the raw data, selection causes the treatment and control groups to be mismatched in terms of age and pretreatment earnings and education. The selection problem is corrected here using the estimated propensity score $\hat{p}(x)$. In columns (2) to (5), this correction is achieved by weighting the treatment observations by $1/\hat{p}(x)$ and the control observations by $1/(1 - \hat{p}(x))$. In columns (6) to (9), we limit the sample to control and treatment observations with $0.2 < \hat{p}(x) < 0.8$. In other words, we block on the propensity score.

Variable	Raw Means			Weighting by the Score		
	Control	Treat.	p-value for Diff.	Control	Treat.	p-value for Diff.
Nominal prize value (mil.)	0.4	1.8	0.0	0.4	1.8	0.0
Age of winner	53.2	56.1	0.07	53.2	56.0	0.06
Male	0.5	0.6	0.00	0.6	0.6	1.00
Yrs of high school	3.5	3.6	0.19	3.5	3.6	0.23
Spouse's yrs of high school	3.4	3.9	0.00	3.4	3.9	0.00
Yrs of college	1.0	1.2	0.36	1.1	1.2	0.54
Spouse's yrs of college	1.5	1.4	0.59	1.5	1.4	0.63
#tix/wk, pretreat.	5.9	5.9	0.99	6.0	5.7	0.70
Working at time won?	0.8	0.8	0.82	0.8	0.8	0.32
Earnings 1978 (thous.)	5.7	7.4	0.04	6.2	6.9	0.40
Earnings 1979 (thous.)	6.5	8.6	0.03	7.0	7.9	0.33
Earnings 1980 (thous.)	7.6	9.6	0.07	8.2	8.8	0.57
Earnings 1981 (thous.)	8.1	10.8	0.02	8.8	10.0	0.30
Earnings 1982 (thous.)	8.7	12.4	0.00	9.5	11.4	0.13
Earnings 1983 (thous.)	9.5	12.9	0.01	10.3	11.9	0.22
Year won prize	1986.0	1986.1	0.49	1986.0	1986.1	0.42
N	137	154		137	154	

Table 6.4 Balance in pretreatment covariates: winners of big versus smaller prizes. This table shows pretreatment variables for two groups: (1) people who won prizes with nominal amounts of $22,000–650,000 (the Control Group) and (2) people who won nominal prizes of $651,000–$10,000,000 (the Treatment Group). Above, we argue that assignment to treatment group is random *conditional* on gender. Men are more likely to win larger prizes. Covariates in the treatment and control group are not balanced in the raw data, but are somewhat more balanced when we weight by the propensity score. The indicator for male is the only covariate used in the propensity score.

wealth of the lottery prize causes marital problems. The vast majority of people in both groups report being happy and in most specifications any differences are not statistically significant. However, when weighting by the score, we find that the treatment people are statistically significantly happier.

In Tables 6.4 and 6.5, we proceed to analyze the treatment effects of winning a large (>650,000) jackpot prize versus a smaller prize. Table 6.4 shows that weighting by the propensity to win the larger prize balances the differences between the two groups.[5] These are likely not the most efficient weights, because we only included the indicator for male in the propensity score. Our working assumption for this exercise is that simple differences between men's and women's responses to advertised jackpots causes the pretreatment imbalances between the large and small jackpot winners. Table 6.4 reinforces this claim.

---

[5]Again, we weight by $1/\hat{p}(x)$ for the treatment and $1/(1 - \hat{p}(x))$ for the control.

| Variable | Weighting by the Propensity Score | | | |
	Mean Control	Mean Treated	Treat. Effect	(t-stat)
Earnings 1989 (thous.)	14.6	10.4	−4.1	(−2.31)
Earnings 1990 (thous.)	14.0	10.4	−3.5	(−1.97)
Earnings 1991 (thous.)	15.0	10.1	−4.9	(−2.64)
Earnings 1992 (thous.)	15.3	9.8	−5.6	(−2.90)
Earnings 1993 (thous.)	15.8	9.7	−6.1	(−3.07)
Earnings 1994 (thous.)	15.8	9.4	−6.3	(−3.17)
Earnings 1995 (thous.)	13.4	8.3	−5.0	(−2.66)
Working now?	0.6	0.4	−0.2	(−3.84)
Value of cars owned (thous.)	15.4	22.9	7.6	(3.46)
Sum of all savings (thous.)	126.3	230.3	104.0	(3.57)
IRA value (thous.)	35.5	34.7	−0.7	(−0.10)
Savings account value (thous.)	21.5	53.4	31.9	(2.57)
Value of small businesses (thous.)	15.5	70.0	54.5	(3.01)
Value of insurance policies (thous.)	72.9	102.2	29.4	(1.43)
Value other major assets (thous.)	7.8	21.4	13.6	(1.91)
Home value (thous.)	126.6	177.4	50.8	(4.16)
Divorced since won?	0.1	0.2	0.1	(1.65)
Generally happy? (1 = yes)	1.0	1.0	0.0	(0.35)
N	137	154		

Child outcomes

Spent on child's private hs (thous.)	2.2	2.1	−0.2	(−0.15)
Spent on child's college ed. (thous.)	8.5	11.8	3.2	(1.16)
Spent on child's grad school ed. (thous.)	0.3	2.2	1.9	(1.75)
Child's yrs of college	1.4	1.8	0.4	(1.81)
Child has 4+ yrs of college?	0.2	0.3	0.1	(1.46)
N	217	259		

Table 6.5   Treatment effects from winning a big versus a smaller prize. Here we examine the treatment effects from winning a big versus a smaller prize (on average $1.8 million versus 0.36 million). The propensity score is used to weight observations to correct for selection into treatment group on the basis of gender. The mean child outcomes reported use the data at the child level, counting each child in the family as an observation, rather than each winner as a single observation. Standard errors are corrected for within family correlation.

One strategy would be to simply block on the indicator for male as in Rubin (1977) and Fisher (1925), and we did this in results not reported here. However, for illustrative purposes, here we estimate instead the propensity score with a single variable (gender).

Table 6.5 shows the estimated treatment effects from winning a large, rather than a small jackpot. Many of the results are similar to the earlier comparison of winners and nonwinners. For example, divorce rates are 9% higher for the large winners. The chief differences between the original treatment effects and those in Table 6.5 are the value of savings and homes. Large jackpot winners report total savings (IRA, stocks, bonds, mutual funds, savings accounts) that are larger than the total savings of small jackpot winners by $104,000. Furthermore, the large jackpot winners have homes worth $50,000 more.

We also explore the effects of winning a large prize on children's educational expenses and attainment. In the second panel of Table 6.5, we examine the data at the level of each individual child within the families that won large and smaller jackpots. Children in families that won a large jackpot have about 0.4 years more of college education and this result is significant at the 10% level. Similarly, children from families that win large jackpots receive about $2,000 more to cover costs of graduate education.

6.5 Concluding remarks

This chapter shows the value of the propensity score in fixing broken experiments. Frequently, natural experiments in social science will not have perfectly matched treatment and control groups. The presence of human factors almost guarantees that some selection into or out of the treatment group will occur. However, if we know the process by which this selection occurs, we can often undo the selection bias by blocking on or weighting by the propensity score.

In this example, there is a seemingly large pretreatment imbalance between lottery winners and nonwinners, and even between large and small jackpot winners. However, all of these differences are due to differences in lottery ticket buying behavior and propensity score adjustment allows us to reduce, and hopefully correct for, selection bias.

The propensity score allows researchers to adjust for a single variable (the estimated propensity score) and is extremely useful in cases like the lottery study, where there is a modest sample size. The lottery data example illustrates the value of the propensity score to social science and medical research. The propensity score is a great tool for enabling a valid comparison between treatment and control groups in order to identify causal effects.

7

The propensity score with continuous treatments

Keisuke Hirano and Guido W. Imbens[1]

7.1 Introduction

Much of the work on propensity score analysis has focused on the case in which the treatment is binary. In this chapter, we examine an extension to the propensity score method, in a setting with a continuous treatment. Following Rosenbaum and Rubin (1983a) and most of the other literature on propensity score analysis, we make an unconfoundedness or ignorability assumption, that adjusting for differences in a set of covariates removes all biases in comparisons by treatment status. Then, building on Imbens (2000) we define a generalization of the binary treatment propensity score, which we label the generalized propensity score (GPS). We demonstrate that the GPS has many of the attractive properties of the binary treatment propensity score. Just as in the binary treatment case, adjusting for this scalar function of the covariates removes all biases associated with differences in the covariates. The GPS also has certain balancing properties that can be used to assess the adequacy of particular specifications of the score. We discuss estimation and inference in a parametric version of this procedure, although more flexible approaches are also possible.

We apply this methodology to a data set collected by Imbens, Rubin, and Sacerdote (2001). The population consists of individuals winning the Megabucks

[1]Department of Economics, University of Arizona, Tuscon, Ariz., and Department of Economics and Department of Agricultural and Resource Economics, University of California, Berkeley. Financial support for this research was generously provided through NSF grants SES-0226164 and SES-0136789.

Applied Bayesian Modeling and Causal Inference from Incomplete-Data Perspectives.
Edited by A. Gelman and X-L. Meng © 2004 John Wiley & Sons, Ltd ISBN: 0-470-09043-X

lottery in Massachusetts in the mid-1980s. We are interested in the effect of the amount of the prize on subsequent labor earnings. Although the assignment of the prize is obviously random, substantial item and unit nonresponse led to a selected sample in which the amount of the prize is no longer independent of background characteristics. We estimate the average effect of the prize adjusting for differences in background characteristics using the propensity score methodology, and compare the results to conventional regression estimates. The results suggest that the propensity score methodology leads to credible estimates that can be more robust than simple regression estimates.

7.2   The basic framework

We have a random sample of units, indexed by $i = 1, \ldots, N$. For each unit i, we postulate the existence of a set of potential outcomes, $Y_i(t)$, for $t \in \mathcal{T}$, referred to as the unit-level dose–response function. In the binary treatment case, $\mathcal{T} = \{0, 1\}$. Here we allow $\mathcal{T}$ to be an interval $[t_0, t_1]$. We are interested in the average dose–response function, $\mu(t) = \mathrm{E}[Y_i(t)]$. For each unit i, there is also a vector of covariates X_i, and the level of the treatment received, $T_i \in [t_0, t_1]$. We observe the vector X_i, the treatment received T_i, and the potential outcome corresponding to the level of the treatment received, $Y_i = Y_i(T_i)$.

To simplify the notation, we will drop the i subscript in the sequel. We assume that $\{Y(t)\}_{t \in \mathcal{T}}$, T, X are defined on a common probability space, that T is continuously distributed with respect to Lebesgue measure on $\mathcal{T}$, and that $Y = Y(T)$ is a well-defined random variable (this requires that the random function $Y(\cdot)$ be suitably measurable).

Our key assumption generalizes the unconfoundedness assumption for binary treatments made by Rosenbaum and Rubin (1983), to the multivalued case:

Assumption 1 (WEAK UNCONFOUNDEDNESS) $Y(t) \perp T \,|\, X$ for all $t \in \mathcal{T}$.

We refer to this as weak unconfoundedness, as we do not require joint independence of all potential outcomes, $\{Y(t)\}_{t \in [t_0, t_1]}$. Instead, we require conditional independence to hold for each value of the treatment.

Next, we define the generalized propensity score.

Definition 1 (GENERALIZED PROPENSITY SCORE) *Let $r(t, x)$ be the conditional density of the treatment given the covariates:*

$$r(t, x) = f_{T|X}(t|x).$$

Then the generalized propensity score is $R = r(T, X)$.

This definition follows Imbens (2000). For alternative approaches to the case with multivalued treatments, see Joffe and Rosenbaum (1999a, 1999b), Lechner (2001), and Imai and van Dyk (2004).

The function r is defined up to equivalence almost everywhere. By standard results on conditional probability distributions, we can choose r such that $R = r(T, X)$ and $r(t, X)$ are well-defined random variables for every t.

The GPS has a balancing property similar to that of the standard propensity score. Within strata with the same value of $r(t, X)$, the probability that $T = t$ does not depend on the value of X. Loosely speaking, the GPS has the property that

$$X \perp 1\{T = t\} | r(t, X).$$

This is a mechanical implication of the definition of the GPS, and does not require unconfoundedness. In combination with unconfoundedness, this implies that assignment to treatment is unconfounded given the generalized propensity score.

Theorem 1 (WEAK UNCONFOUNDEDNESS GIVEN THE GENERALIZED PROPENSITY SCORE) *Suppose that assignment to the treatment is weakly unconfounded given pretreatment variables X. Then, for every t,*

$$f_T(t | r(t, X), Y(t)) = f_T(t | r(t, X)). \tag{7.1}$$

Proof. Throughout the proof, equality is taken as a.e. equality. Since $r(t, X)$ is a well-defined random variable, for each t we can define a joint law for $(Y(t), T, X, r(t, X))$. We use $F_X(x|\cdot)$ to denote various conditional probability distributions for X, and we use $f_T(t|\cdot)$ to denote conditional densities of T. Note that $r(t, X)$ is measurable with respect to the sigma-algebra generated by X. This implies, for example, that $f_T(t|X, r(t, X)) = f_T(t|X)$.

Using standard results on iterated integrals, we can write

$$f_T(t | r(t, X)) = \int f_T(t | x, r(t, X)) \, dF_X(x | r(t, X))$$

$$= \int f_T(t | x) \, dF_X(x | r(t, X))$$

$$= \int r(t, x) \, dF_X(x | r(t, X)) = r(t, X).$$

The left side of equation (7.1) can be written as:

$$f_T(t | r(t, X), Y(t)) = \int f_T(t | x, r(t, X), Y(t)) \, dF_X(x | Y(t), r(t, X)).$$

By weak unconfoundedness, $f_T(t | x, r(t, X), Y(t)) = f_T(t | x)$, so

$$f_T(t | r(t, X), Y(t)) = \int r(t, x) \, dF_X(x | Y(t), r(t, X))$$

$$= r(t, X).$$

Therefore, for each t, $f_T(t | r(t, X), Y(t)) = f_T(t | r(t, X))$. $\square$

When we consider the conditional density of the treatment level at t, we evaluate the generalized propensity score at the corresponding level of the treatment. In that sense, we use as many propensity scores as there are levels of the treatment. Nevertheless, we never use more than a single score at one time.

7.3 Bias removal using the GPS

In this section, we show that the GPS can be used to eliminate any biases associated with differences in the covariates. The approach consists of two steps. First, we estimate the conditional expectation of the outcome as a function of two scalar variables, the treatment level T and the GPS R, $\beta(t, r) = E[Y|T = t, R = r]$. Second, to estimate the dose–response function at a particular level of the treatment we average this conditional expectation over the GPS at that particular level of the treatment, $\mu(t) = E[\beta(t, r(t, X))]$. We do not average over the GPS $R = r(T, X)$; rather we average over the score evaluated at the treatment level of interest, $r(t, X)$.

Theorem 2 (BIAS REMOVAL WITH GENERALIZED PROPENSITY SCORE) *Suppose that assignment to the treatment is weakly unconfounded given pretreatment variables X. Then*
(i) $\beta(t, r) = E[Y(t)|r(t, X) = r] = E[Y|T = t, R = r]$.
(ii) $\mu(t) = E[\beta(t, r(t, X)]$.

Proof. Let $f_{Y(t)|T, r(t, X)}(\cdot|t, r)$ denote the conditional density (with respect to some measure) of $Y(t)$ given $T = t$ and $r(t, X) = r$. Then, using Bayes rule and Theorem 1,

$$f_{Y(t)|T, r(t, X)}(y|t, r) = \frac{f_T(t|Y(t) = y, r(t, X) = r) f_{Y(t)|r(t, X)}(y|r)}{f_T(t|r(t, X) = r)}$$

$$= f_{Y(t)|r(t, X)}(y|r)$$

Hence,

$$E[Y(t)|T = t, r(t, X) = r] = E[Y(t)|r(t, X) = r].$$

But we also have

$$E[Y(t)|T = t, R = r] = E[Y(t)|T = t, r(T, X) = r]$$

$$= E[Y(t)|T = t, r(t, X) = r]$$

$$= E[Y(t)|r(t, X) = r] = \beta(t, r)$$

Hence, $E[Y(t)|r(t, X) = r] = \beta(t, r)$, which proves part (i). For the second part, by iterated expectations, $E[\beta(t, r(t, X))] = E[E[Y(t)|r(t, X)]] = E[Y(t)]$. $\square$

It should be stressed that the regression function $\beta(t, r)$ does not have a causal interpretation. In particular, the derivative with respect to the treatment level t does not represent an average effect of changing the level of the treatment for any particular subpopulation.

Robins (1998, 1999) and Robins, Hernan, and Brumback (2000) use a related approach. They parameterize or restrict the form of the $Y(t)$ process (and hence the form of $\mu(t)$), and call this a marginal structural model (MSM). The parameters of the MSM are estimated using a weighting scheme based on the GPS. When the treatment is continuous these weights must be "stabilized" by the marginal probabilities of treatment. In the approach we take here, we would typically employ parametric assumptions about the form of $\beta(t, r)$ instead of $\mu(t)$, and do not need to reweight the observations.

Two artificial examples

EXAMPLE 1: Suppose that the conditional distribution of $Y(t)$ given X is

$$Y(t)|X \sim N(t + X \exp(-tX), 1).$$

The conditional mean of $Y(t)$ given X is $t + X \exp(-tX)$. Suppose also that the marginal distribution of X is unit exponential. The marginal mean of $Y(t)$ is obtained by integrating out the covariate to get

$$\mu(t) = E[t + X \exp(-tX)] = t + \frac{1}{(t+1)^2}.$$

Now consider estimating the dose–response function using the GPS approach. We assume that the assignment to treatment is weakly unconfounded. For illustrative purposes, we also assume that the conditional distribution of the treatment T given X is exponential with hazard rate X. This implies that the conditional density of T given X is

$$f_{T|X}(t, x) = x \exp(-tx).$$

Hence the generalized propensity score is $R = X \exp(-TX)$.

Next, we consider the conditional expectation of Y given the treatment T and the score R. By weak unconfoundedness, the conditional expectation of Y given T and X is

$$E[Y|T = t, X = x] = E[Y(t)|X = x].$$

Then by iterated expectations

$$E[Y|T = t, R = r] = E[E[Y|T = t, X]|T = t, R = r]$$
$$= E[E[Y(t)|X]|T = t, R = r]$$
$$= E[t + X \exp(-tX)|T = t, R = r] = t + r.$$

As stressed before, this conditional expectation does not have a causal interpretation as a function of t. For the final step, we average this conditional expectation over the marginal distribution of $r(t, X)$:

$$E[Y(t)] = E[t + r(t, X)] = t + \frac{1}{(1+t)^2} = \mu(t).$$

This gives us the dose–response function at treatment level t.

EXAMPLE 2: Suppose that the dose–response function is $E[Y(t)] = \mu(t)$. Also suppose that X is independent of the level of the treatment so that we do not actually need to adjust for the covariates. Independence of the covariates and the treatment implies that the GPS $r(t, x) = f_{T|X}(t|x) = f_T(t)$ is a function only of t. This creates a lack of uniqueness in the regression of the outcome on the level of the treatment and the GPS. Formally, there is no unique function $\beta(t, r)$ such that $E[Y|T = t, R = r] = \beta(t, r)$ for all (t, r) in the support of $(T, r(T))$. In practice, this means that the GPS will not be a statistically significant determinant of the average value of the outcome, and in the limit we will have perfect collinearity in the regression of the outcome on the treatment level and the GPS. However, this does not create problems for estimating the dose–response function. To see this, note that any solution $\beta(t, r)$ must satisfy

$$\beta(t, r(t)) = E[Y|T = t, r(T) = r(t)] = E[Y|T = t] = \mu(t).$$

Hence, the implied estimate of the dose–response function is

$$\int_x \beta(t, r(t, x)) f_X(x) \, dx = \beta(t, r(t)) = \mu(t),$$

equal to the dose–response function.

7.4 Estimation and inference

In this section, we consider the practical implementation of the generalized propensity score methodology outlined in the previous section. We use a flexible parametric approach. In the first stage, we use a normal distribution for the treatment given the covariates:

$$T_i|X_i \sim N(\beta_0 + \beta_1' X_i, \sigma^2).$$

We may consider more general models such as mixtures of normals, or heteroskedastic normal distributions with the variance being a parametric function of the covariates. In the simple normal model, we can estimate β_0, β_1, and σ^2 by maximum likelihood. The estimated GPS is

$$\hat{R}_i = \frac{1}{\sqrt{2\pi\hat{\sigma}^2}} \exp\left(-\frac{1}{2\hat{\sigma}^2}(T_i - \hat{\beta}_0 - \hat{\beta}_1' X_i)^2\right).$$

In the second stage, we model the conditional expectation of Y_i given T_i and R_i as a flexible function of its two arguments. In the application in the next section, we use a quadratic approximation:

$$E[Y_i|T_i, R_i] = \alpha_0 + \alpha_1 T_i + \alpha_2 T_i^2 + \alpha_3 R_i + \alpha_4 R_i^2 + \alpha_5 T_i R_i.$$

We estimate these parameters by ordinary least squares using the estimated GPS $\hat{R}_i$.

Given the estimated parameter in the second stage, we estimate the average potential outcome at treatment level t as

$$\widehat{E[Y(t)]} = \frac{1}{N} \sum_{i=1}^{N} \left(\hat{\alpha}_0 + \hat{\alpha}_1 t + \hat{\alpha}_2 t^2 + \hat{\alpha}_3 \hat{r}(t, X_i) + \hat{\alpha}_4 \hat{r}(t, X_i)^2 + \hat{\alpha}_5 t \hat{r}(t, X_i) \right).$$

We do this for each level of the treatment we are interested in, to obtain an estimate of the entire dose–response function.

Given the parametric model we use for the GPS and the regression function one can demonstrate root-N consistency and asymptotic normality for the estimator. Asymptotic standard errors can be calculated using expansions based on the estimating equations; these should take into account estimation of the GPS as well as the α parameters. In practice, however, it is convenient to use bootstrap methods to form standard errors and confidence intervals.

7.5   Application: the Imbens–Rubin–Sacerdote lottery sample

The data

The data we use to illustrate the methods discussed in the previous section come from the survey of Massachusetts lottery winners, which is described in further detail in the chapter by Sacerdote in this volume, and in Imbens, Rubin, and Sacerdote (2001). Here we analyze the effect of the prize amount on subsequent labor earnings (from social security records), without discretizing the prize variable.

Although the lottery prize is obviously randomly assigned, there is substantial correlation between some of the background variables and the lottery prize in our sample. The main source of potential bias is the unit and item nonresponse. In the survey unit, nonresponse was about 50%. In fact, it was possible to directly demonstrate that this nonresponse was nonrandom, since for all units the lottery prize was observed. It was shown that the higher the lottery prize, the lower the probability of responding to the survey. The missing data imply that the amount of the prize is potentially correlated with background characteristics and potential outcomes. In order to remove such biases, we make the weak unconfoundedness assumption that conditional on the covariates the lottery prize is independent of the potential outcomes.

The sample we use in this analysis is the "winners" sample of 237 individuals who won a major prize in the lottery. In Table 7.1, we present means and standard deviations for this sample. To demonstrate the effects of nonresponse, we also report the correlation coefficients between each of the covariates and the prize, with the t-statistic for the test that the correlation is equal to zero. We see that many of the covariates have substantial and significant correlations with the prize.

Variable	Mean	S.D.	Corr. w/Prize	t-stat	GPS Est.	GPS SE
Intercept					2.32	(0.48)
Age	47.0	13.8	0.2	2.4	0.02	(0.01)
Years high school	3.6	1.1	−0.1	−1.4	0.02	(0.06)
Years college	1.4	1.6	0.0	0.5	0.04	(0.04)
Male	0.6	0.5	0.3	4.1	0.44	(0.14)
Tickets bought	4.6	3.3	0.1	1.6	0.00	(0.02)
Working then	0.8	0.4	0.1	1.4	0.13	(0.17)
Year won	1986.1	1.3	−0.0	−0.4	−0.00	(0.05)
Earnings year−1	14.5	13.6	0.1	1.7	0.01	(0.01)
Earnings year−2	13.5	13.0	0.1	2.1	−0.01	(0.02)
Earnings year−3	12.8	12.7	0.2	2.3	0.01	(0.02)
Earnings year−4	12.0	12.1	0.1	2.0	0.02	(0.02)
Earnings year−5	12.2	12.4	0.1	1.1	−0.02	(0.02)
Earnings year−6	12.1	12.4	0.1	1.1	−0.01	(0.01)

Table 7.1 Summary statistics and parameter estimates of generalized propensity score.

Modeling the conditional distribution of the prize given covariates

The first step is to estimate the conditional distribution of the prize given the covariates. The distribution of the prize is highly skewed, with a skewness of 2.9 and a kurtosis of 15.0. We therefore first transform the prize by taking logarithms. The logarithm of the prize has a skewness of −0.02 and a kurtosis of 3.4. We then use a normal linear model for the logarithm of the prize:

$$\log T_i | X_i \sim N(\beta_0 + \beta_1' X_i, \sigma^2).$$

The estimated coefficients from this model are presented in Table 7.1.

To see whether this specification of the propensity score is adequate, we investigate how it affects the balance of the covariates. This idea is again borrowed from the analysis of binary treatment cases, in which Rosenbaum and Rubin (1983) stress the balancing properties of the propensity score. We divide the range of prizes into three treatment intervals, [0, 23], [23, 80], and [80, 485], with 79 observations in the first group, 106 in the second, and 52 in the last treatment group. For each of the thirteen covariates, we investigate the balance by testing whether the mean in one of the three treatment groups was different from the mean in the other two treatment groups combined. (Alternatively, we could carry out various joint tests to assess the covariate balance.) In Table 7.2, we report the t-tests for each of the thirteen covariates and each of the three groups. The results show a clear lack of balance, with 14 (17) of 39 t-statistics greater than 1.96 (1.645) in absolute value.

Variable	Unadjusted			Adjusted for the GPS		
	[0, 23]	[23, 80]	[80, 485]	[0, 23]	[23, 80]	[80, 485]
Age	−1.7	−0.1	2.0	0.1	0.3	1.7
Years high school	−0.9	1.7	−0.7	−0.5	0.8	−1.0
Years college	−1.2	0.7	0.5	−0.5	0.7	−0.7
Male	−3.6	0.5	4.0	−0.4	0.2	0.1
Tickets bought	−1.1	0.5	0.6	−0.7	0.7	−0.2
Working then	−1.1	−0.3	2.0	−0.0	−0.2	0.3
Year won	−0.6	2.0	−1.6	−0.1	1.1	−1.0
Earnings year−1	−1.8	−0.5	2.3	−0.3	−0.7	0.5
Earnings year−2	−2.3	−0.4	2.6	−1.0	−0.4	0.5
Earnings year−3	−2.7	−0.6	3.1	−1.4	−0.6	1.2
Earnings year−4	−2.7	−0.7	3.1	−0.9	−0.6	1.7
Earnings year−5	−2.2	−0.3	2.4	−1.1	−0.0	2.1
Earnings year−6	−2.1	−0.1	2.3	−1.5	0.4	2.2

Table 7.2 Balance given the generalized propensity score: t-statistics for equality of means.

Next, we report GPS-adjusted versions of these statistics. Take the first covariate (age), and the test whether the adjusted mean in the first group (with prizes less than 23 K) is different from the mean in the other two groups. Recall that we should have

$$X_i \perp 1\{T_i = t\}|r(t, X_i).$$

We implement this by discretizing both the level of the treatment and the GPS. First, we check independence of X_i and the indicator that $0 \le T_i \le 23$, conditional on $r(t, X_i)$. To implement this we approximate $r(t, X_i)$ by evaluating the GPS at the median of the prize in this group, which is 14. Thus, we test

$$X_i \perp 1\{0 \le T_i \le 23\} \mid r(14, X_i).$$

We test this by blocking on the score $r(14, X_i)$. We use five blocks, defined by quintiles of $r(14, X_i)$ in the group with $1\{0 \le T_i \le 23\}$. The five groups are defined by the GPS values for $r(14, X_i)$ in the intervals [0.06, 0.21], [0.21, 0.28], [0.28, 0.34], [0.34, 0.39], and [0.39, 0.45]. (The full range of values for the GPS $r(T, X)$ evaluated at received treatment and covariates is [0.00, 0.45], but the range of $r(14, X)$ is [0.06, 0.45].) For example, the first of these five groups, with $r(14, X_i) \in [0.06, 0.21]$ has a total of 84 observations (16 with $T_i \in [0, 23]$ and 68 with $T_i \notin [0, 23]$). Testing for equality of the average age in the first versus the other two prize groups in this GPS group gives a mean difference of −5.5 with a standard error of 2.2. In the second GPS group, with $r(14, X_i) \in [0.21, 0.28]$ there are 39 observations (16 with $T_i \in [0, 23]$ and 23 with $T_i \notin [0, 23]$), leading to a mean difference of -3.2 (SE 5.3). In the third GPS group, with $r(14, X_i) \in [0.28, 0.34]$ there are 53 observations (15 with $T_i \in [0, 23]$ and 38 with $T_i \notin [0, 23]$), leading to a mean difference of 8.2 (SE 4.4). In the fourth GPS group, with $r(14, X_i) \in [0.34, 0.39]$ there are 36 observations (16 with $T_i \in [0, 23]$ and 20 with $T_i \notin [0, 23]$), leading to a mean difference of 4.7 (SE 3.0). In the fifth GPS group, with $r(14, X_i) \in [0.39, 0.45]$ there are 25 observations (16 with $T_i \in [0, 23]$ and 9 with $T_i \notin [0, 23]$), leading to a mean difference of 0.4 (SE 4.0). Combining these five differences in means, weighted by the number of observations in each GPS group, leads to a mean difference of 0.1 (SE 0.9), and thus a t-statistic of 0.1, compared to an unadjusted mean of -3.1 (SE 1.8) and t-statistic of -1.7.

The adjustment for the GPS improves the balance. After the adjustment for the GPS, only 2 t-statistics are larger than 1.96 (compared to 16 prior to adjustment) and 4 out of 39 are larger than 1.645 (compared to 17 prior to adjustment). These lower t-statistics are not merely the result of increased variances. For example, for earnings in year -1, the mean difference between treatment group $[0, 23]$ and the other two is -3.1 (SE 1.7). After adjusting for the GPS, this is reduced to -0.3 (SE 0.9).

Estimating the conditional expectation of outcome given prize and generalized propensity score

Next, we regress the outcome, earnings six years after winning the lottery, on the prize T_i, and the logarithm of the score R_i. We include all second-order moments of prize and log score. The estimated coefficients are presented in Table 7.3. Again, it should be stressed that there is no direct meaning to the estimated coefficients in this model, except that testing whether all coefficients involving the GPS are equal to zero can be interpreted as a test of whether the covariates introduce any bias.

Estimating the dose-response function

The last step consists of averaging the estimated regression function over the score function evaluated at the desired level of the prize. Rather than report the dose–response function, we report the derivative of the dose–response function.

Variable	Est.	SE
Intercept	9.68	3.34
Prize	-0.03	0.03
Prize-squared/1,000	0.40	0.20
Log(score)	-3.33	3.41
Log(score)-squared	-0.28	0.46
Log(score) × prize	0.05	0.02

Table 7.3 Parameter estimates of conditional distribution of prize given covariates.

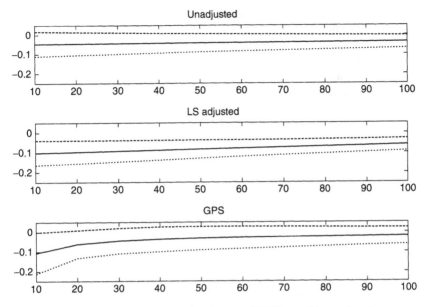

Figure 7.1   Estimated derivatives and 95% confidence bands.

In economic terminology, this is the marginal propensity to earn out of unearned income. (The yearly prize money is viewed as unearned income, and the derivative of average labor income with respect to this is the marginal propensity to earn out of unearned income.) We report the value of the derivative at $10,000 increments for all values of the yearly lottery prize between $10,000 and $100,000. The results are shown in Figure 7.1, along with pointwise 95% confidence bands. The bands are based on 1,000 bootstrap replications, taking into account estimation of the GPS.

The GPS-based estimates are compared to linear regression estimates based on a regression function that is quadratic in the prize, either without additional covariates ("unadjusted") or with the additional covariates included linearly ("LS adjusted").

The GPS estimates imply that the absolute value of the propensity to earn out of unearned income goes down sharply with the level of unearned income, from −0.10 at $10,000 to −0.02 at $100,000, suggesting that those with lower earnings are much more sensitive to income changes than those with higher earnings. The linear regression estimates suggest a much smaller change, with the derivative at a prize of $100,000 equal to −0.04, compared to −0.05 at $10,000.

7.6   Conclusion

Propensity score methods have become one of the most important tools for analyzing causal effects in observational studies. Although the original work of Rosenbaum and Rubin (1983) considered applications with binary treatments, many of the ideas readily extend to multivalued and continuous treatments. We have discussed some of the issues involved in handling continuous treatments, and emphasized how the propensity score methodology can be extended to this case. We applied these ideas to a data set previously studied by Imbens, Rubin, and Sacerdote (2001). We expect that coming years will see further work applying the Rubin causal model approach to a range of settings.

8

Causal inference with instrumental variables

Junni L. Zhang[1]

8.1 Introduction

The Rubin causal model (RCM) has been a well-established framework of causal inference in comparison and evaluation of treatments and programs in economics, medicine, and other fields. In a series of influential papers (Imbens and Angrist, 1994a; Angrist and Imbens, 1995; Angrist, Imbens, and Rubin, 1996; Imbens and Rubin, 1997a, 1997b), Don Rubin and his colleagues reinterpreted the instrumental variable (IV) estimand in the RCM framework and laid out assumptions under which this estimand has causal interpretation as a *local average treatment effect* (LATE), without requiring functional form or constant effect assumptions traditionally employed in econometric IV analyses. This interpretation of IV estimands has been discussed and explored in many studies (e.g., Gerber and Green, 1999, 2000a, 2000b; Ichino and Winter-Ebmer, 1998; Angrist and Evans, 1998; Cruces and Galiani, 2003; Contoyannis and Rice, 2001). A related context arises with studying noncompliance in randomized experiments, where treatment assignment serves as the IV for the effect of treatment receipt on the outcome, and LATE is often termed as CACE (complier average causal effect).

More traditionally, following Haavelmo (1943, 1944), the IVs are used to identify causal parameters in structural equation models, which specify a system of autonomous equations that attempt to capture the causal relationship between

[1]Department of Business Statistics and Econometrics, Peking University.

Applied Bayesian Modeling and Causal Inference from Incomplete-Data Perspectives.
Edited by A. Gelman and X-L. Meng © 2004 John Wiley & Sons, Ltd ISBN: 0-470-09043-X

variables. Often linearity constant causal effects are assumed such that the IV estimand can be interpreted as *average treatment effect* (ATE).

Consider a simple but important case with binary treatment and binary instrument. For illustration, we will consider an example of inferring the effect of college degree on income, using families' proximity to colleges as the IV (see Card, 1995). For individual i, let D_i be an indicator of whether he gets a college degree, let Y_i be his income, and let Z_i be the dichotomized IV: $Z_i = 1$ if family is close to college and $Z_i = 0$ if family is far from college. An example of a structural equation model for inferring the treatment effect is the *dummy endogenous variable model* (see, e.g., Heckman and Robb, 1985; Heckman and Hotz, 1989)

$$Y_i = \beta_0 + \beta_1 \cdot D_i + \varepsilon_i,$$
$$D_i^* = \alpha_0 + \alpha_1 \cdot Z_i + v_i, \tag{8.1}$$

and

$$D_i = \begin{cases} 1 & \text{if } D_i^* > 0, \\ 0 & \text{if } D_i^* \leq 0. \end{cases} \tag{8.2}$$

In this model, β_1 represents the causal effect of college degree on income. ε_i and v_i are disturbance terms. Since people choose to attend colleges on the basis of unobserved factors (such as ability) that tend to be also related to their income, usually ε_i and v_i are correlated, and hence D_i is correlated with ε_i. This has been known as *self-selection problem* or selection bias problem in econometrics.

For Z_i to be a valid IV that can be used to identify β_1, two assumptions are typically made:

$$E[Z_i \cdot \varepsilon_i] = 0, \quad E[Z_i \cdot v_i] = 0 \tag{8.3}$$

and

$$\text{cov}(D_i, Z_i) \neq 0. \tag{8.4}$$

The zero correlation between Z_i and ε_i in (8.3) and the absence of Z_i in (8.1) implies that any effect of families' proximity to colleges on income must be through an effect on college degree. With the second assumption in (8.4), the effect of families' proximity to colleges on college degree is nonzero. The IV estimand (Durbin, 1954) is then defined as the ratio of the covariances

$$\beta^{IV} = \frac{\text{cov}(Y_i, Z_i)}{\text{cov}(D_i, Z_i)}, \tag{8.5}$$

which is equal to β_1 in (8.1) given the two assumptions. Since β_1 as the causal effect of college degree on income is assumed to be constant in the population, the IV estimand can be interpreted as the ATE.

When the causal effect is heterogeneous, however, the IV estimand is no longer equal to ATE. In this chapter, we discuss reinterpretation of the IV estimand and estimation of causal effects using IVs. The materials are mainly drawn from

Angrist, Imbens, and Rubin (1996) and Imbens and Rubin (1997a, 1997b). In Section 8.2, we give the assumptions for the LATE interpretation of the IV estimand. We then discuss estimation of causal effects under these assumptions and sensitivity to these assumptions in Section 8.3. Section 8.4 reviews some recent applications. Section 8.5 then concludes with discussion on IV choice and extensions of the work.

8.2 Key assumptions for the LATE interpretation of the IV estimand

Let Z_i be a binary instrument. To represent the potential outcomes, we use notation different from the previous section. For $z = 0, 1$, let $D_i(z) = 0$ or 1 be the treatment that would be obtained by individual i given instrument z; for $z = 0, 1$ and $d = 0, 1$, let $Y_i(z, d)$ be the outcome that would be observed for individual i given instrument z and treatment d, respectively. An implicit assumption in this notation is the Stable Unit Treatment Value Assumption (SUTVA, Rubin, 1980b, 1990), which requires that individual i is not affected by the values of the instrument and the treatment for other individuals. For individual i, we observe the triple

$$\left(Z_i^{obs} = Z_i, \ D_i^{obs} = D_i(Z_i^{obs}), \ Y_i^{obs} = Y_i(Z_i^{obs}, D_i^{obs}) \right).$$

We make a second assumption that the instrument Z_i is randomly assigned, independent of all the potential outcomes $D_i(z)$ and $Y_i(z, d)$, or ignorable more generally (Rubin, 1978a). The third assumption is that the average effect of Z on D is nonzero, or $E[D_i(1) - D_i(0)] \neq 0$. The fourth assumption is the exclusion restriction for IV: the potential outcomes $Y_i(z, d)$ do not depend on the instrument z; so for $d = 0, 1$, $Y_i(d)$ denotes the outcome that would be observed for individual i under treatment d. Finally, we assume monotonicity such that $D_i(1) \geq D_i(0)$ for all individuals; this assumes that no one would receive the treatment when given instrument $z = 0$ but not when given $z = 1$.

In the context of our example of evaluating the effect of a college degree on income, the five assumptions described above are as follows:

- SUTVA: whether individual i gets a college degree and his subsequent income are unrelated to whether families of other individuals are close to colleges and whether they get college degrees. This assumption might fail, for example, if individuals living far from college but close to each other tend to go to college together.

- Ignorable assignment of the instrument: families' proximity to colleges is exogenous for the potential outcomes of individuals in the study. This assumption might fail, for example, if families choose to live close to college if their kids promise to go to college.

- Exclusion restriction: after taking into account whether an individual gets a college degree, his income is not affected by whether his family lives close to colleges. This assumption might fail, for example, if individuals are more constrained when their family lives closer to college, and thus lose some opportunities to become more self-supportive.

- Nonzero average effect of Z on D: the distance of family from colleges changes the likelihood that individuals get college degrees. Card (1995) presented evidence that supports this assumption.

- Monotonicity: no one would get a college degree when his family lives far from colleges but not when his family lives close to colleges. This assumption might fail, if individuals with families far from colleges tend to work harder in order to go to college.

We will address violation of the exclusion restriction or the monotonicity assumption in the next section. At present, we will assume that all five assumptions are satisfied.

Under the SUTVA assumption, we can define the causal effects of Z on D and on Y.

Definition 1 *Causal Effects of Z on D and Z on Y.*
The causal effect for individual i of Z on D is $D_i(1) - D_i(0)$.
The causal effect for individual i of Z on Y is $Y_i(1, D_i(1)) - Y_i(0, D_i(0))$.

If exclusion restriction is further assumed, we can define the causal effect of treatment D on outcome Y, which is the causal effect of ultimate interest.

Definition 2 *Causal Effect of D on Y.*
The causal effect for individual i of D on Y is $Y_i(1) - Y_i(0)$.

Under the SUTVA assumptions, ignorable assignment of the instrument and nonzero average effect of Z on D, the units can be partitioned according to the four possible joint values of $(D_i(0), D_i(1))$, as shown in Table 8.1. This partition is an example of principal stratification (Frangakis and Rubin, 2002). For individual

	$D_i(0)$	$D_i(1)$	Causal Effect of Z on Y	
			No Exclusion	With Exclusion
Never-takers	0	0	$Y_i(1, 0) - Y_i(0, 0)$	0
Compliers	0	1	$Y_i(1, 1) - Y_i(0, 0)$	$Y_i(1) - Y_i(0)$
Defiers	1	0	$Y_i(1, 0) - Y_i(0, 1)$	$-(Y_i(1) - Y_i(0))$
Always-takers	1	1	$Y_i(1, 1) - Y_i(0, 1)$	0

Table 8.1 Partition of the population by $D_i(0)$ and $D_i(1)$, and the corresponding causal effect of Z on Y, with or without exclusion restriction.

i, let C_i denote the principal stratum indicator. In the context of our example, the four principal strata are as follows:

- The individuals who would not get a college degree regardless of whether their families live close to colleges, $\{i : D_i(0) = D_i(1) = 0\}$. We label them never-takers ($C_i = n$). Under exclusion restriction, the causal effects of Z on Y are 0 for never-takers.

- The individuals who would get a college degree regardless of whether their families live close to colleges, $\{i : D_i(0) = D_i(1) = 1\}$. We label them always-takers ($C_i = a$). Under exclusion restriction, the causal effects of Z on Y are 0 for always-takers.

- The individuals who would get a college degree when their families live close to colleges but not otherwise, $\{i : D_i(0) = 0, D_i(1) = 1\}$. We label them compliers ($C_i = c$). Under exclusion restriction, the causal effects for compliers of Z on Y are equal to those of D on Y.

- The individuals who would not get a college degree when their families live close to colleges but would otherwise, $\{i : D_i(0) = 1, D_i(1) = 0\}$. We label them defiers ($C_i = d$). Under exclusion restriction, the causal effects for defiers of Z on Y are opposite to those of D on Y.

We refer to never-takers, always-takers, and defiers together as noncompliers.

Under the SUTVA assumptions, ignorable assignment of the instrument, and nonzero effect of Z on D, one can show that

$$\beta^{IV} = \frac{E[Y_i(1, D_i(1)) - Y_i(0, D_i(0))]}{E[D_i(1) - D_i(0)]}, \tag{8.6}$$

thus the IV estimand is equal to the ratio of the average causal effect of Z on Y and the average causal effect of Z on D.

By virtue of the exclusion restriction, the causal effects of Z on Y are 0 for never-takers and always-takers. By virtue of the monotonicity assumption, there are no defiers. Further, by assuming nonzero effect of Z on D, the proportion of compliers is nonzero and is equal to the average causal effect of Z on D. Combing all five assumptions, we can see that the average causal effect of Z on Y is proportional to the average causal effect of D on Y for compliers, and the proportional factor is equal to the proportion of compliers. Therefore, the IV estimand can be written as

$$\beta^{IV} = E[(Y_i(1) - Y_i(0))|D_i(1) - D_i(0) = 1]. \tag{8.7}$$

The IV estimand thus identifies a LATE, the average treatment effect for compliers. In our example, it is equal to the average causal effect of college degree on income for those who would get a college degree only because their families live close to colleges.

In the context of randomized studies with noncompliance, if we use treatment assignment Z as the IV for the effect of treatment receipt D on outcome Y, the same five assumptions can be made. In particular, the second assumption is automatically satisfied, the third assumption (exclusion restriction) assumes that treatment assignment affects outcome only through treatment receipt, and the fifth assumption (monotonicity) assumes that there is only one-sided noncompliance such that no one would always go against the treatment assignment. Under these assumptions, as shown in (8.7), the IV estimand identifies the complier average causal effect (i.e., CACE).

8.3   Estimating causal effects with IV

Since we cannot observe both $D_i(0)$ and $D_i(1)$, we cannot directly observe the principal strata for the individuals. What we can observe are the four groups based on Z_i^{obs} and D_i^{obs}. In the context of our example, these four observed groups are:

- OBS$(0, 0) = \{i : Z_i^{obs} = 0, D_i^{obs} = 0\}$, those whose families live far away from college and do not get college degrees;

- OBS$(0, 1) = \{i : Z_i^{obs} = 0, D_i^{obs} = 1\}$, those whose families live far away from college and get college degrees;

- OBS$(1, 0) = \{i : Z_i^{obs} = 1, D_i^{obs} = 0\}$, those whose families live close to college and do not get college degrees;

- OBS$(1, 1) = \{i : Z_i^{obs} = 1, D_i^{obs} = 1\}$, those whose families live close to college and get college degrees;

The data pattern and latent principal strata associated with each observed group are shown in Table 8.2. Without monotonicity assumption, each of the four observed groups is a mixture of two latent principal strata. With monotonicity assumption, we can identify the OBS$(1, 0)$ group as never-takers, and OBS$(0,1)$ group as always-takers.

| | Latent Types | |
OBS(Z_i^{obs}, D_i^{obs})	No Monotonicity	With Monotonicity
OBS(1,1)	Always-takers, compliers	Always-takers, compliers
OBS(1,0)	Never-takers, defiers	Never-takers
OBS(0,1)	Always-takers, defiers	Always-takers
OBS(0,0)	Never-takers, compliers	Never-takers, compliers

Table 8.2   Classification of the population by Z_i^{obs} and D_i^{obs}, and the latent types of individuals belonging to each observed group, with and without monotonicity assumption.

Since we never observe never-takers with $D_i^{obs} = 1$ or always-takers with $D_i^{obs} = 0$, the treatment effect of D on Y for never-takers and always-takers cannot be estimated from the observed data, so we cannot estimate the ATE. As shown in the previous section, a LATE can be estimated by assuming exclusion restriction and monotonicity. These two assumptions, however, are not directly verifiable from the observed data. In the rest of this section, we will discuss what can be estimated from the observed data given these two assumptions, and sensitivity to these two assumptions.

The IV estimand

As discussed in Section 8.2, with exclusion restriction and monotonicity assumption, the IV estimand is equal to LATE, or the average causal effect of D on Y for the compliers.

Suppose we want to relax only the exclusion restriction. Angrist, Imbens, and Rubin (1996) showed if we assume that there is a direct effect of the instrument on outcome Y for each individual, and the instrument and the treatment have additive effects on outcome Y for each complier, then

$$\text{bias of IV estimand for LATE} = \frac{E(H_i)}{Pr(i \text{ is a complier})},$$

where

$$H_i = Y_i(1, d) - Y_i(0, d), \quad d = 0, 1,$$

is the direct effect of the instrument on outcome. The higher the correlation between the instrument and the treatment (i.e., the stronger the instrument), the higher the proportion of compliers, and therefore the less sensitive the IV estimand is to violations of the exclusion restriction.

If we relax the monotonicity assumption only, Angrist, Imbens, and Rubin (1996) showed that the bias of the IV estimand for LATE is

$$\text{bias} = -\frac{Pr(i \text{ is a defier})}{Pr(i \text{ is a complier}) - Pr(i \text{ is a defier})}$$
$$\times (E[Y_i(1) - Y_i(0)|i \text{ is a defier}] - E[Y_i(1) - Y_i(0)|i \text{ is a complier}]).$$

The smaller the proportion of defiers, or the stronger the instrument, or the less variation there is in the causal effect of D on Y between defiers and compliers, the less sensitive the IV estimand is to violations of the monotonicity assumption.

Estimating outcome distributions

Before proceeding, it is helpful to have more notation. Let ϕ_n, ϕ_a, ϕ_c, and ϕ_d denote the population proportions of never-takers, always-takers, compliers, and defiers respectively. Let $g_{tz}(y)$ denote the distribution of $Y_i(z, D_i(z))$ for individuals of type t ($t = n, a, c, d$) and for $z = 0, 1$. Let $f_{zd}(y)$ denote the directly estimable distribution of Y_i^{obs} for the OBS(z, d) group.

Imbens and Rubin (1997b) showed that under the exclusion restriction and the monotonicity assumption, one can in principle estimate the entire marginal outcome distributions $g_{tz}(y)$. The exclusion restriction implies that $g_{n0}(y) = g_{n1}(y) = g_n(y)$ and $g_{a0}(y) = g_{a1}(y) = g_a(y)$. Under the monotonicity assumption, there is no defier, so $\phi_d = 0$. As shown in the last column of Table 8.2, the OBS(1, 0) group and the OBS(0, 1) group can be identified as never-takers and always-takers respectively, so $g_n(y) = f_{10}(y)$ and $g_a(y) = f_{01}(y)$. At the same time, since the instrument is independent of $D_i(0)$ and $D_i(1)$, it is also independent of the principal stratum indicator C_i. So in large samples we can obtain the proportions of the principal strata: $\phi_n = \Pr(D_i^{obs} = 0|Z_i^{obs} = 1)$, $\phi_a = \Pr(D_i^{obs} = 1|Z_i^{obs} = 0)$ and thus $\phi_c = 1 - \phi_n - \phi_a$.

Since the OBS(0, 0) group is a mixture of never-takers and compliers, the sampling distribution of $f_{00}(y)$ is a mixture distribution of $g_n(y)$ and g_{c0}. The OBS(1, 1) group is a mixture of always-takers and compliers, so the sampling distribution of $f_{11}(y)$ is a mixture distribution of $g_a(y)$ and $g_{c1}(y)$.

$$f_{00}(y) = \frac{\phi_c}{\phi_c + \phi_n} g_{c0}(y) + \frac{\phi_n}{\phi_c + \phi_n} g_n(y),$$

$$f_{11}(y) = \frac{\phi_c}{\phi_c + \phi_a} g_{c1}(y) + \frac{\phi_a}{\phi_c + \phi_a} g_a(y).$$

We can invert the relations to get $g_{c0}(y)$ and $g_{c1}(y)$ as:

$$g_{c0}(y) = \frac{\phi_n + \phi_c}{\phi_c} f_{00}(y) - \frac{\phi_n}{\phi_c} f_{10}(y), \tag{8.8}$$

$$g_{c1}(y) = \frac{\phi_a + \phi_c}{\phi_c} f_{11}(y) - \frac{\phi_a}{\phi_c} f_{01}(y). \tag{8.9}$$

Estimation of the marginal outcome distributions can help the policy-makers understand the distributional effects of treatments, and thus is more desirable than a simple LATE estimate provided by the IV estimand. These estimates, however, depend heavily on the assumptions and large sample size. If the sample size is small, there is no guarantee that the sample estimates $\widehat{g}_{c0}$ and $\widehat{g}_{c1}$ will be nonnegative, as required by a probability density function. This is similar to the case that unbiased estimators for the variances can lead to negative estimates. Similar to the solution there, we need to constrain the estimates $\widehat{g}_{c0}$ and $\widehat{g}_{c1}$ to be nonnegative.

Imposition of nonnegativity can be done through nonparametric or parametric method. In nonparametric method, histogram estimates $\widehat{g}_{c0}$ and $\widehat{g}_{c1}$ can first be obtained in correspondence to (8.8) and (8.9), and then nonnegativity is imposed by revising the estimates to be

$$\widehat{g}_{cz}^{pos}(y) = \frac{\max(0, \widehat{g}_{cz}(y))}{\int \max(0, \widehat{g}_{cz}(y)) dy},$$

for $z = 0, 1$. In parametric method, it is assumed that $g_n(y)$, $g_a(y)$, $g_{c0}(y)$, and $g_{c1}(y)$ follow some parametric distributions, such as multinomial distribution (with cells defined similar to the histogram estimates) or normal distribution, then find the maximum likelihood estimates (MLEs). Through a Monte Carlo simulation example, Imbens and Rubin (1997b) showed that if exclusion restriction and monotonicity hold, although imposition of nonnegativity makes the estimators biased for LATE, it decreases the root-mean-squared error and median-absolute error. This is again similar to the result from imposing nonnegativity on variance estimates.

If we relax the exclusion restriction only, the distribution estimands obtained from (8.8) and (8.9) are actually

$$\widetilde{g}_{c0}(y) = g_{c0}(y) + \frac{\phi_n}{\phi_c}[g_{n0}(y) - g_{n1}(y)],$$

$$\widetilde{g}_{c1}(y) = g_{c1}(y) + \frac{\phi_a}{\phi_c}[g_{a1}(y) - g_{a0}(y)].$$

The bias depends on how much g_{n0} is different from g_{n1} and g_{a0} is different from g_{a1}. Imposing nonnegative on the density estimates cannot let the bias go away.

If we want to relax the monotonicity assumption but keep the exclusion restriction, the situation becomes more complex, since the proportions are estimated assuming monotonicity, and they will be biased for the true population proportions. In general, the bias of the estimands $\widetilde{g}_{c0}(y)$ and $\widetilde{g}_{c1}(y)$ obtained from (8.8) and (8.9) will be related to the true proportions of the principal strata and the true marginal outcome distributions.

Bayesian analysis

More principled inferences come from Bayesian analysis. First, consider the case without exclusion restriction or monotonicity assumption. Let η_{tz} denote the parameters for $g_{tz}(y)$, and let $g_{tz}^i = g_{tz}(Y_i^{\text{obs}})$. Let

$$\pi = (\phi_n, \phi_a, \phi_c, \phi_d, \eta_{n0}, \eta_{n1}, \eta_{a0}, \eta_{a1}, \eta_{c0}, \eta_{c1}, \eta_{d0}, \eta_{d1})$$

be the vector of parameters, and let $p(\pi)$ be its prior distribution. Let $\mathbf{Z}^{\text{obs}}$ be the N-vector of Z_i^{obs}, $\mathbf{D}^{\text{obs}}$ be the N-vector of D_i^{obs}, and $\mathbf{Y}^{\text{obs}}$ be the N-vector of Y_i^{obs}. The posterior distribution is

$$p(\pi | \mathbf{Z}^{\text{obs}}, \mathbf{D}^{\text{obs}}, \mathbf{Y}^{\text{obs}}) \propto p(\pi)$$

$$\times \prod_{i \in OBS(0,0)} (\phi_n g_{n0}^i + \phi_c g_{c0}^i) \prod_{i \in OBS(0,1)} (\phi_a g_{a0}^i + \phi_d g_{d0}^i)$$

$$\times \prod_{i \in OBS(1,0)} (\phi_n g_{n1}^i + \phi_d g_{d1}^i) \prod_{i \in OBS(1,1)} (\phi_a g_{a1}^i + \phi_c g_{c1}^i)$$

Taking the view of data augmentation (Tanner and Wong, 1987), we can augment the observed data by $\mathbf{C}$, the N-vector of C_i, and thus the complete data are

Z_i^{obs}	D_i^{obs}	$\Pr(C_i = n)$	$\Pr(C_i = a)$	$\Pr(C_i = c)$	$\Pr(C_i = d)$
0	0	$\dfrac{\phi_n g_{n0}^i}{\phi_n g_{n0}^i + \phi_c g_{c0}^i}$	0	$\dfrac{\phi_c g_{c0}^i}{\phi_n g_{n0}^i + \phi_c g_{c0}^i}$	0
0	1	0	$\dfrac{\phi_a g_{a0}^i}{\phi_a g_{a0}^i + \phi_d g_{d0}^i}$	0	$\dfrac{\phi_d g_{d0}^i}{\phi_a g_{a0}^i + \phi_d g_{d0}^i}$
1	0	$\dfrac{\phi_n g_{n1}^i}{\phi_n g_{n1}^i + \phi_d g_{d1}^i}$	0	0	$\dfrac{\phi_d g_{d1}^i}{\phi_n g_{n1}^i + \phi_d g_{d1}^i}$
1	1	0	$\dfrac{\phi_a g_{a1}^i}{\phi_a g_{a1}^i + \phi_c g_{c1}^i}$	$\dfrac{\phi_c g_{c1}^i}{\phi_a g_{a1}^i + \phi_c g_{c1}^i}$	0

Table 8.3 $\Pr(C_i = t \mid Z_i^{\text{obs}}, D_i^{\text{obs}}, Y_i^{\text{obs}}, \pi)$, conditional probability of individual i being type t given observed data $(Z_i^{\text{obs}}, D_i^{\text{obs}}, Y_i^{\text{obs}})$ and parameters π.

$(\mathbf{Z}^{\text{obs}}, \mathbf{D}^{\text{obs}}, \mathbf{Y}^{\text{obs}}, \mathbf{C})$. The Gibbs sampler (Geman and Geman, 1984; Gelman and Rubin, 1992) can be used to simulate π from its posterior distribution by iteratively imputing $\mathbf{C}$ from its conditional distribution given $(\mathbf{Z}^{\text{obs}}, \mathbf{D}^{\text{obs}}, \mathbf{Y}^{\text{obs}}, \pi)$, and drawing π from its conditional distribution given $(\mathbf{Z}^{\text{obs}}, \mathbf{D}^{\text{obs}}, \mathbf{Y}^{\text{obs}}, \mathbf{C})$.

The conditional distribution $\Pr(C_i = t \mid Z_i^{\text{obs}}, D_i^{\text{obs}}, Y_i^{\text{obs}}, \pi)$ in the first step of each iteration is given in Table 8.3. Let $\Psi(t) = \{i \mid C_i = t\}$ denote the set of units of type t. The conditional distribution in the second step of each iteration is given by

$$p(\pi \mid \mathbf{Z}^{\text{obs}}, \mathbf{D}^{\text{obs}}, \mathbf{Y}^{\text{obs}}, \mathbf{C}) \propto p(\pi) \prod_{i \in \Psi(n) \cap OBS(0,0)} \phi_n g_{n0}^i \prod_{i \in \Psi(c) \cap OBS(0,0)} \phi_c g_{c0}^i$$

$$\times \prod_{i \in \Psi(a) \cap OBS(0,1)} \phi_a g_{a0}^i \prod_{i \in \Psi(d) \cap OBS(0,1)} \phi_d g_{d0}^i$$

$$\times \prod_{i \in \Psi(n) \cap OBS(1,0)} \phi_n g_{n1}^i \prod_{i \in \Psi(d) \cap OBS(1,0)} \phi_d g_{d1}^i$$

$$\times \prod_{i \in \Psi(a) \cap OBS(1,1)} \phi_a g_{a1}^i \prod_{i \in \Psi(c) \cap OBS(1,1)} \phi_c g_{c1}^i$$

For convenience of drawing from this conditional distribution, we can let the prior distribution of π be

$$p(\pi) = p(\phi_n, \phi_a, \phi_c, \phi_d) \prod_{t \in \{n,a,c,d\}} \prod_{z=0,1} p(\eta_{tz}),$$

and choose conjugate priors when possible. For example, $p(\phi_n, \phi_a, \phi_c, \phi_d)$ can be a Dirichlet distribution; if the outcome is binary and binomial distribution is used for g_{tz}, or if the outcome is continuous and normal distribution is used for g_{tz}, the corresponding conjugate priors can be used for the η parameters.

If the exclusion restriction is assumed, then $g_{n0} = g_{n1}$ and thus $\eta_{n0} = \eta_{n1}$, similarly, $g_{a0} = g_{a1}$ and thus $\eta_{a0} = \eta_{a1}$. The only difference from the above procedure is in the second step of each iteration, where only one η parameter needs to be drawn for always-takers, and only one η parameter needs to be drawn for never-takers. If the monotonicity assumption is made, then $\phi_d = 0$, $\Psi(d)$ is empty, and the two distributions g_{d0} and g_{d1} are irrelevant. The above Bayesian analysis can still be performed.

There are several advantages of the principled Bayesian analysis. First, one can investigate sensitivity to the exclusion restriction and the monotonicity assumption easily by examining how the posterior distributions for causal estimands (e.g. LATE) change. Second, as shown by Imbens and Rubin (1997a), even assuming exclusion restriction and monotonicity, Bayesian analysis with full posterior distribution can yield inference differing from MLE and IVE (IV estimator) with normal approximation. In a Monte Carlo simulation, they showed that the central probability intervals from Bayesian analysis have higher coverage rate than those from MLE with normal approximation, and are less wide than those from IVE with normal approximation.

8.4   Some recent applications

In a series of papers, Gerber and Green (1999, 2000a, 2000b) studied the effect of personal canvassing on voter turnout. In some voter mobilization experiments, lists of registered people were randomly assigned to treatment and control groups. Some of the people in the treatment group were successfully contacted for personal canvassing. Using IV analysis with treatment assignment as the IV, Green and Gerber showed that actual canvassing increased voter turnout.

Ichino and Winter-Ebmer (1998) presented evidence based on Germany that supports the existence of heterogeneous returns to schooling and the validity of the LATE interpretation of IV. They used two different IVs, one is the indicator whether the father of individual i served actively in World War II, the other is the indicator whether father of individual i has a degree higher than high school. They argued that the different IV estimates thus obtained should be interpreted as LATE estimates of the returns for different subgroups in the population.

IV estimation together with its LATE interpretation have also been discussed and applied, for example, in evaluating the effect of childbearing on labor supply (Angrist and Evans, 1998; Cruces and Galiani, 2003), the effect of school interruption on earnings (Meng and Gregory, 1999), the impact of welfare benefit denial on future receipt (Green and Warburton, 2001), and the effect of psychological interventions on depression (Dunn et al., 2003).

8.5   Discussion

Rubin (1986) made a point that observational study can only be informative about the causal effect of treatment for those whose treatment status can be thought of having been manipulated in some way. This is exemplified in estimating causal effect with IV: the average causal effect can only be estimated for those who would be induced to take the treatment by changing the value of the instrument. Different instruments will thus lead to estimates of average causal effects for different subpopulations. For example, in estimating the effect of college degree on income, if we use father's education as the instrument, we will estimate the average causal effect for those who would get college degrees only because their fathers have college degrees, which is different from the LATE from using families' proximity as the instrument. Therefore, in choosing IV, we not only want the IV to be valid (so that the five assumptions in Section 8.2 are satisfied), but we also need to consider whether the corresponding LATE is of policy interest.

The extension of the framework in this chapter can be extended for multivalued instrument and multiple instruments along the line in Imbens and Angrist (1994), and for multivalued treatment along the line in Angrist and Imbens (1995). Under assumptions similar to those in Section 8.2, the IV estimands in those cases are equal to weighted averages of LATEs. The marginal outcome distributions cannot be estimated as in the binary-treatment, binary-instrument case, towing to the complex mixture structure. Bayesian analysis can still be performed, but careful modeling should be considered, and the payoff could be substantial.

In real studies, covariates are usually observed together with the outcomes of interest. They can serve several purposes: they can help predict the principal strata and the missing potential outcomes for the individuals, thus making inference more precise; they can make inferences more specific by estimating LATEs for different subpopulation indexed by the covariates. With discrete covariates, one can seek LATE for each joint value of the covariates. Angrist, Graddy, and Imbens (2000) showed that if additive linear structure for the (possibly continuous) covariates is assumed, the IV estimand is equal to a weighted average of LATEs. The Bayesian analysis presented in Section 8.3 can be extended to incorporate covariates by making the proportions ϕ and the outcome distributions g_{tz} depend on the covariates.

9

Principal stratification

Constantine E. Frangakis[1]

9.1 Introduction: partially controlled studies

We often need to evaluate the effects of treatments or other interventions on the outcomes of biologic, clinical, economic, or other behavioral nature. A general type of such studies has some factors of interest that are controlled and others that are not controlled. Such "partially controlled" studies are increasingly met in practice, especially when controlled studies cannot be conducted at all. We argue that in such partially controlled studies, we can use a framework, "principal stratification", for formulating and addressing in a systematic way the problems that arise.

The goal here is to provide a review of principal stratification. The next section provides two examples of partially controlled studies, for demonstration (see also Rubin, 2000; Zhang, 2002). The three main sections discuss the role of principal stratification, respectively, in formulating quantities of interest (estimands), flexible assumptions, and better designs for partially controlled studies.

9.2 Examples of partially controlled studies

Example and goals in the study of surrogate endpoints

When conducting clinical trials to compare treatments on a primary outcome (end-point), we also record variables of the patient's progression after the treatment

[1]Department of Biostatistics, Johns Hopkins University, Baltimore, Md. The discussion at biosun01.biostat.jhsph.edu/~cfrangak/papers/discussion-pstrat.pdf provides additional comments, including points of critique and how they are addressed by principal stratification. This work was supported in part by NEI grant RO1 EY 014314-01 and was completed, while the author was on leave, at the Press Room of the President of the Hellenic Republic.

Applied Bayesian Modeling and Causal Inference from Incomplete-Data Perspectives.
Edited by A. Gelman and X-L. Meng © 2004 John Wiley & Sons, Ltd ISBN: 0-470-09043-X

but before the primary outcome is measured. Such posttreatment variables, called "surrogate endpoints" are of increasing interest (e.g., Prentice, 1989; Freedman, Graubard, and Schatzkin, 1992; Buyse et al., 2000). We focus here on a fundamental question: how to evaluate treatment effects on the outcomes that are associative (i.e., occur together) and that are dissociative (i.e., do not occur together) with effects on the surrogate.

To focus on the main points, we consider a template study with a standard ($z = 1$) and a new ($z = 2$) therapy for cancer patients. For patient i, consider the potential outcomes (Rubin, 1974, 1977, 1978a) of the patient under each treatment $z = 0,1$: $Y_i(z)$ for the survival time (the primary endpoint), and $S_i(z)$ for a measure (L = low, H = high) of cancer response two months after treatment assignment. Also, assume for simplicity that no patient dies before two months so that cancer response is measured, that treatments $\{Z_i\}$ are completely randomized, and that, for subject i, cancer response $S_i^{obs} = S_i(Z_i)$ and, later, survival time $Y_i^{obs} = Y_i(Z_i)$ are measured, thereby creating what we call a "validation" study.

To evaluate the above question, it is important to have a definition of causal effects. A causal effect of the treatment on the outcome Y is defined as a comparison (e.g., difference or ratio of averages) between the ordered sets of potential outcomes on a common set of subjects, for example, the comparison between the ordered sets

$$\{Y_i(1): i \in \text{set}_1\} \text{ and } \{Y_i(2): i \in \text{set}_2\}, \qquad (9.1)$$

if the groups of subjects, set$_1$ and set$_2$, being compared are identical (Neyman, 1923; Rubin, 1978). For example, a comparison of the distribution $\Pr(Y_i(1))$ to $\Pr(Y_i(2))$ describes causal effects for all subjects.

The main goal for the posttreatment variable S here is to evaluate if it possesses the following property:

Causal necessity: S is necessary for the effect of treatment on the outcome Y in the sense that an effect of treatment on Y can occur only if an effect of treatment on S has occurred.

The property of causal necessity relates to the degree to which the treatment acts on the outcome together or separately from acting on the surrogate. This information is central feedback about the mechanisms of treatment action, for example, in guiding pathways of drug development.

To approach quantifying this property, Prentice (1989) defined S^{obs} to be a surrogate if it satisfies certain criteria, the main being that the observed outcome Y_i^{obs} should be conditionally independent of the assigned treatment Z_i given the observed value S_i^{obs} of the posttreatment variable in the validation study. (Prentice (1989), used a hazard regression parameterization for multiple-time measurements on S^{obs}. For clarity, we discuss the single-time measurement case.) Related definitions have been proposed when exact independence is not expected and that compare results (e.g., r-squares) of the regression of the outcome on treatment before and after conditioning on the variable S^{obs} (e.g., Freedman et al., 1997;

Lin, Fleming, and De Gruttola, 1997; Buyse and Molenberghs, 1998; Gail, Pfeiffer, Houwelingen, and Carroll, 2000). All these approaches are based on the general idea to call S^{obs} a surrogate if S^{obs} is a good predictor (relative to treatment Z) of outcome Y^{obs} when conditioning on S^{obs} and Z. Thus, in principle, all current definitions generate from Prentice's main criterion of a statistical surrogate:

Statistical surrogate (Prentice (1989) criterion): S is a *statistical* surrogate for a comparison of the effect of $z = 1$ versus $z = 2$ on Y if, for all fixed s, that comparison of the distributions

$$\Pr(Y_i^{obs}|S_i^{obs} = s, Z_i = 1) \text{ and } \Pr(Y_i^{obs}|S_i^{obs} = s, Z_i = 2). \qquad (9.2)$$

is an equality.

Equality (9.2) is also known as "net-treatment" equality (Rosenbaum, 1984b). It is important to note, however, that the *net-treatment equality does not generally have the interpretation of (causal) treatment effect equality.* Since treatment is randomized, $\Pr(Z_i = 1|S_i(1), S_i(2), Y_i(1), Y_i(2))$ is a common constant across subjects, which implies that assignment is ignorable (Rubin, 1978) and that the comparison in (9.2) is equivalent to the comparison in

$$\Pr(Y_i(1)|S_i(1) = s) \text{ and } \Pr(Y_i(2)|S_i(2) = s). \qquad (9.3)$$

The last comparison is problematic if the treatment has any effect on the posttreatment variable. Then, the groups $\{i : S_i(1) = s\}$ and $\{i : S_i(2) = s\}$ (i.e., who get posttreatment value s under standard and new treatment, respectively) are not the same subjects, and, by (9.1), the equality does not evaluate a treatment effect. Frangakis and Rubin (2002) show that the standard approaches (e.g., Prentice, 1989; "individual-level surrogacy" of Buyse et al., 2000) do not in principle satisfy the property of causal necessity. This problem, although known to epidemiologists (e.g., see Rosenbaum, 1984; Robins and Greenland, 1992) has not been addressed appropriately.

In Section 9.4, we review how principal stratification provides a new criterion that satisfies the property of causal necessity and quantifies associative and dissociative effects.

Example and goals in the study of needle exchange

Needle exchange programs (NEPs) attempt to reduce HIV transmission among injection drug users (IDUs) (e.g., Bastos and Strathdee, 2000). A NEP consists of sites, usually vans, where a user can visit and exchange a used needle for a clean one. However, controversy exists on whether NEPs actually help or not (e.g., Bruneau, Lamothe, and Franco, 1997; Drucker, Lurie, Wodak, and Alcabes, 1998). We consider estimation of the effect on HIV that is attributable to using versus not using the NEP.

Most of the existing methods evaluate the NEPs either by (a) comparing IDUs who use the NEP to IDUs who do not use it, with respect to HIV transmission, a comparison called "as-treated" or by (b) comparing HIV rates before and after the institution of a NEP (e.g., Keende, Stimson, Jones, and Parry-Langdon, 1993; van Ameijden, van den Hoek, and Coutinho, 1994; Drucker, Lurie, Wodak, and Alcabes, 1998). Such methods are limited by the fact that the decision about who uses the NEP, and who provides outcomes (HIV tests) is not controlled by the study. Consequently, even after as-treated analyses stratify on certain measured variables, IDUs who use the NEP can be at higher (or lower) risk for HIV, before the start of NEP, compared to IDUs who do not use the NEP (e.g., Drucker, Lurie, Wodak, and Alcabes, 1998; Schechter et al., 1999). Then any differences in HIV rates observed between NEP users and nonusers can reflect differences between the groups and not the impact of NEP. Analogous complications arise for before-after comparisons because of trends of the HIV epidemic, and from likely differences between IDUs who agree to be tested for HIV (for whom outcomes get measured), and those who do not agree (for whom the outcome is unmeasured) (e.g., Kaplan, 1994).

To address these complications, new research methods must capitalize on the factor that *is controlled* by these studies: the location of the NEP sites (vans) relative to subjects' residences. For example, in the Baltimore NEP (Vlahov et al., 1997), the location of a NEP site was chosen to be within each of a number of broader areas that had exhibited high HIV rates in the years prior to the study. Importantly, however, within those broad areas, the location of the NEP sites was chosen essentially at random. More generally, then, and within broad areas, IDUs who live closer to NEP sites are comparable, before the NEP starts, to IDUs who live farther from the NEP sites. Moreover, it is expected that larger distance of an IDU from the NEP site decreases the likelihood that the IDU exchanges needles at the site and/or accepts to get tested for HIV, as suggested by positive relations between the access to and use of services in other settings (e.g., McClellan, McNeil, and Newhouse, 1994). These points indicate that we can use the controlled factor of location with a method to provide a better evaluation of the NEPs.

Such a method, however, did not really exist. In particular, if there were a single uncontrolled factor, the needed method for NEP evaluation would share some aspects with the more standard method of instrumental variables in some earlier studies for other evaluations; for example, see Card (1986), McClellan, McNeil, and Newhouse (1994), and Angrist, Imbens, and Rubin (1996). The problem in the NEP study, however, is the presence of more than one uncontrolled factor simultaneously, that is, here both exchange of needles at the NEP, and whether or not the subject provides outcomes (Figure 9.1). For more such demanding studies, earlier work has shown that the standard instrumental variables method is not appropriate to estimate the treatment effects (Frangakis and Rubin, 1999).

In Section 9.5, we discuss how principal stratification can be used to better evaluate such studies with combined uncontrolled factors.

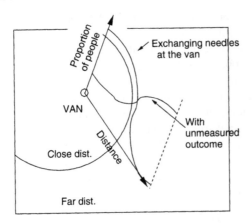

Figure 9.1   Needle exchange program as a partially controlled study. Combined factors of (i) exchanging needles and (ii) measuring outcome, partially controlled through distance of the program from the subjects.

9.3   Principal stratification

In the two partially controlled examples discussed above, a common goal is the estimation of certain causal effects of the controllable factor on the outcome that also takes into consideration the partially controlled factors (the cancer response, for the surrogate endpoint example and the exchange behavior, for the needle exchange example). In this section, which borrows essentially from Frangakis and Rubin (2002), we propose such estimands that use the posttreatment variable and that are always causal effects.

To do this, first consider, more generally, a group of subjects $i = 1, \ldots, n$, where each can be potentially assigned either a standard "treatment" $(z = 1)$ or a new "treatment" $(z = 2)$. Also, let Y denote the outcome at a specific time after assignment of each unit, where we let $Y_i(z)$ be the value of Y if unit i is assigned treatment z, for $z = 1, 2$. Finally, let S denote the partially controlled (posttreatment) factor, where we let $S_i(z)$ be the value of S if unit i is assigned treatment z, for $z = 1, 2$. Consider all the potential values of the posttreatment variable jointly, and construct the following partitions.

  (a) The basic principal stratification P_0 with respect to posttreatment variable S is the partition of units $i = 1, \ldots, n$ such that within any set of P_0, all units have the same vector $(S_i(1), S_i(2))$.

  (b) A principal stratification P with respect to posttreatment variable S is a partition of the units whose sets are unions of sets in the basic principal stratification P_0.

An example of a principal stratification P is the partition of subjects into the set whose posttreatment variable is unaffected by treatment in this study (i.e., with $S_i(2) = S_i(1)$) and into the remaining subjects (i.e., with $S_i(2) \neq S_i(1)$). Generally, we cannot directly observe subjects' principal strata because we cannot directly observe both $S_i(1)$ and $S_i(2)$ for any i. Nevertheless, consideration of the principal strata is important in order to determine which quantities are causal. Generally, a principal stratification generates the following estimands.

Let P be a principal stratification with respect to the posttreatment variable S, and let S_i^P indicate the stratum of P to which unit i belongs. Then, a principal effect with respect to P is defined as a comparison of potential outcomes under standard versus new treatment within a principal stratum ς in P, that is, a comparison between the ordered sets

$$\{Y_i(1): S_i^P = \varsigma\} \text{ and } \{Y_i(2): S_i^P = \varsigma\}. \tag{9.4}$$

Principal effects are important thanks to their conditioning on principal strata. Although the potential variable $S_i(1)$ generally differs from $S_i(2)$, the value of the ordered pair $(S_i(1), S_i(2))$ is, by definition, not affected by treatment, just like the pair (birthdate, gender). Therefore, we have

Property. The stratum S_i^P, to which unit i belongs, is unaffected by treatment for any principal stratification P.

And, by definition (9.1), we have,

Property. Any principal effect, as defined in (9.4), is a causal effect.

Using the examples of surrogate endpoints and needle exchange, we discuss next how these properties of principal stratification help in three important aspects of partially controlled studies. First, in formulating better estimands. Second, in allowing assumptions for more appropriate analysis. Third, in implementing designs that allow more reliable estimation.

9.4 Estimands

Example on surrogate endpoints (continued)

Consider first the four finest principal strata with respect to the binary surrogate of early cancer response in our template study:

1. subjects whose cancer response would be low no matter the treatment, $\{i: S_i(1) = S_i(2) = L\}$, and whom we label "sicker" patients (the terms "sicker" etc., are for convenience, and do not imply knowledge of all the characteristics that underlie the principal strata);

2. subjects whose cancer response would be high no matter the treatment $\{i: S_i(1) = S_i(2) = H\}$, and whom we call "healthier";

3. subjects whose cancer response under new treatment would be better than under standard treatment, $\{i: S_i(1) = L \text{ and } S_i(2) = H\}$, and whom we label "normal";

4. subjects whose cancer response under new treatment would be worse than under standard treatment, $\{i: S_i(1) = H$ and $S_i(2) = L\}$, and whom we label "special."

Then, we propose the following criterion of surrogacy:

Definition 1 S is a principal *surrogate for a comparison of the effect of $z = 1$ versus $z = 2$ on Y if, for all fixed s, that comparison between the ordered sets*

$$\{Y_i(1): S_i(1) = S_i(2) = s\} \text{ and } \{Y_i(2): S_i(1) = S_i(2) = s\}, \qquad (9.5)$$

results in equality.

The above criterion in words is that causal effects of treatment on outcome Y may only exist when causal effects of treatment on the posttreatment variable S exist. Thus, our criterion based on principal stratification immediately satisfies the property of causal necessity of Section 9.2.

To see the contrast with a statistical surrogate, note that, although definition (9.5) does not involve an assumption about the assignment model for Z_i, under randomization, (9.5) implies that the same comparison applied to

$$\Pr(Y_i^{obs}|S_i(1) = S_i(2) = s, Z_i = 1) \text{ and } \Pr(Y_i^{obs}|S_i(1) = S_i(2) = s, Z_i = 2).$$
$$(9.6)$$

also results in equality. Then we have the following:

Result 1. (a) If the posttreatment variable S is a statistical surrogate (equation (9.2)) then it is not, generally, a principal surrogate (equation (9.5)). (b) If the posttreatment variable S is a principal surrogate, then it is not, generally, a statistical surrogate.

To understand better the implications of Result 1, we offer a proof for part (b) by discussing the example in Figure 9.2 for the comparison of averages (to show the result, in the figure we need only consider scenarios with no "special" subjects).

The subgroups of patients who experience no causal effect of treatment on the early cancer response ("sicker" and "healthier") experience no causal effect of treatment on survival. Therefore, by criterion (9.6), early cancer response is a principal surrogate in this study.

However, when $s = L$, the subjects $\{i: S_i^{obs} = L, Z_i = 1\}$ in the left-side conditioning of (9.2) is the mixture of "sicker" and "normal" patients under standard treatment, whereas the subjects $\{i: S_i^{obs} = L, Z_i = 2\}$ are, in fact, a different group of subjects—the "sicker" patients only—under new treatment. Using the numbers of Figure 9.2, the left side of (9.2) has mean 20 months, whereas the right side of (9.2) has mean 10 months. It follows that early cancer response is not a statistical surrogate. Therefore, although the standard interpretation would be that the new treatment decreases survival whenever it cannot change a low value of the surrogate, that conclusion is incorrect, as the principal surrogacy of S clearly

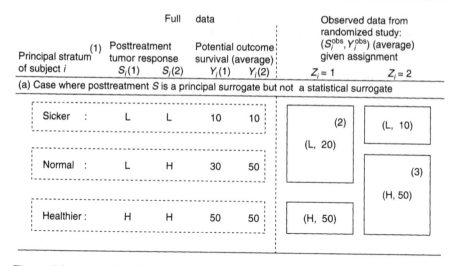

Figure 9.2   Example where early cancer response is a principal but not a statistical surrogate. Notes: (1) We set equal proportions for each principal stratum, for simplicity of demonstration; (2)$\frac{1}{2}(10) + \frac{1}{2}(30)$; (3)$\frac{1}{2}(50) + \frac{1}{2}(50)$.

indicates. Part (a) of Result 1 can also easily be shown. The discrepancy between the two criteria indicated in Result 1 occurs more generally because a statistical surrogate does not generally involve causal effects.

More generally than assessing principal surrogacy, we can evaluate the effects of treatment on outcome that are associative and dissociative with effects on the posttreatment variable in the validation study. An effect on outcome that is dissociative with an effect on surrogate is defined as a comparison between $\{Y_i(1): S_i(1) = S_i(2)\}$ and $\{Y_i(2): S_i(1) = S_i(2)\}$, that is, it occurs without an effect on the surrogate. An effect on outcome that is associative with an effect on the surrogate is defined as a comparison between $\{Y_i(1): S_i(1) \neq S_i(2)\}$ and $\{Y_i(2): S_i(1) \neq S_i(2)\}$. Note that both the associate and the dissociative effects can, in principle, be further stratified on basic principal strata.

9.5   Assumptions

Example on needle exchange (continued)

To discuss the role of principal stratification on flexibility of assumptions, we return to the example on needle exchange, discussed as a simplified context of the Baltimore NEP (Vlahov et al., 1997). Details of evaluating the program can be found in Frangakis et al. (2004).

The Baltimore NEP can be considered as part of a larger cohort study, the ALIVE study (Vlahov et al., 1991), which follows IDU subjects, offering clinical testing for HIV and interviews, independently of attendance to the NEP, although

ALIVE is linked with the NEP component (after appropriate subjects' consent). We consider IDUs in ALIVE who are HIV negative at the beginning of needle exchange, and, focus in an area served by a single NEP van. To formulate the appropriate question and methods for evaluation, consider the following data for each person that would arise if the van were placed at a distance d from person i:

1. $E_i(d)$ for the indicator for whether or not the subject would exchange needles at the van (1 for yes);

2. $Y_i(d)$ for the subject's HIV status five years after the van begins service; and

3. $M_i(d)$ for the indicator for whether or not the HIV status in (ii) would be measured (1 if the subject would consent to testing at the ALIVE or NEP visit).

In the simple setting, distance is dichotomized, $d = 1, 2$ for close and far respectively, and "far" is assumed to be chosen far enough so that no subject living far ($d = 2$) from the van would visit it to exchange needles (i.e., $E_i(2) = 0$), although subjects living close may or may not exchange needles (the condition can be relaxed).

There are two finest principal strata of needle exchange in this study:

1. a "never-exchanger," denoted by $U_i = n$, that is, a subject who, in the context of this study, would never exchange needles whether the van was placed close or not; and

2. a "close-exchanger," denoted by $U_i = c$, that is, a subject who would exchange needles if and only if the van was placed close.

These strata relate to those considered by Angrist, Imbens, and Rubin (1996) for single posttreatment variables. In the NEP, important implications also have the analogous classification of subjects by measurement status $M_i(d), d = 1, 2$.

From the earlier discussion, the estimand of interest for evaluating the effect on HIV attributed to the NEP should be a principal effect of distance on HIV with respect to exchanging needles. To define and estimate such effect, we focus on the main issues related to the points of the chapter, and refer the reader to (Frangakis et al., 2004) for issues of less relevance.

Since never-exchangers are, by definition, people who would not exchange needles at the van, regardless of its location, placing the van far or close to never-exchangers can be assumed to not have an important effect on either their outcome HIV status in five years, or their behavior in allowing the study to measure that status. More explicitly, this is stated as follows.

Compound Exclusion Restriction: if subject i is a never-exchanger, that is, if $E_i(1) = E_i(2)$, then: (a) $Y_i(1) = Y_i(2)$; and (b) $M_i(1) = M_i(2)$.

We now define the estimand of interest as follows. Denote the average outcome among each principal stratum by $\overline{Y}_u(d) := E\{Y_i(d)|U_i = u\}$. By the properties of principal stratification, comparisons of HIV rates by distance within principal strata are well-defined causal effects. One such comparison is

$$\text{ECE} := \overline{Y}_{u=c}(d = 1)/\overline{Y}_{u=c}(d = 2),$$

the proportional effect of close versus far distance on HIV *among* "close-exchangers," which we call the "exchange causal effect" (ECE) (see also Imbens and Rubin, 1997a,b; Frangakis and Rubin, 1999). In the above effect, all close-exchangers do exchange needles when close to the NEP (numerator), and no close-exchanger exchanges needles when far from the NEP (denominator). In addition, by compound exclusion, distance can only affect HIV status for close-exchangers, in other words, there is no effect of distance on HIV that is dissociative with exchange. For these reasons, we take the principal effect ECE to represent the effect of distance on HIV and that is attributable to exchange.

We consider two additional assumptions that can help estimation. First, vans are placed by study staff, so, on the basis of earlier arguments, it is reasonable to assume that placement of the vans is independent of the potential outcomes of the variables measured after placement of the vans, conditionally on the observed covariates X_i available to the study staff. This essentially implies that the mechanism of determining actual distance D_i of the nearest needle exchange site to subject i is ignorable.

Owing to the second uncontrolled factor, we do not observe all outcomes Y. It is, therefore, necessary to connect information from HIV statuses that are measured to HIV statuses that are not measured. Quite generally, in order to avoid confounding, such connections first need to balance (e.g., match) important covariates between subjects with and without HIV measures. Exchange of principal stratum U_i, as a pretreatment characteristic, is likely a predictor of outcome, in the sense that close-exchangers can be at very different risk for HIV than never-exchangers even in the absence of the NEP, and also a predictor of who has HIV status measured. Therefore, if we know the exchange stratum for all subjects, we should first stratify subjects by it, before connecting those without measured HIV status to those with measured HIV status. One way to formalize this is with the following assumption.

Latent ignorability: Among subjects who have the same principal stratum U_i, the same other observed covariates X_i, and the same distance from the van d, the potential outcomes are independent of whether or not those outcomes are measured:

$$\Pr(Y_i(d), M_i(d) \mid X_i, U_i) = \Pr(Y_i(d) \mid X_i, U_i) \Pr(M_i(d) \mid X_i, U_i). \qquad (9.7)$$

Under ignorability of distance of the vans from the subjects, latent ignorability, and compound exclusion for missing outcomes and needle exchange, and together with certain additional conditions, Frangakis and Rubin (1999) obtained results that, in this application, amount to the following:

Result 2. The effect on HIV attributable to exchange is estimable, but it is not estimable by standard instrumental variables methods.

Because measuring HIV data is not controlled, of course, latent ignorability, as other assumptions, should be judged, not by whether or not it is certain to be correct, but *in comparison to other approaches*, on the inferences and the sensitivity analyses it can produce. The point is that a framework that allows that the mechanism of who provides measurements can depend on the latent exchange status U, also allows inferences to be closer to the true NEP effects *than* inferences that do not allow such possibility (e.g., Gilbert, Bosch, and Hudgens, 2003).

For the NEP, these points are demonstrated practically in Frangakis et al. (2004). Their results point to a reduction of up to 90% in HIV incidence that is attributable to needle exchange, a benefit that is substantially larger than the one indicated by the standard method. Importantly, they also show that the *comparison* between the standard and the new method gives additional insights into the characteristics of those who exchange needles more often, and provides additional support for the benefit of NEPs.

9.6   Designs and polydesigns

The new ways of formulating questions and analyzing data to address them, also suggest that we should start using different designs, specific to the new questions and analyses of principal stratification. Better designs can play an important role when either the cost is differential for different measurements (e.g., on outcome versus controlled versus partially controlled factors) or when there is concern in the robustness of the answers to model specification. Designs addressing cost issues are discussed by Jo (1999) and Frangakis and Baker (2001). The case is more demanding when we consider designs to address possible misspecification of the model for principal stratification.

Suppose that, to do this, instead of considering the "full design" (all the data), we create a "reduced design," that is, a subset of the full data together with the rule that created this subset. An example of this in the context of the NEP study is discussed in Li, Frangakis, and Varadhan (2004). Over the duration of six years, a total of only 54 subjects were diagnosed as new HIV cases. In the NEP, then, a "reduced design" can be one where we match each new HIV case with a subject who was non-HIV (control) at the same time, and so that the control is "close" to the case in some metric of covariates measured before the program starts. The "reduced design" has the advantage that it explicitly focuses on controls that are similar to the cases in the covariates, and, therefore, as in other situations with "case-control" designs, the "reduced design" avoids extrapolation of the model that can have unwanted consequences if the model is misspecified.

It is, therefore, noteworthy why "reduced designs" in partially controlled studies have not been used frequently for estimation. One can see two reasons for this.

The first reason is that, in the more traditional, instrumental variables framework for special cases of partially controlled studies, the properties of the model tie together assumptions specific to the "full design" as well as assumptions on potential outcomes and principal strata (even if such quantities are not stated explicitly in that framework). This, of course, is expected, because the equations of the IV framework have as reference the *observed* outcomes and *observed* exposures to the uncontrolled treatment, which are the result of a *combination* of potential outcomes, principal strata, and mechanisms of selection and assignment of subjects. This issue, originally pointed out by Imbens and Rubin (1994) and Angrist, Imbens, and Rubin (1996) in relation to appreciating the role of different assumptions, also has important consequences for designs. In particular, in a design that differs from the "full design," such as the "reduced design," it becomes quite difficult for the standard framework to distinguish between the two types of assumptions, in the sense of replacing those specific to the "full design" by those of the "reduced design," and keeping the ones specific to the potential outcomes and principal strata. This task is, nevertheless, straightforward within the framework of principal stratification, which emphasizes making each assumption explicit, as pointed out in the previous section.

The second reason for the infrequent use of reduced designs in partially controlled studies is more subtle. From the likelihood of a reduced design as induced from the model on the principal stratification in the full design, we can find that the estimand of interest (e.g., the associative effect ECE of (9.7)) is not necessarily identifiable, even if it is identifiable in the full design. This is because partially controlled studies are more complex in structure than more familiar studies, such as a simple case-control modeled by conditional logistic regression. It would appear, therefore, that addressing robustness is in conflict with identifiability in a reduced design.

To address this conflict, Li, Frangakis, and Varadhan (2004) propose a class of "polydesign methods", which are methods that use a combination of designs, specifically here the "full design" and the "reduced design." The key idea of polydesign methods is to use the reduced design to estimate only certain parameters identifiable by it, and use the full design to estimate the remaining parameters, in a way that keeps identifiability of the main estimand of interest. Li, Frangakis, and Varadhan (2004) show that polydesign methods can combine both, identifiability of the estimands and more robust estimation for possible model misspecifications that relate to the data that are present in the full design but not in the reduced design. Operational details of polydesign methods are a subject for further work.

Part II

Missing data modeling

10

Nonresponse adjustment in government statistical agencies: constraints, inferential goals, and robustness issues

John Eltinge[1]

10.1 Introduction: a wide spectrum of nonresponse adjustment efforts in government statistical agencies

I thank the editors of this volume for the invitation to present brief comments on some issues in nonresponse modeling and adjustment encountered by government statistical agencies. The general missions of statistical agencies vary widely. At

[1]Bureau of Labor Statistics, Washington, D.C. The author thanks Tom Belin, David Binder, Shail Butani, Steve Cohen, Mark Crankshaw, Don Dillman, Bob Fay, Wayne Fuller, Darrell Green, Bob Groves, Rachel Harter, Susan Hinkins Graham Kalton, Dan Kasprzyk, Rod Little, Sharon Lohr, Mike Macaluso, Van Parsons, Jon Rao, Don Rubin, Fritz Scheuren, Joe Sedransk, Sandy West, and Kirk Wolter for many helpful discussions of nonresponse adjustment and statistical agencies. The views expressed in this chapter are those of the author and do not necessarily reflect the policies of the U.S. Bureau of Labor Statistics.

Applied Bayesian Modeling and Causal Inference from Incomplete-Data Perspectives.
Edited by A. Gelman and X-L. Meng © 2004 John Wiley & Sons, Ltd ISBN: 0-470-09043-X

one extreme, some agencies, or programs within agencies, focus on production of estimates for relatively simple population aggregates. At the other extreme, other agencies or programs emphasize scientific research based on extensive modeling and a correspondingly high degree of conditioning in analyses and subsequent formal inferences. Still other agencies and programs collect data intended to meet needs at both ends of this spectrum. Within the general survey context, Skinner, Holt, and Smith (1989) discuss this spectrum of survey goals in some depth. For the current discussion of nonresponse modeling and adjustment, it may be especially useful to consider three factors that may display quite distinct characteristics at different points along this spectrum: constraints, inferential goals, and robustness.

10.2   Constraints

In many cases, statistical agency decisions to develop and implement a given non-response adjustment procedure are influenced by fairly strong constraints. Some examples include compatibility with legacy production systems, incremental costs of changes in these production systems, availability of personnel with specific types of training, and timely availability of relevant auxiliary data. In some cases, these constraints are well defined and readily admit a formal mathematical characterization. In such cases, one could consider extension of classical efforts at survey optimization under constraints (e.g., Cochran, 1977) to optimize (approximately) nonresponse adjustment procedures. In other cases, the constraints are clearly important, but do not readily admit a formal deterministic characterization. It would be of interest to study the extent to which the uncertainty of some of these operational constraints could be characterized realistically in a stochastic form. For such cases, a Bayesian approach might be useful in subsequent development of approximately optimal procedures, and in comparison of these new procedures with current agency practice.

10.3   Complex estimand structures, inferential goals, and utility functions

Historically, much of the research literature on nonresponse modeling and adjustment has focused on cases in which principal interest resides in a small or moderate number of estimands that are identified *a priori*. See, e.g., Rubin (1996) and references cited therein). However, statistical agencies are often responsible for production of estimates for a very large number of population parameters; or for production of public-use datasets that in turn may be used in production of many parameter estimates. In many cases, some of the parameters of interest may not have been anticipated when the agency designed its nonresponse adjustment procedure. In addition, the presence of a large number of estimands and potential data users will entail a wide range of priorities among inferential goals, and a corresponding disparity in data users' utility functions. In many cases, agency decisions on implementation of a given nonresponse adjustment procedure may depend on trade-offs among these competing utility functions.

In addition, when a survey involves a large vector of observations on an individual unit, or a large vector of estimands, the underlying observational structure and associated estimand structure may involve important complex conditional relationships which in turn are important in the implementation of a given nonresponse adjustment procedure. Some important issues associated with complex estimand structure have been considered in some special cases (e.g., Heeringa, Little, and Raghunathan, 2002), but there are a large number of open questions in this area that would benefit from additional study.

10.4   Robustness

In work with nonresponse adjustment procedures, statistical agencies encounter several important sets of robustness issues. First, in keeping with the comments in Section 10.3, one often has a large number of estimands, and the sample size associated with some of these estimands may be relatively small. In such cases, one may have concerns regarding influential observations, especially in establishment surveys with highly skewed underlying populations. These problems can be exacerbated when patterns of nonresponse are uneven across subpopulations. For such cases, it is beneficial for agencies to have available diagnostics to identify observations that are influential for specific subsets of the large number of potential estimands (see, e.g., Zaslavsky, Schenker, and Belin, 2001 and references cited therein).

Second, the setting described in Section 10.3 can lead agencies to be especially concerned about traditional issues associated with omission of important predictor variables from nonresponse models and associated adjustment procedures. Variants on this issue, potential solutions, and associated critiques, have been considered in general survey, observational study and nonresponse contexts for many years. See, e.g., Rosenbaum and Rubin (1983b), Hansen, Madow, and Tepping (1983), and references cited therein. However, agencies may benefit from additional consideration of these issues in a framework that emphasizes a large number of estimands and associated competing utility functions.

10.5   Closing remarks

In the past, nonresponse adjustment methods used by some government statistical agencies have been occasionally described as baroque bordering on rococo; and some incremental efforts to update agency practice have been compared to the Ptolemaic theory of epicycles awaiting a Copernican revolution. Such critiques, taken in good humor, can be useful if they spur the statistical community to a broader and deeper consideration of the factors that give rise to general agency practice for nonresponse adjustment. The preceding remarks have highlighted three factors:

(a) The often dominant role of constraints.

(b) The complexity of estimand structures, inferential goals, and utility functions.

(c) Robustness.

Each of these factors has received attention in both the Bayesian and randomization-based literature on nonresponse. However, additional systematic study could shed a considerable amount of additional light on approximate optimization of nonresponse adjustment methods for statistical agencies. Two areas are of special interest.

First, although factors (a) and (b) above have been studied in depth for specific cases, general attempts to develop a systematic approach to nonresponse adjustment have tended to treat (a) and (b) somewhat as side conditions. In some cases, this is reasonable either because factors (a) and (b) truly are of secondary interest or because a systematic characterization of (a) and (b) is simply not feasible. However, in many practical applications, factors (a) and (b), and related implications for (c), can have a major—and sometimes dominant—effect on the efforts to optimize agency practice. Consequently, to the extent that we can formally characterize (a) and (b) in a realistic way, it would be of interest to incorporate these factors more systematically into the development, implementation, and evaluation of agency nonresponse adjustment methods. This would help ensure a stronger match between methodological development and agency practice, and has the potential to lead to several very rich classes of statistical research problems.

Second, the preceding sections noted that nonresponse modeling and adjustment efforts by government statistical agencies encompass a wide range of users, with correspondingly wide-ranging inferential goals and utility functions. Many of these users are likely to approach nonresponse adjustment methods as consumers of a technology, even though these adjustment methods may have been originally developed on the basis of first-principles statistical science. The general process of conversion of scientific results into a broadly applicable technology, and subsequent adoption of that technology by a relatively wide range of users, has been studied in some depth in the sociology, engineering, and business literature; see, e.g., Rogers (1995), Drucker (1985), and references cited therein. Much of this literature is quite controversial, and one would naturally be cautious about wholesale application of that literature to the adoption and diffusion processes observed in statistical science and statistical technology. Nonetheless, several general themes of this literature (e.g., differences in utility functions and risk/reward profiles; expectations on robustness and observable feedback loops; and degrees of customization) do appear to have close parallels in the development and implementation of nonresponse adjustment methods. Careful consideration of these themes by statistical researchers, by managers of government statistical agencies, and by users of specific survey datasets may be very helpful in calibrating statistical research work with the needs of agencies and users, and in accelerating the adoption and diffusion of improved nonresponse adjustment methods.

11

Bridging across changes in classification systems

Nathaniel Schenker[1]

11.1   Introduction

A common practice in data collections is to classify responses into categories for analysis. For example, narrative responses on jobs can be classified into industry and/or occupation categories, people's descriptions of their races can be classified into race categories, and medical diagnoses can be classified into disease categories. When a data collection is repeated over time, but the system for classifying responses into categories changes, problems of noncomparability can arise, especially if data classified using an earlier system are to be compared or combined with data classified using a later system. This noncomparability can be viewed as an issue of missing data: Either the units classified using the earlier system can be viewed as missing the categories under the later system, or vice versa.

In this chapter, I discuss two projects, each of which had the goal of bridging the transition between classification systems, that is, handling the missing-data problem caused by the transition. In the first project, Rubin's (1978b, 1987b) multiple imputation was used to achieve comparability of industry and occupation codes in public-use files from the 1970 and 1980 censuses. This was, I believe, the first application of multiple imputation to a large public-use database, and I had the good fortune of working with Rubin on the project when I was his doctoral student

[1]National Center for Health Statistics, Centers for Disease Control and Prevention, Hyattsville, Md. The views expressed in this chapter are those of the author and do not necessarily reflect the views of the United States government.

Applied Bayesian Modeling and Causal Inference from Incomplete-Data Perspectives.
Edited by A. Gelman and X-L. Meng   © 2004 John Wiley & Sons, Ltd   ISBN: 0-470-09043-X

as well as early in my postdoctoral years. In the second project, methods that can be viewed as imputation were used by the National Center for Health Statistics of the Centers for Disease Control and Prevention, with assistance from the Bureau of the Census, to bridge the transition from single-race reporting to multiple-race reporting in the census. Analytic approximations to multiple imputation were then used to assess the variability due to race bridging. This was one of the most recent efforts to bridge across a change in a classification system. The two bridging problems, while similar in their basic structures, had many differences both in their features and in the methods used to solve them.

11.2   Multiple imputation to achieve comparability of industry and occupation codes

Overview

In each decennial census, employment information is obtained from individuals using open-ended descriptions of occupations, which are then coded into several hundred specific categories for both occupation and industry.

To provide relatively up-to-date information, the classification scheme is revised somewhat by the Bureau of the Census for each census. Major changes, however, were made for the 1980 census. For example, fewer than one-third of the 1970 occupation categories mapped into single categories in the 1980 classification. As a consequence, public-use databases from the 1980 census had industry and occupation codes that were not directly comparable to those in public-use databases from previous censuses, and in particular the 1970 census. The lack of comparability of codes across time made it difficult to study such topics as occupation mobility and labor force shifts by demographic characteristics.

As discussed in Section 11.1, the industry and occupation coding problem could be viewed as a problem of missing data. Because the 1980 classification system was thought to be more broadly used and superior to earlier systems, it was chosen as the standard, and thus, the units in the 1970 public-use files were viewed as missing 1980 codes.

The 1970 public-use databases contained over one million records, with written descriptions of occupations only in physical storage, and therefore it would have been prohibitively expensive to recode the 1970 data according to the 1980 scheme. There existed, however, a double-coded sample of about 127,000 units from the 1970 census, that is, a sample with occupations coded using both the 1980 and 1970 schemes, which was created by the Bureau of the Census for purposes other than those described here.

A project to multiply impute 1980 industry and occupation codes to public-use samples from the 1970 census was carried out with funding from the National Science Foundation and support from the Bureau of the Census. The double-coded sample from the 1970 census was used to build models predicting 1980 codes from the 1970 codes and covariates. The models were then used to create multiple imputations of 1980 codes for two public-use samples from 1970 with a combined total of about 1.6 million units.

Descriptions and evaluations of this project can be found in Treiman and Rubin (1983), Rubin (1983), Rubin and Schenker (1987a), Treiman, Bielby, and Cheng (1988), Weidman (1989), Clogg et al. (1991), and Schenker, Treiman, and Weidman (1993). At the time of the project, which was carried out early in the development of multiple imputation, many of the issues, techniques, and results were quite novel. Although about two decades have gone by since the project took place, many aspects of the project are still useful and important. In the following subsections, I highlight some of the details.

Special features of this missing-data problem

The industry and occupation coding problem had three features that were unusual for missing-data problems. The first, which made this problem easier than many, was that the reasons for "nonresponse" were known, since the double-coded sample from the 1970 census was a probability sample. The second, which made this problem harder than many, was that the level of missing data in the 1970 public-use files was very high. In fact, 1980 codes could be viewed as missing for all of the records in the 1970 public-use files.

Finally, the double-coded sample was drawn independently of the 1970 public-use samples. Thus, the "observed" data used to fit the imputation model were not part of the data set to which multiple imputation was to be applied. Rubin and Schenker (1987a) argued that this feature would result in multiple-imputation inferences that are conservative. This was, I believe, the first published discussion of the conservatism that can result when the imputer uses information that is unavailable to the analyst of the multiply imputed data. Issues of this type have been discussed further in Fay (1991, 1992, 1993), Kott (1992), Meng (1994a), and Rubin (1996).

Methods used

Models predicting the 1980 codes for each 1970 code, given covariates, were fitted to the double-coded sample. The models were used to impute five sets of 1980 codes ($M = 5$, in the traditional notation of multiple imputation) for two 1970 public-use samples, each with about 800,000 records. To simplify modeling and to avoid imputing impossible 1980 industry and occupation pairs, the 1980 industry code was imputed first, and then the 1980 occupation code was imputed conditional on the 1980 industry code.

Modeling a polytomous outcome as a sequence of binary outcomes

The basic imputation model used was binary logistic regression. If a 1970 code mapped into more than two 1980 codes, a sequence of binary logistic regressions was used. Consider, for example, the situation in which a single 1970 code had four possible 1980 codes associated with it, say codes A through D in order of sample size in the double-coded sample. Separate binary logistic regression models were fitted to predict the dichotomies A versus (B, C, or D), B versus (C or D), and C versus D. Imputation of the 1980 code for a given unit in the 1970 public-use sample was carried out in the same sequence, with the first model used to impute A versus (B, C, or D), the second model used to impute B versus (C or D) if A was not imputed in the first step, and the third model used to impute C versus D if B was not imputed in the second step. Modeling a polytomous outcome as a sequence of binary outcomes had two benefits: (1) it allowed the use of simpler software and (2) it prevented the quality of the fitted models for 1980 codes with more data available (e.g., code A) to be affected by lack of data for less populous codes (e.g., code D).

Including many substantively important variables

It is standard advice that, when using multiple imputation, it is beneficial to include as many variables as possible in the imputation model; see, for example Meng (1994a) and Rubin (1996). In the industry and occupation coding project, in addition to including variables in the logistic regression models that were thought to be good predictors of the 1980 codes, an effort was made to include as many substantively important variables as possible, even if such variables were not statistically significant predictors of the 1980 codes. This was done to ensure that if the variables were used in subsequent analyses of the multiply imputed data, then the analyses would reflect the uncertainty about the relationships of the variables to the 1980 codes. (Not including a variable in an imputation model implies that it is known with certainty that the variable is unrelated to what is being imputed, which is usually not the case.) As part of the modeling process, knowledgeable social scientists were asked which variables were most important to and most often used by analysts of data on industries and occupations. The models for 1980 industry codes included categorical variables for age, race, sex, race-by-sex and age-by-sex interactions, class of worker (private industry, government, or self-employed), amount of work, and geography. The models for 1980 occupation codes included the same variables that were used for industry codes, as well as categorical variables for earnings and 1980 industry code.

Simple Bayesian methods for logistic regression

Since the predictors used were categorical, the data used to fit each logistic regression model could be represented in terms of a contingency table. Because an effort was made to include a large number of predictors, the contingency table often had sparse data, especially when the outcomes involved were less populous industries or occupations. For example, each contingency table for a 1980 industry code dichotomy had 4,608 cells, while the number of observations per table ranged from 4 to 3,500.

To handle the sparse data in estimation, simple Bayesian methods for logistic regression, as developed and evaluated in Rubin and Schenker (1987b) and Clogg et al. (1991), were used. Briefly, for a contingency table with $2c$ cells being used to fit a logistic regression model with p parameters for predicting, say, 1980 code A versus code B, the methods added fp/c "prior" observations to each cell of the contingency table corresponding to code A, and $(1 - f)p/c$ prior observations to each cell corresponding to code B, where f was the marginal fraction of sample units with code A. Thus, the total number of observations added to the entire table was equal to the number of parameters being estimated in the logistic regression model. After the prior observations were added, traditional maximum likelihood methods for fitting logistic regression models were applied to the augmented sample.

The simple Bayesian methods used in the project can be thought of as an extension of the use of the Jeffreys prior for estimating a binomial proportion to the problem of logistic regression (Rubin and Schenker, 1987b). They also have the flavor of empirical Bayes estimation (Morris, 1983). Use of the methods in the project ensured that the posterior distributions of the logistic regression parameters were unimodal and easy to maximize, and it facilitated the reflection of uncertainty in the multiple imputations, regardless of the sample sizes involved in fitting the logistic regressions.

Properly reflecting uncertainty and using the SIR algorithm

To properly reflect the uncertainty due to estimating the parameters of a logistic regression model in the multiple imputations for a dichotomy, say 1980 code A versus code B, the following two steps were followed for each of the $M = 5$ sets of imputations. First, a random vector of the logistic regression parameters was drawn from the approximate posterior distribution of the parameters. Second, for each unit in the public-use sample needing imputation of the dichotomy in question, (1) the drawn vector of parameters and the unit's covariate values were used to compute the probability of code A, and (2) a random imputation of code A or B was created using the probability computed (and its complement, the probability of code B). The two-step procedure of first drawing parameter values and then imputing given the drawn parameter values is a standard technique for properly reflecting the uncertainty due to estimating the parameters of an imputation model when the pattern of missing data is simple. For more complicated patterns, advances in computational statistics, such as Markov chain Monte Carlo methods (Gilks, Richardson, and Spiegelhalter, 1996; Kass, Carlin, Gelman, and Neal, 1998; Gelman, Carlin, Stern, and Rubin, 2003; Chapter 11), are useful.

In the first step of the two-step procedure, an approximation to the posterior distribution of the logistic regression parameters was needed. An obvious candidate was a normal approximation, with mean equal to the posterior mode and variance/covariance matrix equal to the negative of the inverse of the second-derivative matrix of the log-posterior. This approximation was relatively easy to obtain after application of the simple Bayesian methods described earlier. However, because of the small sample sizes as well as unequal splits in the categories of the outcomes often occurring in the logistic regressions, the actual posterior distributions were often not approximated well by normal distributions. To obtain better approximations, the sampling/importance resampling (SIR) algorithm (Rubin, 1983, 1987a, 1988) was used; see K. H. Li's Chapter 24 in this book for a detailed discussion of the SIR algorithm. First, a large sample was drawn from the normal approximation. Then, for each vector in the sample, the ratio of the posterior density to the normal density, evaluated at the vector, was calculated. Finally, $M = 5$ vectors, to be used for the five sets of imputations, were resampled from the original sample, with probabilities proportional to the ratios of densities. The size of the original sample drawn from the normal approximation varied across the logistic regression problems, with large samples drawn for problems with more unequal splits in the categories of the outcomes.

Evaluations and lessons learned

In addition to the methodological innovations in the industry and occupation coding project, and the general idea of treating noncomparability as a missing-data problem, there were several lessons learned from the evaluations of data from the project.

Rubin and Schenker (1987a) conducted a Monte Carlo study, using data from the double-coded sample on the agriculture industry, to investigate the properties of the methods used in the project. They found that on average, the actual coverage rates of intervals based on the multiply imputed data were close to the nominal levels. In contrast, single imputation was found to result in actual coverage rates that were substantially below the nominal levels. They also found that, owing to the high fractions of missing information in the industry and occupation coding problem, it was important to properly reflect the uncertainty due to estimating the parameters of the imputation model as discussed earlier. Not doing so resulted in actual coverage rates that were well below the nominal levels, although still better than those resulting from single imputation.

Treiman, Bielby, and Cheng (1988) conducted an evaluation of the data resulting from using the project's imputation models to multiply impute 1980 industry codes for the units in the double-coded sample (treating the actual 1980 codes as unknown). They found that the imputed codes approximated the actual codes well. Their analysis was an intermediate step in the industry and occupation coding project, and at the time of the analysis, multiple imputations for the 1970 public-use files had not yet been created. They also considered the question of whether the use of a multiply imputed public-use file from 1970 (a larger file, but with multiply imputed codes rather than directly assigned codes) would yield more precise results than the use of the double-coded file (a smaller file, but with directly assigned codes rather than multiply imputed codes). As an example, they examined estimates (and standard errors) of the mean years of school completed by workers in various industry categories. By extrapolating the results on the extra variability due to imputation from analyses of the multiple imputations for the double-coded sample to the situation of a file of the size of the public-use file, they concluded that the standard errors obtained from the multiply imputed public-use file would be substantially smaller than those obtained from analyses of the double-coded sample with directly assigned codes.

The latter question considered by Treiman, Bielby, and Cheng (1988) was also addressed by Schenker, Treiman, and Weidman (1993), who compared analyses of an actual multiply imputed public-use file with analyses of the double-coded sample with directly assigned codes. They examined estimates (and standard errors) of changes between 1970 and 1980 in the percentage of workers in 12 different occupations who were female. Consistent with the results of Treiman, Bielby, and Cheng (1988), they found that smaller standard errors were usually obtained with the multiply imputed public-use file. Schenker, Treiman, and Weidman (1993) also estimated the fraction of missing information for their analyses of the multiply imputed public-use file, using a formula suggested by Rubin (1987b; Section 3.1). This fraction differs from the simple missing-data rate, which can be considered to be 100% for the industry and occupation coding problem, since it measures how much information is lost by having to use multiply imputed codes rather than known codes, and thus accounts for the predictive power of the covariates in the imputation models. For the 12 occupations considered, the fraction of missing information varied from 2 to 83%, demonstrating how much the fraction can depend on the specific estimation problem being addressed. This is a good counter-argument to the idea of estimating a single inflation factor to correct standard errors in the presence of missing data rather than using a technique such as multiple imputation.

11.3 Bridging the transition from single-race reporting to multiple-race reporting

Overview

In 1997, the Office of Management and Budget issued revised standards for the collection of race information within the Federal statistical system (Office of Management and Budget, 1997). One major revision allows each respondent to a Federal data collection to choose more than one race category in describing the person in question. The prior standards, issued in 1977, had specified that only a single-race category be chosen (Office of Management and Budget, 1977). This change presents challenges for analyses that involve data collected under both the 1977 and 1997 race reporting systems, since the data on race are not comparable.

There were four race categories under the 1977 standards: American Indian or Alaska Native (AIAN); Asian or Pacific Islander (API); Black; and White. Under the 1997 standards, the four single-race categories from 1977 have been expanded to five: AIAN; Asian; Black or African American; Native Hawaiian or Other Pacific Islander (NHOPI); and White. In addition, any combination of these five categories may be used. The five 1997 single-race categories can be collapsed into the four

1977 race categories (by collapsing the Asian and NHOPI categories to form the API category).

As most people still report only a single race under the 1997 system, a common proposed solution is to try to bridge the transition by assigning a 1977 race category to each multiple-race report under the 1997 system, and to conduct analyses using just the observed and assigned 1977 race categories; see, for example, Office of Management and Budget (2000). Thus, analogous to the industry and occupation coding problem, the problem of noncomparability of race reporting can be viewed as a missing-data problem, in which 1977 race categories are missing and need to be imputed for people assigned a multiple-race race category under the 1997 system.

A specific issue that sparked interest in race bridging at the National Center for Health Statistics involves the calculation of vital rates by race. Such rates are frequently used in epidemiologic and other studies. Beginning in 2000, data from the decennial census, which are used to calculate the denominators for rates, were collected under the 1997 race reporting system. In contrast, vital event (e.g., birth and death) record systems, which provide the data for the numerators, are implementing the change to the new standards over the next several years. Thus, numerators will often be available under the 1977 race categories, whereas denominators will be available under the 1997 categories.

To enhance comparability between the 2000 census and data classified according to the 1977 standards, in particular data on vital events, the National Center for Health Statistics, with assistance from the Bureau of the Census, has produced estimates of the population counts that would have been obtained had the 1977 standards been used. The estimates, which have been released publicly, result from bridging the data in the Census 2000 Modified Race Data Summary File, which contains the counts of the resident population for each of the (single- or multiple-) race categories under the 1997 standards, by county, age, sex, and Hispanic origin, to the four categories specified in the 1977 standards.

Descriptions and evaluations of this project, which is still ongoing, can be found in Ingram et al. (2003), Schenker (2003), and Parker et al. (2004); and precursors to the methods used in the project are discussed in Schenker and Parker (2003). In the following subsections, I highlight some of the aspects of the project and offer some contrasts to the industry and occupation coding project discussed in Section 11.2.

Special features of this missing-data problem

In contrast to the industry and occupation coding problem, the race bridging problem had very low rates of missing data overall, although the rates were higher for some specific race groups and geographical areas. For example, analysis of the Census 2000 Modified Race Data Summary File indicates that only 1.3% of the overall population was classified into multiple-race categories. In contrast to this overall low rate, there is a much higher rate of multiple-race reporting involving the 1977 race group AIAN. While 0.9% of the population was classified into the AIAN category, 0.4% was classified as AIAN/White; thus, over three-tenths (i.e., over $0.4/(0.4 + 0.9)$) of the reporting involving the AIAN race is accounted for by multiple-race reports.

Also, in contrast to the industry and occupation coding problem, the reasons for "nonresponse" in the race bridging problem are not well understood; that is, the factors leading up to someone being described by a multiple-race category, so that they cannot be straightforwardly assigned to a 1977 race category, are not fully known.

A final feature of the race bridging problem that deserves mention is that the variable in question, race, is sometimes reported inconsistently. For example, the race of a person is often reported by proxy: In the census, it might be reported by a family member other than the person in question; in another survey, it might be reported by a different family member; and on a death certificate, it might be reported by a funeral director on the basis of information from an informant or on observation. This inconsistency is an issue to consider in race bridging, but it is also an important issue even without the problem of race bridging.

Methods used

The National Health Interview Survey (NHIS) is an ongoing household-based survey of the civilian noninstitutionalized population, with about 40,000 households containing about 100,000 persons surveyed per year. The NHIS has allowed multiple-race reporting for all persons being surveyed since 1982. In a follow-up question, for each person classified into a multiple-race category, it asks for the "primary" race, that is, the single-race category that best describes the person.

Categorical regression models predicting primary race were fitted to data on the roughly 4,000 persons from the 1997–2000 NHIS who were classified into multiple-race groups but also had accompanying primary-race descriptions. For each of the 11 multiple-race groups formed by combinations of AIAN, API, Black, and White, by county and person-level covariate combination, the count in the Census 2000 Modified Race Data Summary File was distributed into the applicable 1977 race groups in proportion to the estimated probabilities from the appropriate fitted regression model.

Bridging backward rather than forward

Whereas the industry and occupation coding project bridged from the older classification system to the newer one, the race bridging project bridged in the other direction. There were two main reasons for this. First, as mentioned earlier, only 1.3% of the people in the Census 2000 Modified Race Data Summary File were classified into multiple-race groups. Thus, predicting a 1977 race category for each multiple-race person effectively defined a problem with 1.3% of the values missing. In contrast, predicting how races classified under the 1977 system, such as for vital events, would have been classified under the 1997 system effectively defines a problem with 100% of the data missing, since each person would have the possibility of being put into a multiple-race category. Second, the modeling required is much simpler when bridging from the 1997 system to the 1977 system than vice versa. For example, a person classified as Black/White under the 1997 system has two possibilities under the 1977 system: Black or White. In contrast, a person classified as White under the 1977 system has several multiple-race possibilities under the 1997 system, including any multiple-race group for which White is a component.

Separate versus combined models

For each of the six largest multiple-race groups in the NHIS (AIAN/Black, AIAN/White, API/Black, API/White, Black/White, AIAN/Black/White), a separate logistic or multinomial logit model was fitted. Since the remaining five multiple-race groups had small sample sizes, a single multinomial logit model was fitted for these groups, based on the combined data for all 11 groups. The use of separate models for larger groups and a combined model for smaller groups was roughly in the spirit of Bayesian or empirical Bayesian procedures that compromise between separate estimates and pooled estimates based in part on their relative precisions (Morris, 1983; Gelman, Carlin, Stern, and Rubin, 2003; Chapter 5). Further research will be aimed at refining the modeling, especially for the small groups.

Selection of predictors

As discussed in Section 11.2, when creating imputations for a public-use file, it is often desirable to include more predictors than necessary rather than fewer predictors than necessary. Since the fitted models in the race bridging project were to be applied to the Census 2000 Modified Race Data Summary File, the person-level predictors that could be used were limited to those in the summary file, that is, age, sex, and Hispanic origin. However, since the counts in the summary file were given by county, there is the potential to include a number of contextual, mainly county-level, predictors. At the time that the bridging of the summary file was carried out, most county-level data from the 2000 census were not yet available, so the contextual variables were limited to region of the country, urbanicity, summaries of the single-race distribution in the 2000 census, and the prevalence of multiple-race reporting in the 2000 census. As more contextual variables become available, a goal in the race bridging effort is to determine which, if any, additional variables should be included in the bridging models.

Imputing "best" estimates

The primary need of users of the Census 2000 Modified Race Data Summary File was to have point estimates of the counts by 1977 race. Moreover, most such users would not be familiar with multiple imputation. Hence, the race bridging project created only a single file of bridged counts, analogous to the Census 2000

Modified Race Data Summary File, with "best" estimates of the counts under the 1977 system (i.e., point estimates based on the estimated probabilities from the regressions). In contrast, the use of multiple imputation would produce multiple versions of the file of bridged counts, each version representing a random draw from the predictive distribution of the counts.

Assessing variability due to race bridging

Often, when there is no need for race bridging, census counts are treated as non-random population quantities. In contrast, bridged census counts such as those produced in the race bridging project, are estimates and thus have random variability. Schenker and Parker (2003) suggested the use of multiple imputation to assess the variability due to race bridging.

For the race bridging project, in which only a file of "best" estimates had been created, Schenker (2003) assessed the variability due to race bridging by adapting the methods of Schafer and Schenker (2000) for inference with imputed conditional means to the race bridging problem. These methods can be viewed as a first-order approximation to multiple imputation with an infinite number of versions of the file of bridged counts. Schenker (2003) gave a detailed discussion contrasting the use of the Schafer/Schenker methods with the use of multiple imputation. Briefly, however, the Schafer/Schenker methods are more efficient (since they approximate multiple imputation with "$M = \infty$"), and the production of a single file of "best" estimates of the counts simplifies point estimation. On the other hand, for variance estimation, the Schafer/Schenker methods involve more complicated formulas than does multiple imputation, and the Schafer/Schenker formulas need to be customized for each inference problem, as illustrated by Schenker (2003). Moreover, the Schafer/Schenker methods involve more linear approximations than does multiple imputation, and thus could be less effective when sample sizes are small.

Evaluations and lessons learned

As mentioned earlier, the race bridging project is ongoing. Future efforts will focus on refining the methods used to bridge the counts from the 2000 census, and on applying the methods to other years of data and perhaps in other contexts.

Ingram et al. (2003) provided detailed evaluations of the bridged census counts, and Parker et al. (2004) provided further evaluations of the methodology, with an emphasis on vital rates. As might be expected, the largest impacts of bridging are for the races that have the largest amounts of multiple-race reporting relative to their populations. Overall, the largest relative amount of multiple-race reporting involves the AIAN race, followed by, in decreasing order, API, Black, and White. The allocations of people from multiple-race groups into 1977 race categories vary among states, reflecting the effects of the covariates in the bridging models. Moreover, the estimated parameters differ between the bridging models for the multiple-race groups, suggesting that the race reporting mechanisms differ across groups.

As for the assessments of variability due to race bridging, Schenker (2003) found that the relative standard errors of the bridged census counts tend to be higher for finer geographic levels and lower for coarser geographic levels. For each state or the District of Columbia, the relative standard error of the bridged count for any race is no larger than 0.05. At the national level, for birth and death rates by age group and 1977 race, use of bridged census counts in the denominators does not add substantially (on an absolute basis) to the relative standard errors of the rates.

11.4 Conclusion

The problem of bridging across a change in a classification system is in some senses an "artificial" type of missing-data problem, since the missing data are caused by a change in the system rather than by nonresponse per se. Nevertheless, it is a problem that is not uncommon, since classification systems are often changed to provide up-to-date information on the current characteristics of the population. As illustrated by the two bridging problems discussed in this chapter, treating a bridging problem as a missing-data problem can be very useful, as it facilitates bringing lessons from the large amount of research on methods for handling missing data to bear.

12

Representing the Census undercount by multiple imputation of households

Alan M. Zaslavsky[1]

12.1 Introduction

It has been known at least since 1950 that the decennial US census undercounts the population (Citro and Cohen, 1985). Some households are omitted entirely from the census roster, while others are included but with one or more members omitted. Also, some households and individuals are counted more than once. Such errors in coverage are extremely contentious due to the role of census population estimates in the allocation of political power and resources (Anderson and Fienberg, 1999; Choldin 1994).

Major efforts were devoted to estimation of the undercount in the 1990 and 2000 censuses in the United States. Estimation classes were defined by the characteristics of persons, such as age or sex, and of households or blocks (small geographic areas), such as type of dwelling, tenure (owner-occupied or rented housing unit) or urbanicity. For each class, an undercount rate was calculated using data from a second survey conducted independently of the census, the Post Enumeration Survey (PES). These rates expressed estimated net omissions as a fraction of enumerated persons in that class. Although the 1990 census had a net undercount of 0.5 to 1%, some classes had a larger estimated net undercount,

[1] Department of Health Care Policy, Harvard Medical School, Boston, Mass.

Applied Bayesian Modeling and Causal Inference from Incomplete-Data Perspectives.
Edited by A. Gelman and X-L. Meng © 2004 John Wiley & Sons, Ltd ISBN: 0-470-09043-X

exceeding 8%, and some had a small net overcount (Hogan, 1993). The 2000 census presented a more confusing picture, with inconsistent coverage estimates from different methodologies (Citro, Cork, and Norwood, 2004).

"Adjustment" is any process that corrects the enumeration for estimated coverage error. Coverage rates for adjustment classes (groups of observations regarded as exchangeable for adjustment) can be used to calculate adjusted counts of individuals, using weights that incorporate inverse probabilities of enumeration. For more complex analyses, however, it would be desirable to place the added persons into households in microdata rosters or samples. Household microdata are used for tabulations by household characteristics and for a wide range of research purposes. The adjusted records are only useful if the composition of the adjusted households (represented here by a vector giving the number of household members from each adjustment class) and the relationships of its individual members are logically consistent and typical of the types of households found in their area.

To describe abstractly the patterns of plausible households and create new households that fit them is a daunting task. Imputing persons into a special category of unrelated individuals to represent omissions, as in the 1990 undercount estimates, sidesteps this problem at the cost of creating a skewed picture of relationships in the block.

One solution to this problem is to assign weights to the households enumerated in the census lists for the block, making the weighted counts of households and of persons in each adjustment class agree with the corresponding adjusted totals (Zaslavsky, 1988). This methodology, a generalization of raking ratio estimation, changes the proportionate composition of the block, but all of the households are real. However, the weighting methodology is not based on a model of the underlying processes of census undercoverage, in particular of the within-household omissions, which cause households to appear to be smaller than they actually are. A simulation of the reweighting procedure examined the effect of household reweighting on the distribution of the number of adults in households with children, under simple models of simulated undercount. Owing to the simulated undercount, some two-adult households were observed as having a single adult member. Two-adult households are unusually numerous, so their prevalence was underestimated even after weighting, whereas one-adult households were overestimated owing to the large number of two-adult households that are erroneously classified as one-adult households. Thus, the limitation of weighting strategies is that it is difficult to incorporate complex information either about the prevalence of different types of households or about the undercount mechanism into the adjustment methodology.

This chapter describes an alternative approach, in which households are imputed into the census roster to represent omitted households, while persons are imputed into enumerated households to represent persons omitted from those households, under explicit probability models for the misenumeration processes and the distribution of household types. This approach lends itself to multiple imputation (Rubin, 1978b, 1987b), in which the entire imputation process is repeated several times to represent the variability introduced by estimation of the underenumeration.

Applying standard combining rules to repeated analyses of the imputed data sets, a census data user would calculate valid standard errors for estimates of interest (be they simple tabulations or complex model parameters) reflecting all known sources of variability. This unifying scheme facilitates analyses of any degree of complexity that use familiar complete-data methods, at the same time properly representing bias and variance introduced by all known forms of nonresponse and error in the census.

The remainder of this chapter develops methods for imputing households from the posterior distribution of true household types given the observed roster, emphasizing the overall framework rather than the details of models and simulation results. More extensive results appear in Zaslavsky (1989), where these methods were first presented.

12.2   Models

Notation and overall model structure

Define a "household type" by the vector of counts of persons from each adjustment class (for example, white women over 65 years old in urban areas) making up the household. Let T_{bh} and U_{bh} be the true and observed types respectively of the h-th household in census block b, N_{Tb} and N_{Ub} be the true and observed number of households in block b, and $\mathbf{T} = \{T_{bh}\}$, $\mathbf{U} = \{U_{bh}\}$, $\mathbf{N}_T = \{N_{Tb}\}$ and $\mathbf{N}_U = \{N_{Ub}\}$ the arrays of true and observed values. The following hierarchical model specifies distributions of hyperparameters Φ, block-level parameters $\{\omega_b\}$, true values $\mathbf{N}_T, \mathbf{T}$ and observed data $\mathbf{N}_U, \mathbf{U}$.

General hyperparameter: a general hyperparameter $\Phi = (\Phi_\mu, \Phi_\omega, \Phi_\alpha)$ (with some prior distribution) governs all of the other distributions.

Block-level parameters and independence of blocks: Conditional on Φ, the distributions of the data and parameters for different blocks are independent. For each block, a block-level parameter ω_b represents the particular characteristics of that block. The number of households N_{Tb} is also an unobserved block-level parameter. The distribution of (ω_b, N_{Tb}) is governed by the hyperparameter component Φ_ω.

$$(\omega_b, N_{Tb}) \mid \Phi \sim P_\omega(\cdot, \Phi_\omega), \text{ i.i.d. for } b = 1, \ldots B. \tag{12.1}$$

Distribution of true and observed household types: The pairs (T_{bh}, U_{bh}) representing true and observed household types are drawn independently, conditional on the block and general parameters. We factor their distribution as the product of the probability of the true type in that block, $\mu_b(t) = \Pr(T_{bh} = t \mid \Phi_\mu, \omega_b)$ and the probability that the true type is recorded as a particular observed type, $\alpha_b(t, u) = \Pr(U_{bh} = u \mid T_{bh} = t, \Phi_\alpha, \omega_b)$. The former distribution describes the distribution of true household types in various blocks, while the latter describes the process that causes households to be misclassified (i.e., the undercounting process). If $\alpha_b(t, u) > 0$ (t could be observed as u), we say that "t is a cover for u" or $t \succ u$.

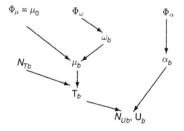

Figure 12.1   Overall structure of the hierarchical model.

The factorization $p(U, T) = p(T)p(U \mid T)$ characterizes the joint distribution of (T, U) by submodels for two scientifically distinct processes, the "distributional parameterization" of Rubin and Zaslavsky (1989). An alternative "direct predictive parameterization" models $P(T \mid U, \theta)$, thus specifying a predictive model from which T might be drawn directly. An advantage of our distributional parameterization, besides its greater scientific interpretability, is that it separates the distribution of T, which might be expected to vary across small areas, from the undercoverage process $U \mid T$, which is estimated from a relatively small sample (the PES), and therefore must be modeled as relatively consistent across areas. Thus, this factorization is better adapted to the structure of the available data.

With this notation, the distribution of observed and true household types in block b is

$$p(T_{bh}, U_{bh} \mid \omega_b, \Phi) = \alpha_b(T_{bh}, U_{bh})\mu_b(T_{bh}), \quad \text{i.i.d. for } h = 1, \dots N_{Tb}. \quad (12.2)$$

The hierarchical model for $(\mathbf{N}_T, \mathbf{T}, \mathbf{U})$ is summarized by Figure 12.1 and the following equation:

$$p(\mathbf{N}_T, \mathbf{T}, \mathbf{U}) = \int \left[\prod_b \int \prod_{h=1}^{N_{Tb}} p(U_{bh} \mid T_{bh}, \omega_b, \Phi_\mu)p(T_{bh} \mid \omega_b, \Phi_\mu) \right.$$

$$\left. \times \ dP(N_{Tb}, \omega_b \mid \Phi_\omega) \right] dP(\Phi). \quad (12.3)$$

Models for the local distribution of household types

Households are of types drawn from the discrete set of possible household compositions, which is very large. There is no simple or reliable way of characterizing the possible household types (compositions) that includes all the plausible types and excludes the impossible ones. Rather, we assume that the set of possible household types equals the set of observed types. This is a reasonable approach in this setting because most households are enumerated accurately, so any type that actually occurs more than a few times is highly likely to be observed.

The distributions of individuals by race and age, and of households by family size and other variables, vary substantially across blocks. This variability is reflected in (12.3) by the dependency of μ_b on the block parameter ω_b.

We model prevalences of household types in various blocks as:

$$\mu_b(t) = \frac{\mu_0(t)\exp(r(t)'\omega_b)}{\sum_{t'}\mu_0(t')\exp(r(t')'\omega_b)}, \tag{12.4}$$

where $r(t)$ is a low-dimensional function that summarizes the composition of household type t and ω_b is a parameter for block b. The general parameter Φ_μ enters through μ_0 and possibly through the definition of $r(t)$ as a function of household characteristics. Because $\mu_0(t)$ assumes a distinct value for each type t ($\mu_0(t)$ is specified nonparametrically), the model is semiparametric, like the proportional hazards model for survival times (Cox, 1972), which it resembles in form.

This model has the following interpretation: in block b, the probability that each household will be of type t is proportional to the product of a factor $\mu_0(t)$ corresponding to the prevalence of type t generally and a factor $\exp(r(t)'\omega_b)$ corresponding to the relative prevalence of households like t (having similar values of $r(t)$) in a block like b. Thus, it is unnecessary to model or describe all possible household types. Only the interblock differences need be modeled; these differences are much better understood than the probabilities of the particular types. The components of $r(t)$ should be functions of household characteristics whose distributions differ most among the blocks; these can be selected by multiple discriminant analysis.

In the simulations of Section 12.4, $r(t)$ is taken to be one-dimensional. In this case, the joint distribution of t and b is the log-multiplicative model for two-way contingency tables with ordered categories (Agresti, 1984; Goodman, 1981), with known scores $r(t)$ on the t dimension and estimated scores ω_b on the b dimension.

Models for errors in enumerating households

If enumeration errors are always undercounts, misclassifications take the form of omission of one or more persons from the household. Then $\{u : t \succ u\}$ consists of subsets of the members of t, that is, households obtained by dropping one or more members from t; conversely, $\{t : t \succ u\}$ can be obtained by adding members to u. The α_b matrix is sparse because only misclassifications due to omissions have positive probability. These features may be exploited in data structures and algorithms.

The structure of the models is similar when persons are also overcounted or are misclassified owing to misrecording of characteristics; then U is obtained by adding members to T or changing the classification of one or more members, respectively. These extensions are not considered further in this chapter but with adequate data could be explored to build more realistic models for misclassifications.

Households that are completely missed in the census constitute a truncated type $U = 0$. These differ from households that are counted but for which there is no information on their composition: for example, when a neighborhood informant indicates that a housing unit is occupied but the number or characteristics of occupants cannot be determined. The latter constitutes special observed types that are included in the count N_{Ub}, but "type 0" households are not. Hence $N_{Ub} \leq N_{Tb}$.

Whole-household omission probabilities $\alpha_b(t, 0)$ might be modeled by logistic regression on summaries of household composition (number of members, number of adults, proportions by race, and so forth). If there is another special status (such as "occupied but no interview obtained"), the regression is polytomous.

Specific models for within-household omissions, $\alpha_b(t, u)$ for $u \neq 0$, can be built upon models for omissions of persons with appropriate modeling of dependency between persons within the same household. For example, an omission model with random household effects can be approximated by a simple additive loglinear model for independent omissions with an extra nonlinear (e.g., quadratic) term for dependency (Zaslavsky, 1989, Section 12; Darroch, Fienberg, Glonek, and Junker, 1993, Section 4.3). Either of these models requires processing PES data differently than in past censuses, since previous efforts have focused on marginal omission probabilities for classes of individuals, ignoring household structure.

12.3   Inference

Overview

Proper multiple imputations (Rubin, 1987) may be obtained as draws of completed data from the posterior distribution of the complete data given the observed data. In this context, the complete data correspond to the full population including all members of all households; the completed data are imputed "true" rosters.

The data are observed households U_b for every block, other block-level covariate information $\mathbf{x}_b$, and the true rosters (N_{Tb}, $\mathbf{T}_b$) for a sample of blocks that fall into the PES. To impute we must draw from the joint distribution $p(\mathbf{N}_T, \mathbf{T}, \Phi, \omega \mid \mathbf{U}, \mathbf{T}_{\text{PES}})$. We outline here the steps of this full Bayesian inference, recognizing that for some parameters maximum likelihood estimation might be an adequate approximation and computationally cheaper in large data sets.

Only in the PES blocks are complete pairs (T_{bh}, U_{bh}) observed; thus, in accordance with PES procedures, we assume that Φ_α, the general parameter controlling α_b, is estimated from PES data alone and hence is independent of the remaining parameters. We can sample from the posterior distributions of parameters of these loglinear and logistic regression models using standard approximations for exponential family models.

The remaining parameters and imputed values are the type prevalence parameter μ_0 (equivalent to Φ_μ), the block composition parameters ω, and the general parameter Φ_ω of their prior distribution, the true block counts N_{Tb}, and the true household types $\mathbf{T}$. Using a Gibbs sampler, each of these can be drawn in turn conditional on all of the others; the stationary distribution after many such cycles is the desired posterior distribution. The following sections present the specific algorithms for sampling these variables and suggest plausible specifications of their prior distributions.

Inference for μ_0

As described in Section 12.2, we assume that the support of μ_0 is the set of observed types $\mathbf{U}$. Conditional on other parameters, the likelihood of μ_0 is

$$\prod_t \mu_0(t)^{n(t)} \bigg/ \prod_b \left(\sum_t \mu_0(t) e^{r(t)'\omega_b} \right)^{N_{Tb}}, \tag{12.5}$$

where $n(t)$ is the number of households of type t. With a proper Dirichlet prior $p(\mu_0) \propto \prod_t \mu_0(t)^{\delta_t - 1}$, the posterior distribution can be approximated by a scaled Dirichlet distribution $(c_1\mu_0(1), c_2\mu_0(2), \ldots) \sim \text{Dirichlet}(n(1) + \delta_1, n(2) + \delta_2, \ldots)$, where c_t is a suitably weighted average of $e^{r(t)'\omega_b}$. (The approximation is obtained by Taylor linearization of the logarithm of the denominator, from which $c_t \approx (En(t))/\mu_0(t)$, where the expectation is under current values μ_0 and ω and includes a prior "pseudo-data" block 0 with $n_0 = \sum \delta_t$ and $\omega_0 = 0$.) In simulations, this approximation appears to be a good proposal distribution in a Metropolis-Hastings step, with adequate acceptance rates.

A proper prior for μ_0 ($\delta_t > 0$) must be assumed even though it introduces a small bias in the estimation of μ_0. With an improper prior ($\delta_t = 0$), the Gibbs sampler is nonrecurrent and will converge to a μ_0 with support on a minimal set of household types. With a uniform prior $\delta_t \equiv \delta$, δ should be small to avoid biasing the posterior draws of μ_0 toward low-prevalence types. Bias can be further reduced by specifying a simple model for $(\delta_1, \delta_2, \ldots)$ reflecting general relationships of prevalence to characteristics of types, for example, a loglinear model with predictors like the number of household members and the prevalences of the adjustment classes to which the members belong.

Inference for block parameters ω and Φ_ω

Conditional on $\mathbf{N}_{Tb}$ and μ_0, the likelihoods of each of the ω_b are distinct, given by

$$L(\omega_b) = \prod_t e^{n_b(t)r(t)'\omega_b} \bigg/ \left(\sum_t \mu_0(t) e^{r(t)'\omega_b} \right)^{N_{Tb}}, \tag{12.6}$$

where $n_b(t)$ is the number of type t households in block b. This is the likelihood of a polytomous regression model with an offset.

A flexible prior distribution for ω_b is the familiar multivariate normal regression model $\omega_b \mid \Phi_\omega \sim N(\beta \mathbf{x}_b, \Sigma)$, where $\Phi_\omega = (\beta, \Sigma)$, possibly with some prior constraints on the coefficient matrix β. The regression model allows us to incorporate observable block characteristics $\mathbf{x}_b$ into the predicted distribution for each block. Methods for drawing from such normal-exponential-family hierarchical models are well known (e.g., Gelman, Carlin, Stern, and Rubin, 2003; Section 16.4).

Inference for block counts N_{Tb}

Although (12.1) assumes a distribution for block sizes, in practice block definitions are often quite arbitrary. Census blocks in urban areas largely correspond to city blocks bounded by streets or other natural features, but almost half of the census blocks cover areas with no population at all, such as highway median strips, wilderness areas, or bodies of water. Furthermore, blocks that are anticipated before the census to have little or no population sometimes are found at census time to be occupied by new housing developments. Therefore, unbiasedness in estimation of N_{Tb} is more important than incorporating unreliable prior information on the empirical size distribution of blocks. Any estimator that systematically under- or overestimated the population of a certain class of blocks (e.g., small blocks) might be regarded as arbitrary and unfair (since block boundaries are themselves arbitrary) and therefore legally and politically unacceptable.

Instead of pursuing a posterior inference for the distribution of N_{Tb} from (12.1) and (12.2), a single vague prior distribution for N_{Tb} is posited in all blocks. Once N_{Tb} is drawn from its posterior distribution, $\mathbf{U}_b^*$ is $\mathbf{U}_b$ augmented with $N_{0b} = N_{Tb} - N_{Ub}$ households with "missing" status.

Unit or *whole-household* omissions exemplify *truncated* observations; the number of unobserved households is unknown. This is a distinct case from *censored* observations that are known to be present although there is no information about the type, as when a housing unit (physical structure) is determined to be occupied but no response can be obtained. (Such households constitute another set of observed types.) Once the number of truncated households N_{0b} is determined, these households are assigned a distinct "observed" type $U = 0$. Hence we now consider imputation of N_{Tb} when the number of observed households N_{Ub} is known.

Let $\bar{\alpha}_{0b} = \sum \mu_b(t)\alpha_b(t, 0)$ be the expected probability that a household will be missed altogether ("type 0") in block b. The inference conditional on the observed households $\mathbf{U}$ is the same as that conditional only on the number of observed households N_{Ub}; since all the households are drawn from the same distribution, the observed households give no additional information on the unobserved household (except perhaps through estimation of μ_b). Hence we write $N_{Ub} \mid N_{Tb} \sim \mathrm{Bin}(N_{Tb}, 1 - \bar{\alpha}_0)$. Then $\mathrm{E}[N_{Ub} \mid N_{Tb}] = (1 - \bar{\alpha}_{0b})N_{Tb}$, so a natural unbiased (conditional on $\bar{\alpha}_{0b}$) estimator of N_{Tb} is $N_{Ub}/(1 - \bar{\alpha}_{0b})$.

A simple prior specification preserves the key properties of this unbiased estimator in multiple imputation of N_{Tb} and hence of $N_{0b} = N_{Tb} - N_{Ub}$, the number of omitted households. Suppose that N_{Tb} has the improper prior distribution $\Pr(N_{Tb} = n) \propto \frac{1}{n}$. Then $N_{Tb} \mid U$ has a (proper) negative binomial posterior distribution with expectation $N_{Ub}/(1 - \bar{\alpha}_0)$, so the posterior expectation is an unbiased estimator of the total number of households. Meng and Zaslavsky (2002) show that this single observation unbiased prior (SOUP) is the only prior distribution having this property under the binomial model. They also show that inference under this prior is insensitive to the boundaries drawn (somewhat arbitrarily) between blocks with similar population distributions μ_b and coverage properties α_b.

Although the SOUP distribution is improper, by restricting the number of observed households to some broad range (say, $1 \leq N_{Ub} \leq 10^{10}$) we obtain a proper prior with no appreciable effect on the inference. (The "infinite" mass at 0 means only that no households will be imputed into blocks in which none were observed, the only prudent course in the absence of substantive information about the size and characteristics of the block.) Therefore, the procedure is legitimately Bayesian. The motivation, however, lies in the frequency properties of the procedure, that is, its unbiasedness, rather than in credibly representing prior beliefs about the distribution of household sizes. Unbiasedness is often considered a major criterion for "fairness" of census methodology, since a procedure that tended to under- or overestimate the population of blocks with certain characteristics would share many of the defects of using uncorrected census data. A procedure that yields roughly unbiased estimates of population in each class of blocks might be more acceptable than a biased one even if it has slightly larger mean squared error. Furthermore, if the population estimate for each block is unbiased, the population estimates for larger areas made up of many blocks are consistent as the area size increases.

Drawing true types T_b

Conditional on N_{Tb}, μ_b, and α_b, the posterior distribution of T_{bh}, by Bayes' theorem, is

$$\Pr(T_{bh} = t \mid U_{bh} = u, N_{Tb}, \mu_b, \alpha_b) = \alpha_b(t, u)\mu_b(t) \Big/ \sum_{t'} \alpha_b(t', u)\mu_b(t'),$$

(12.7)

Directly applying (12.7) could be computationally expensive when the number of types is large. Several algorithms for imputation of $\mathbf{T}$ that do not require summing over all possible values of T are described in Zaslavsky (1989, Section 11.1). These include a sampling procedure to estimate the denominator of (12.7) and several rejection sampling procedures.

The types of the N_{0b} unobserved households can also be drawn under (12.7), that is, with probability proportional to $\alpha_b(t, 0)\mu_b(t)$, in the same manner as for households of observed types.

Imputation of covariates

Households may be associated with covariate information in addition to their type. Some such data, such as characteristics of housing units, are asked of all respondents. Other data, such as income, are collected only from the sample of households that complete the long-form census questionnaire. Issues regarding covariates are discussed in the context of weighting in Zaslavsky (1989, Section 8). When imputing the roster, explicit models may be formulated relating true covariate values Y_{Tbh} to the true and observed types T_{bh} and U_{bh}, observed covariates values Y_{Tbh}, and the distribution of covariate values for households of type T.

Taking income as an illustrative covariate, a range of models might be considered. At one extreme, we might assume that reported income is equally valid regardless of whether or not the household type was observed correctly. Alternatively, the distribution of true income might be assumed to be similar for a given *true* type, regardless of whether or not it was observed correctly. In the latter case, reported income in households with omitted members would have to be (stochastically) adjusted for the difference between the income distributions for the observed and imputed types, perhaps reflecting underreporting of income for the omitted member. More complex models might be required if households with omissions (or unenumerated households) differ systematically from fully enumerated households with similar composition, for example, if they tend to be poorer.

In an imputation framework, covariates can also be multiply imputed under models that specify $p(Y_T \mid Y_U, T, U)$. There is currently little empirical basis for fitting such models, however, since PES implementations have not included collection of long-form data.

12.4   Simulation evaluations

We conducted a simulation (Zaslavsky, 1989; Section 14) to evaluate the bias of estimates of some population summaries under the imputation procedure, and to evaluate the usefulness of our block distribution model for describing real census data. We used the 1% Public Use Microdata Sample for the state of California from the 1980 census, comprising 231,459 persons in 86,447 households. The simulated blocks are the 111 "subcounty group units", each of which covers a part of a large city or one or more smaller cities and towns, with an average of 2,085 persons (779 households) per block.

Persons were classified into 60 classes by sex, age (5 levels), race (3 levels), and form of tenure (renter or owner). Of the 7,812 distinct household types, 60.9% appeared only once, while 5.8% of the types (453 types) comprised over 80% of the households (69,737 households).

The median number of covering types per type (not counting a type as covering itself) is 4, and 24.2% of the types have no covers. However, the most common household types are the smaller households, which tend to have more covers. Thus, most of the types with no covers are represented by a single household, over 90% of households have at least 11 covering types, and the median number of covering types per household is 141. For most observed households, the data represent a rich distribution of possible "true" types.

Models for μ_b of the form described in Section 12.2 were fitted, where $r(t)$ weighted together the fractions of household members who were Black or Hispanic and seven other variables. The fit of the block-level model is highly significant (likelihood ratio test, $\Delta G = 6628$ on 110 d.f.).

Because no data set was available for fitting the omission models of Section 12.2, plausible specifications were constructed for use in simulations, using marginal undercount rates from the 1985 Test of Adjustment Related Operations (TARO), approximately doubled by using the same rates for within- and whole-household omissions (Diffendal, 1988).

Within-household omissions were assumed independent. Whole-household omission rates were posited to follow a logistic model of the form logit $\alpha(t, 0) =$ logit $a(t) - 0.2 s(t) + 0.6$, where $s(t)$ = size (number of persons) of household t, and $a(t)$ = mean value of adjustment class omission rates for persons in household t. This specification reflects evidence that whole-household omission rates are higher for smaller households and for households whose individual members are from classes that have high omission rates (Fein and West, 1988).

The California data set and variously constructed subsets of it were treated as representing the "truth" for a single large block. Simulated "observed" data sets with omissions were drawn. Then μ_0 and N_{Tb} were estimated from the "observed" data by maximum likelihood (using an EM procedure); this calculation is equivalent to mean-imputation of adjusted data sets.

In each case, the "true", "observed", and "adjusted" data sets were summarized by total number of households and total population, population counts by race (Asian, Hispanic, and Other), sex, and age (five levels), and fraction of households by total size and by number of adults for households with children. For these household size measures, substantial bias remained in data sets adjusted by reweighting (as described in the Introduction to this chapter), with almost twice the correct number of single-adult families with children and substantial differences in other categories (Zaslavsky, 1988).

We evaluated how well the imputation procedure corrected biases in the simulated observed data by averaging over imputations. The simulated adjusted household and population counts are all very close to the "true" values, within 1.6% of the correct values, despite the large simulated undercounts of population (10.6%) and households (6.2%) and the large differentials in simulated undercount (ranging from 4.4 to 16.7% for various groups).

Adjusted proportions of households by size were also generally close to the truth, despite large biases (from -20.0 to $+53.6\%$ relative to the true proportion) in the observed data. The largest remaining biases were for the largest size categories (by either measure), which were underestimated by about 5.5% of their true shares. This might be due to the paucity of covers for the larger household types, which made it hard to impute additional persons to large households. To solve this problem it might be necessary to abandon the purely nonparametric approach used here to define possible household types, at least insofar as the more complex (larger) household types are concerned, or to enrich the model for misclassification of these larger types to create more covers.

Nonetheless, the bias of the estimation procedure is at most modest, and this holds for a wider range of measures than for the reweighting methodology.

12.5 Conclusion

The work described here addresses a challenging problem in inference from data under misclassification, complicated by the large number of possible classifications, the complex structure of both the original classifications and the classification errors, the truncation of some observations, and the importance of small-area variation to the required estimates. Although many problems remain to be solved, the models and computational methods described here might prove useful in other applications sharing some of these features. A helpful feature of the explicit hierarchical modeling approach is that alternative components can be substituted where desired, while retaining the overall structure of the model. In particular, Schafer (1995) proposed a parametric alternative for modeling household types and their distributions across blocks that would be less affected by the sparseness of the larger household types. Schafer's model could replace our nonparametric model μ_0 wholly, or as a supplement for the range of types (particularly larger households) in which sparseness is an issue.

This research reflects the contributions of Don Rubin in a number of spheres. Don led a long-term collaboration with the Census Bureau, directing his own talents and those of his students, including many represented in this volume, to solving complex methodological problems arising in the Census Bureau's work, particularly in the areas of missing data, undercount estimation, record linkage, and nondisclosure.

The use of imputation as a unifying strategy for expressing complex models was prescient; Rubin (1978) anticipates the widespread adoption of data augmentation (Tanner and Wong, 1987) and the Gibbs sampler (Gelfand and Smith, 1990). More broadly, Don Rubin has been among the leading proponents of using Bayesian hierarchical models to describe complex phenomena while obtaining valid measures of uncertainty. Frequency evaluation of Bayesianly derived procedures (Rubin, 1984) is critical to their acceptance in applied practice, especially for closely scrutinized applications like the census.

The methods described here, or indeed any form of adjustment for undercount, have not been adopted in the Census Bureau's practice, for both technical and policy reasons. In complex, high-profile applications like these, however, the statistician's role is not only to solve an immediate problem, but to widen the range of alternatives, extending the policymaker's methodological horizon and eventually improving statistical practice.

13

Statistical disclosure techniques based on multiple imputation

Roderick J. A. Little, Fang Liu, and Trivellore E. Raghunathan[1]

13.1 Introduction

Statistical disclosure control (SDC) is the modification of statistical data to prevent third parties from revealing sensitive information about the respondents (such as persons, households, businesses, etc.) Easy access to data via the Internet and electronic media has increased concerns about respondent privacy. SDC methods provide tools for modifying data to maintain wide dissemination of information while reducing the risk of disclosure of the identity of respondents. Ideally, an SDC tool should protect the identity of respondents in a data set, allow valid statistical inferences from the modified data with minimum information loss, and should achieve a reasonable balance between the competing goals of protection of confidentiality and dissemination of information. This chapter summarizes a cluster of SDC techniques that protect against disclosure by deleting the original values of variables in the data set and replacing them by $D > 1$ sets of values drawn from their predictive distributions. The imputation uncertainty is reflected

[1] Department of Biostatistics and Institute for Survey Research, University of Michigan, Ann Arbor, Mich. This research was supported by a grant from the National Science Foundation. We also thank Dr. Ramesh Dandekar for providing the data set analyzed in Section 13.5.

Applied Bayesian Modeling and Causal Inference from Incomplete-Data Perspectives.
Edited by A. Gelman and X-L. Meng © 2004 John Wiley & Sons, Ltd ISBN: 0-470-09043-X

by the method of multiple imputation (MI, Rubin, 1987b), applied to the set of D imputed data sets.

We first discuss *full synthesis*, which replaces the entire data file by predicted records from a model fitted to the observed data. This method provides essentially perfect protection from disclosure, since no actual records need to be released. However, it requires a full model for the joint distribution of the survey variables, and the quality of inferences from the synthetic data sets depends on how well this large model is specified. Also, disclosure protection comes at the expense of a considerable loss of information, although this loss can be appropriately taken into account in the inference by applying the MI combining rules described below.

We then discuss methods of *partial synthesis* that modify a part of the observed data. We assume the variables can be divided into two sets: key variables X (such as age, sex, and locality) that provide identifying information to intruders from publicly available sources, and nonkey variables that are not available from public databases, including potentially sensitive information like HIV status or income. Partial synthesis methods can be divided into methods that synthesize sensitive nonkey variables, so that values of these variables are not available to intruders who identify individuals in the database (Raghunathan, Reiter, and Rubin, 2003; Reiter, 2003; Rubin, 1993), and methods that synthesize key variables, so that individuals in the database cannot be identified. We focus here on the latter approach, which provides the potential to protect a large number of nonkey variables by synthesizing a modest number of keys (Little, 1993), though it does require knowledge of the key variables available to potential intruders. In particular, we consider Multiple Imputation of Keys (MIKe), which synthesizes all the values of the key variables, and Selective Multiple Imputation of Keys (SMIKe), which synthesizes the values of key variables for a selected subset of the cases, including the cases deemed most likely to be identified by the intruder. As with full synthesis, the information loss can be propagated in statistical inferences by MI, although, as discussed below, the combining rules are different from those for full synthesis or for MI for missing data. An attractive feature of partial synthesis is that the loss of information is much lower than with full synthesis, and in fact can be reduced to negligible levels by increasing the number D of multiply-imputed data sets.

The synthesized values in these methods are predictions based on a model for the population values. SDC methods not based on formal statistical models include global recoding (Willenborg and De Waal, 1996), local suppression (Willenborg and De Waal, 1996), data swapping (Dalenius and Reiss, 1982; Greenberg, 1987), microaggregation (Defays and Anwar, 1998), and post randomization (PRAM) (Gouweleeuw, Willenborg, and de Wolf, 1998). An important weakness of these model-free procedures is that they do not ensure valid statistical inferences based on modified data sets. In particular, point estimates based on the released data may be biased; furthermore, the incurred modification uncertainty is not taken into account in statistical analyses, so standard errors are too small and confidence intervals do not have their nominal level of coverage. The MI procedures discussed here do provide valid inferences, provided the imputation model is correctly specified.

Sections 13.2 and 13.3 provide overviews of full synthesis and partial synthesis, and Section 13.4 describes simple measures of protection and information loss. Section 13.5 describes an application of these methods to data from the 1995 panel of the Commercial Building Energy Consumption Survey. Section 13.6 provides concluding remarks and some topics for future research.

13.2 Full synthesis

Rubin (1993) first proposed full synthesis for disclosure control, where the original data set is replaced by D multiple synthetic data sets with all the variables imputed from their joint posterior predictive distribution. The method follows the basic principles of a Bayesian analysis of survey data (Rubin, 1987; Chapter 2), where the nonsampled portion of the population is treated as "missing data", and a Bayesian model is used to construct the posterior predictive distribution of the nonsampled values conditional on the observed sample values. Inferences are based on MI and can be viewed as approximations of Monte Carlo Bayesian inference. Although Bayesian in etiology, these inferences have desirable repeated sampling properties under a well-specified model.

Multiple synthetic populations can be constructed by appending sets of predictions of the nonsampled values to the sampled values, but they are often large and unwieldy. Rubin (1993) proposed addressing this by releasing a simple random sample from each synthetic population instead of the full population. The sampling affords maximum protection against disclosure if none of the original sample records are released, but introduces an additional source of variability that needs to be incorporated in the analysis.

The imputer model should be chosen to provide valid inferences for a wide variety of analyst models. Biased inferences can result when the imputer model is a submodel of the one used by an analyst (see for example the discussions of uncongeniality in Meng (1994a) and Rubin (1996)). Hence, the imputer model should be expansive, in the sense of including as submodels models that may be fitted by analysts. For this reason, Raghunathan, Reiter, and Rubin (2003) considered a nonparametric approach for creating multiple synthetic samples using the approximate Bayesian bootstrap (Rubin, 1987; Rubin and Schenker, 1986).

Suppose that the population consists of N subjects and a simple random sample of size n is drawn from this population. Suppose that $Y = (Y_1, Y_2, \ldots, Y_p)$ are the p survey variables of interest with multivariate distribution $p(y_1, y_2, \ldots, y_p)$. The objective is to simulate from the posterior predictive distribution of Y given the observed data $Y_{\text{obs}} = \{(y_{i1}, y_{i2}, \ldots, y_{ip}), i = 1, 2, \ldots, n\}$. The following approximate Bayesian bootstrap (Rubin, 1981b) simulates approximate draws from this posterior distribution for a likelihood based on the empirical distribution and a noninformative Dirichlet prior distribution:

1. Draw $n - 1$ uniform random numbers and order them, yielding the ordered sequence $a_0 = 0 < a_1 < a_2 < \cdots < a_{n-1} < a_n = 1$.

2. Draw N uniform random numbers, $u_1, u_2, \ldots, u_N$. Select sample unit i as unit j of the synthetic population if $a_{i-1} \le u_j < a_i$, for $j = 1, 2, \ldots, N$.

3. A synthetic sample is a simple random sample from the population simulated in Step 2.

4. Repeat steps 1 to 3 D times to create D synthetic samples.

Confidentiality protection in this approach is limited by the fact that actual data from the sampled units are replicated in the synthetic samples. To improve protection but maintain the nonparametric nature of the synthetic data creation, we suggest replacing the values in step 2 by draws from the predictive distribution corresponding to, for example, a sequence of semiparametric regression models (Hastie and Tibshirani, 1990; Green and Silverman, 1994). This approach is a generalization of Abowd and Woodcock (2001), who apply the sequential regression approach to multiple imputation of missing data (Kennickell, 1991; Raghunathan, Lepkowski, Van Hoewyk, and Solenberger, 2001).

Specifically, we obtain predictions $\hat{Y}_j$ of a variable Y_j from the generalized additive model,

$$g(E(Y_j)) = f_o + \sum_{k \neq j}^{P} f_j(Y_k) + \sum_{\ell=1}^{L} f_\ell(Z_\ell) \qquad (13.1)$$

where Z_ℓ are interactions or nonlinear functions of $\{Y_k, k = 1, 2, \ldots, p; k \neq j\}$ and g is a suitable link function, for example, the identity function for continuous variables, logistic for binary, log for counts, and so on. For a continuous variable, define the perturbed values as $Y_j^* = \hat{Y}_j + e_j^*$, where e_j^* are the residuals drawn at random from the observed residuals, $Y_j - \hat{Y}_j$. In general, the perturbations of the data are draws from the distribution centered at (13.1). For binary variables, the perturbations are created by drawing an independent uniform random number and setting the value to 1 if it is smaller than the corresponding predicted value in (13.1), and 0 otherwise. For count variables, the predictions may be drawn from a Poisson distribution with the mean equal to the predicted value in (13.1) or by adding randomly drawn residuals to the predictions, as in the case of continuous variables. The terms on the right side of (13.1) can be empirically determined on the basis of the data analysis of the original data, or lower-order interaction terms can be routinely included. The order of the sequential regressions can be randomized for each synthetic sample. In each sequence, previously perturbed Y's should be used as predictors in subsequent regression models to model association between the Y's. Smoothing parameters or roughness penalty parameters can be chosen to control the smoothness of the regression function.

13.3   SMIKe and MIKe

We now describe SMIKe, an SDC technique in which keys of sensitive cases and a selected *mixing set* of nonsensitive cases are multiply imputed from their posterior predictive distributions, and each set of imputed keys is released to the public with the rest of the data. We first describe SMIKe for categorical key variables, and then outline extensions to data where the key variables are both categorical and continuous. What follows can be applied for MIKe by considering the mixing set to be all cases in the data set.

The steps of SMIKe are (a) selection of sensitive cases and mixing sets of nonsensitive cases; (b) construction of an imputation model and MI of the values of key variables for the sensitive cases and their mixing sets; (c) assessment of disclosure risk and information loss for selected analyses involving the key variables; and (d) release of the imputed data along with tools for valid inferences. The amount of information synthesized can be varied through the size of the mixing sets to achieve a good balance between protection and information loss.

Suppose in a data set with n cases, there is a set of categorical key variables X, the cross-tabulation of which forms a contingency table with K categories; let Y denote a vector containing q nonkey variables, which may be continuous, categorical, or a mixture of both. Cells in the contingency table based on X are defined as *sensitive* if they contain less than s cases, where s is a sensitivity threshold chosen by the analyst. The cases in sensitive cells are called *sensitive cases*, and the objective of the SMIKe is to reduce the chance that an intruder will identify these cases.

Selection of nonsensitive cases

Each sensitive case i ($i = 1, 2, \ldots, n_s$), where n_s is the number of sensitive cases) is associated with a *mixing set* M_i of insensitive cases. In general, it is advantageous to include cases in the mixing set that are similar to the sensitive case with respect to Y, since imputing keys with relatively homogeneous sets of cases tends to distribute cases over the set of sensitive and nonsensitive cells, thus promoting the mixing of sensitive and nonsensitive cases and increasing protection. Thus, it is suggested that cases in M_i are chosen from donor key cells that are close to the sensitive cell according to some metric. For example, if Y is continuous with components transformed to be approximately normal and y_i is the value of Y for sensitive case i, then cases in the mixing set might be selected from a cell or cells j that are close as measured by the Mahalanobis distance $(y_i - \overline{y}_j)^T S^{-1} (y_i - \overline{y}_j)$, where $\overline{y}_j$ is the mean of Y in cell j and S is the pooled within-cell covariance matrix of Y. Mixing sets can be chosen by randomly sampling cases within cells (random selection), or by selecting cases that have values of Y close to y_i (purposive selection). These choices have implications for the imputation model, as discussed below. The mixing sets for different sensitive cases may overlap. Also, the mixing sets can be further constrained to avoid information loss for particular analyses. For example, if we want to preserve the row margins of a table formed by a subset of key variables, we may only allow sensitive cases to be mixed with cases from the same row in that table.

Let M denote the union of sensitive cases and nonsensitive cases in the mixing sets $\left(M = \bigcup_{i=1,2,\ldots,n_s}(i \bigcup M_i)\right)$, and let n_M denote the number of cases in M. The fraction of cases with keys imputed is thus n_M/n. The size of n_M determines how many key values are imputed; the larger the mixing sets, the greater the gains in protection and losses in information. M is a subset of the union of sensitive cases and all the nonsensitive cases in donor cells, which we denote as C. MIKe is a special case of SMIKe, where $n_M = n$ and the keys for all the cases are imputed.

Construction of an imputation model for keys

Let (x_M, y_M) denote the values of X and Y in M. The values x_M are deleted, and then multiply imputed from the predictive distribution $p(x_M|y_M, M)$ of x_M given y_M based on an imputation model estimated using the original data. The predictive distribution $p(x_M|y_M, M)$ is written conditional on M, since for valid inferences we need to take into account the method of selection of the mixing set M, which differs from the full sample. In particular, the distribution of the key variables in M differs because the definition of sensitive cases and their mixing sets is based on the distribution of X. The selection of cases in the donor cells also depends on Y under purposive selection, but does not depend on Y under random selection of cases from a given donor cell. These differences have implications for the modeling of $p(x_M|y_M, M)$. The most straightforward approach is to model the distribution of X and Y using only the data in M. However, this is inefficient if the set of cases in M is small, particularly if Y is high dimensional. Factor the joint distribution of (x_M, y_M) as:

$$p(x_M, y_M, M) = p(x_M|M)p(y_M|x_M, M).$$

If the selection of the mixing set is based only on X, as in random selection, then

$$p(y_M|x_M, M) = p(y_M|x_M, C),$$

and this conditional distribution can be modeled using the larger set of cases C. In fact, the entire data set could be used to model this distribution, but restricting the model to the donor cells limits the modeling task to the cells that are relevant to the imputation. Under purposive selection based on a subset of the Y-variables, say Y_1, let y_{M1} denote the values of Y_1 in the mixing set and y_{M2} the values of the other Y-variables; then the predictive distribution

$$p(y_{2M}|y_{1M}, x_M, M) = p(y_{2M}|y_{1M}, x_M, C)$$

can be estimated on the basis of the cases in C and $p(y_{1M}, x_M|M)$ estimated using the cases in M.

As with full synthesis, the imputation model can be parametric, semiparametric or nonparametric; a parametric model may be adequate here since the impact of model misspecification is limited to the subset of values being imputed. In our applications, imputes of the keys are drawn from the predictive distribution

$$p(x_i|y_i, M) = \int p(x_i|y_i, M, \theta) p(\theta \mid M)\, d\theta,$$

which can be accomplished by drawing $\theta^{(d)}$ from the posterior distribution of θ, and then drawing x_i from $p(x_i|y_i, M, \theta^{(d)})$ for cases in the mixing set.

For continuous keys, all cases will tend to be deemed sensitive based on a cross-classification, leading to the MIKe, where all key values are imputed. When keys include continuous and categorical variables, MIKe can be applied to the continuous keys and SMIKe can be applied to the categorical keys, treating the continuous keys like nonkey variables. To reduce information loss in the continuous keys, SMIKe can be applied to a coarsened version of the continuous variables, and the continuous variables are then imputed for all cases, conditioning on the imputed coarsened variables.

13.4 Analysis of synthetic samples

Suppose for completed data set d ($d = 1, \ldots, D$), $\hat{\phi}^{(d)}$ is an estimate of ϕ, a scalar parameter of interest, and $V^{(d)}$ is an estimate of the variance of $\hat{\phi}^{(d)}$. The synthetic estimate for both full and partial synthesis is

$$\hat{\phi}_{\text{syn}} = \sum_{d=1}^{D} \hat{\phi}^{(d)}/D. \tag{13.2}$$

The variance estimate for full synthesis (as described in Section 13.2) is

$$T_{\text{syn}} = (1 + D^{-1})B - W, \tag{13.3}$$

where $B = \sum_{d=1}^{D} (\hat{\phi}^{(d)} - \hat{\phi}_{\text{syn}})^2/(D - 1)$ and $W = \sum_{d=1}^{D} V^{(d)}/D$ (Raghunathan, Reiter, and Rubin, 2003). For small D, say $D < 10$, the variance estimate in (13.3) can occasionally be negative. Raghunathan and Rubin (2000) suggest replacing negative values by $T_{\text{syn}} = kW/n$, where k is the size of the synthetic sample. This *ad hoc* fix gave well-calibrated confidence intervals in simulations by Reiter (2002). Raghunathan and Rubin (2000) and Raghunathan, Reiter, and Rubin (2003) provide analytical results demonstrating the validity of inferences from a repeated sampling perspective, and show for various regression models that inferences from the original data and multiple synthetic samples are comparable. Reiter (2002) also demonstrates that this approach results in valid inferences for a complex survey design, if the model used to construct the predictive inference incorporates these complex design features.

For the partial synthesis methods of Section 13.3, the variance associated with $\hat{\phi}_{\text{syn}}$ is given by

$$T_{\text{syn}} = W + B/D, \tag{13.4}$$

(Reiter, 2004; Little and Liu, 2003). The combining rules in (13.3) and (13.4) both differ from that for missing data problems, namely $T_{\text{mis}} = W + (1 + 1/D)B$ (Rubin, 1987).

The associated fraction of information loss for inferences for scalar ϕ is $\gamma = B/(DT_{\text{syn}})$, where T_{syn} is given by (13.3) for full synthesis and (13.4) for partial synthesis. The former is larger than the information loss for missing data, and the latter is smaller, and tends to zero as the number of MI data sets D increases. To assess information loss for a given amount of imputation, γ might be computed for a range of analyses of interest.

Assessment of disclosure risk

The assessment of disclosure risk is difficult since it requires conjectures about the behavior of a data intruder, and we provide a brief discussion here. We first describe a measure of disclosure risk $R(\text{orig})$ for the original data set. If the data are a census of the population, and cases in key cells with more than s cases are assumed to have negligible disclosure risk, then equating risk with the probability of disclosure yields the measure

$$R(\text{orig}) = \sum_{i=1}^{n} r_i(\text{orig}), \tag{13.5}$$

where

$$r_i(\text{orig}) = \begin{cases} 1/n_i, & \text{if } n_i \leq s \\ 0, & \text{otherwise} \end{cases}, \tag{13.6}$$

n_i is the number of sample cases of the key cell containing unit i. A simple modification of (13.6) for an equal probability sample of the population with sampling fraction f is

$$r_i = \begin{cases} f/n_i, & \text{if } n_i \leq fc \\ 0, & \text{otherwise} \end{cases}, \tag{13.7}$$

c is a prespecified cutoff value for the population.

A simple measure of the disclosure risk for the dth synthesized data set is

$$R^{(d)} = \sum_{i=1}^{n} r_i^{(d)},$$

where

$$r_i^{(d)} = \delta_i f/n_i^{(d)}, \tag{13.8}$$

where $\delta_i = 1$ if the case i is imputed to its original key cell and $n_i^{(d)}$, the number of cases in that cell after imputation, is less than or equal to fc; otherwise $\delta_i = 0$. Then

$$R_1(\text{smike}) = \sum_{d=1}^{D} R_1^{(d)}/D, \tag{13.9}$$

averaging the risk measure of the MI data sets. This simple measure does not account for the fact that additional information is available if the D data sets are considered in aggregate rather than one at a time. For a more complex measure that addresses this issue, see Liu and Little (2002). Absolute disclosure risk is important in applications, but the relative reduction in disclosure risk is useful for measuring the trade-off between the information loss and protection P, which is given by

$$P_1 = 1 - \frac{R_1(\text{smike})}{R(\text{orig})}. \tag{13.10}$$

13.5   An application

In this section, we apply full synthesis, MIKe and SMIKe to 13 variables from the Energy Information Administration's 1995 Commercial Building Energy Consumption Survey (CBECS), a survey ($n = 5,655$) that provides information concerning building characteristics and electricity, natural gas, and fuel oil consumption for a national sample of commercial buildings. The 13 variables comprised two categorical key variables, namely principal building activity (pba) and year of construction (year), which formed a table with $K = 171$ cells, and two continuous key variables, namely space and number of employees. The remaining nine variables are the non-key variables Y, consumption of electricity, natural gas and major fuel (3 variables), the associated expenditures (3 variables), and percent of floor space heated, cooled or lit (3 variables). All these variables were logtransformed to reduce skewness.

Model for full synthesis: The procedure outlined in Section 13.2 was implemented, creating $D = 10$ imputations. The projection pursuit regression package in S-plus was used to perturb the values, with a maximum of 10 terms in the model. Projection pursuit regression is a generalized additive model for a linear combination of the predictor variables, as in (13.1). These linear combinations are carefully chosen to succinctly extract the variation in the predictor space. We included all two-factor interaction terms as predictors as well.

Model for partial synthesis: For unit i, denote the two categorical keys as X_{0i}, the two continuous keys as X_{1i}, and the nine continuous nonkeys after transformation as Y_i, and all the continuous variables as $Z_i = (X_{1i}, Y_i)$. Partial synthesis was based on the following general location model:

$$X_{0i} \sim_{ind} \text{Multinomial } (\pi_1, \ldots, \pi_K), \sum_{k=1}^{K} \pi_k = 1,$$

$$Z_i \mid X_{0i} = k \sim_{ind} N(\mu_k, \Sigma_k). \tag{13.11}$$

The covariance matrix in the multivariate normal distribution was initially assumed to be a constant over cells k, but diagnostics revealed that it was important to allow this matrix to vary over k. Some cases with gross outlying values of Z_i

were also excluded in fitting the imputation model. A standard Bayesian analysis of this model assuming standard noninformative priors yields the following posterior distribution of the parameters given the data prior to masking:

$$\pi_1, \ldots, \pi_K | \text{data} \sim \text{Dirichlet}(n_1 + 1/2, \ldots, n_K + 1/2)$$

$$\mu_k | \pi \, \Sigma_k, \text{data} \sim N(\bar{z}_k, \Sigma_k/n_k)$$

$$\Sigma_k | \pi, \text{data} \sim \text{inverse-Wishart}((n_k - 1)S_k, n_k - 1), \qquad (13.12)$$

where $(n_k, \bar{z}_k, S_k)$ denote respectively the sample count, the sample mean of Z, and the sample covariance matrix of Z in the inverse-Wishart distribution with degrees of freedom ν (e.g., Schafer, 1997). In MIKe, the keys for all cases i are imputed. For SMIKe, all values of the continuous keys are imputed but the categorical keys are imputed for a subset of cases. Specifically, we set $s = 3$ as the threshold for a sensitive cell in the contingency table formed by X_0, yielding 26 out of 171 sensitive cells and 50 out of 5,655 sensitive cases. We set $n_{\text{mix}} = 4$ as the size of mixing set for each sensitive case. The matching sets consisted of 136 nonsensitive cases in 34 nonsensitive cells, chosen to be close to the sensitive cells in the Mahalanobis distance. The set M with categorical keys imputed includes $50 + 136 = 186$ cases and the set C for modeling consists of 671 cases.

Evaluation: To compare inferences based on the original data and the data modified by SMIKe, MIKe, and full synthesis, estimates of the regression

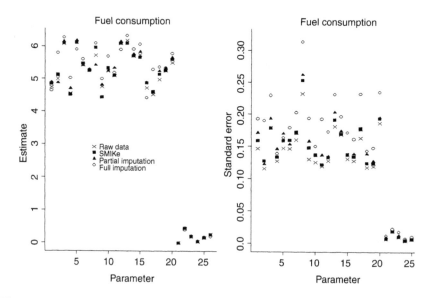

Figure 13.1   Estimates of regression coefficients (26 parameters) and their standard errors from the multiple regression of log(fuel consumption) in the raw data and data sets modified by SMIKe, MIKe, and full synthesis.

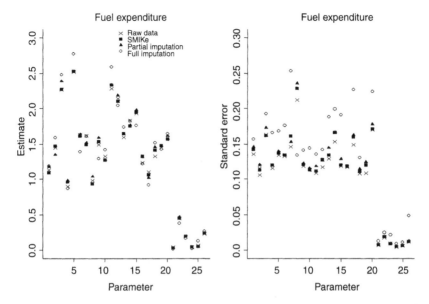

Figure 13.2   Estimates of regression coefficients (26 parameters) and their standard errors from the multiple regression of log(fuel expenditure) in the raw data and data sets modified by SMIKe, MIKe, and full synthesis.

coefficients and their associated standard errors from regressions of log (fuel consumption) and log (fuel expenditure) based on the original and imputed data are plotted in Figures 13.1 and 13.2. Results from various other statistical analyses can be found in Liu (2003). The regressions include 26 predictors and do not include a constant term. The estimates of the regression coefficients based on the imputed data are close to the original ones, and the standard errors of the estimates from SMIKe and MIKe are close to those from the original sample, indicating very little information loss. The standard errors from full synthesis are larger, suggesting more information loss. Measures on protection are given in Table 13.1. Compared to the perfect protection given by full synthesis, those given by SMIKe and MIKe are comparable, especially when h gets larger, P_1 approaches to 1 as well. Liu (2003) provides more discussion on information loss and protection for this application.

SDC Technique	MIKe					SMIKe		
h	10	10	20	50	100	20	50	100
P_1	0.909	0.972	0.991	0.996	0.924	0.976	0.994	0.997

Table 13.1   Protection given by SMIKe and MIKe in the CBECS data.

13.6   Conclusions

The full synthesis method provides the greatest degree of disclosure protection of the methods discussed; indeed, the method can provide inferences that are comparable (if less efficient) than inferences from the original data, without having to disclose any of the values in the original data set. Thus, the method shows promise for situations where very high levels of disclosure protection are sought. On the other hand, the method involves an extensive model-building effort, particularly for large data sets, and models need to be relatively nonparametric to avoid potential distortions from the effects of model misspecification. SMIKe is attractive since it can yield major gains in disclosure protection with much less imputation than is needed for full synthesis. The method is less affected by model misspecification than full synthesis, and involves less onerous imputation tasks.

An alternative to SMIKe that preserves the distribution of the key variables in the original data is multiple and stochastic swapping of keys (MaSSK), which swaps the values of the key variables between paired cases in a data set and releasing multiple swapped data sets. MaSSK can be viewed as a model-based version of data swapping that yields valid inferences. For more information about this method, see Liu (2003).

An important feature of these methods is the ability to create valid inferences from the MI data sets, using the combining rules discussed in Section 13.4. Methods are also available for hypothesis testing and inferences for multiparameter estimands, analogous to the methods for missing data (Rubin, 1987; Little and Rubin, 2002) with the total variances modified as in (13.3) for full synthesis and (13.4) for partial synthesis. Real-data examples such as that in Section 13.5 cannot demonstrate the validity of inferences, but this has been shown in simulation studies described elsewhere (Reiter, 2002; Raghunathan, Reiter, and Rubin, 2003; Little and Liu, 2003). More assessments of the frequency validity of these inferences would be useful on large and more realistic problems. Also, better measures of disclosure risk would be welcome. We look forward to further developments of the methods described here in the future.

14

Designs producing balanced missing data: examples from the National Assessment of Educational Progress

Neal Thomas[1]

14.1 Introduction

The National Assessment of Educational Progress (NAEP) is an ongoing collection of surveys of students and teachers in the United States. Demographic and educational environmental variables are collected from sampled students who are subsequently administered a brief examination. Teachers of sampled classes also complete a questionnaire. The goal of NAEP is to estimate the proficiency of US students in different subject areas, how it varies with demographic and educational environmental conditions, and how it changes over time. The primary reporting of a NAEP survey consists of mean performance and percent above prespecified cutpoints by region (Northeast, Southeast, Midwest, West), ethnicity, and gender on scales described in Section 14.2.

To increase the number of different items (test questions) that can be administered, matrix-sampling designs have been used since 1985 in which sampled students are administered different forms each containing a small subset of a large collection of items. These designs allow many more items to be administered,

[1] Department of Clinical Biostatistics, Pfizer Corp., Groton, Ct.

Applied Bayesian Modeling and Causal Inference from Incomplete-Data Perspectives.
Edited by A. Gelman and X-L. Meng © 2004 John Wiley & Sons, Ltd ISBN: 0-470-09043-X

but they result in a large amount of (planned) missing data because sampled students are not administered most items in the survey. The primary reporting, and secondary analyses based on public-release files, are performed using multiple imputation. Robert Mislevy (personal communication) attributes the idea of using multiple imputation to analyze matrix-sampled NAEP data to Darrell Bock, who proposed it after attending an early seminar on multiple imputation given by Donald Rubin in the Statistics Department at the University of Chicago. NAEP is one of the earliest substantial applications of multiple imputation, and one of the most ambitious uses of statistical methods in large-scale public policy research. A brief overview of the statistical methods utilized in NAEP with an emphasis on the handling of the planned missing item responses is in Section 14.2. Beaton and Zwick (1992), along with accompanying articles, provide a more detailed introduction to the statistical methods used for NAEP.

Each NAEP survey measures a domain such as reading or mathematics. The ability of a student to correctly respond to cognitive items (test questions) within a domain is called their "proficiency". Within a domain such as Mathematics, for example, items are administered in several (5) related subdomains like algebra and geometry. The design issue considered in Section 14.3 is how to allocate items measuring different subdomains, for example, assign some students items measuring algebra, other students mostly geometry items, and some students a very small number of items measuring proficiencies in both subdomains. With very limited time to administer the survey, and several subdomains to measure, it was conjectured that concentrating the items for each student in only one or two subdomains would yield more accurate proficiency estimates because the proficiencies that were well measured could be used to impute proficiencies in the subdomains not measured, taking advantage of the correlation between proficiencies. Such designs are also popular because some "extended" item types, which involve several related items on a common complex task, measure a single subdomain and require too much time to allow items measuring each subdomain. Results summarized in Section 14.3 show that a class of "balanced" designs is optimal.

The simultaneous analysis of data from the multiple correlated proficiencies are contrasted with the results for bivariate normal data subject to missingness in Little and Rubin (2002). The NAEP analyses are a generalization of the results in Little and Rubin, and the balanced designs have the peculiar feature that despite high correlation between proficiencies (such as algebra and geometry), which allow for much more accurate estimation of each individual proficiency using all of the items administered to a student, there is no improvement from multivariate analyses in the estimation of population parameters included in the primary reporting.

In addition to the covariates defining the subpopulations for the primary reporting, such as region, ethnicity, and gender, NAEP collects numerous secondary covariates, including responses to questions such as: "How many math courses have you taken?" and "How many hours do you watch TV each day?". In Section 14.5, the role of secondary covariates in the estimation of primary population parameters is considered. As with cognitive items measuring different proficiencies, the secondary covariates do not contribute to the estimation of the primary population parameters when optimal balanced designs are utilized, despite the fact that using the secondary covariates can substantially improve the estimation of proficiencies for individual students. The practical consequences of this unanticipated result are discussed. The role of secondary covariates in the creation of public-use files is also discussed.

14.2   Statistical methods in NAEP

Notation

The q primary and secondary covariates for the ith student in a sample of size n are denoted by $x_i' = (x_{i1}, \ldots, x_{iq})$. The variables representing the cognitive items for the ith student are denoted by y_i, and are partitioned into p subdomains (e.g., algebra and geometry) $y_i' = (y_{i1}, \ldots, y_{ip})$, corresponding to p latent proficiencies in a multidimensional Item Response (IRT) model. Each y_{ij} is composed of item scores, $y_{ijk}, k = 1, \ldots, s_j$, that are binary or ordinal with values coded as $0, 1, \ldots, m_{jk}$, for the kth item measuring the jth proficiency.

A model representing the data is specified in two stages. First, a latent proficiency vector, $\theta_i = (\theta_{i1}, \ldots, \theta_{ip})$, is hypothesized for the ith student, which determines the distribution of the item scores through logistic IRT models, where conditional on the latent proficiencies, all item responses are assumed independent of each other and the x_i. Second, the θ_i are assumed to follow a multivariate normal distribution conditional on the covariates. Most NAEP reporting is based on summaries of the θ_i.

Models for the cognitive responses

A logistic item response model is used for the cognitive data y_i conditional on x_i, θ_i, and the item parameters $\beta = (\beta_1, \ldots, \beta_p)$, $\beta_j = \{\beta_{jk}, k = 1, \ldots, s_j\}$. The probabilities of binary scored items are modeled by a three-parameter logistic IRT model,

$$\Pr(y_{ijk} = 1 \mid \theta_i, x_i, \beta) = c_{jk} + (1 - c_{jk}) / [1 + \exp\{a_{jk}(\theta_{ij} - b_{jk})\}], \quad (14.1)$$

where $\beta_{jk} = (a_{jk}, b_{jk}, c_{jk})$. The response probabilities of an ordinal item are modeled by a partial credit model,

$$\Pr(y_{ijk} = l \mid \theta_i, x_i, \beta) = \frac{\exp\left\{\sum_{h=0}^{l} a_{jk}(\theta_{ij} - b_{jkh})\right\}}{\sum_{q=0}^{m_{jk}} \exp\left\{\sum_{h=0}^{q} a_{jk}(\theta_{ij} - b_{jkh})\right\}}, \quad l = 0, \ldots, m_{jk},$$

$$(14.2)$$

where $\beta_{jk} = \{a_{jk}, b_{jkh}, h = 1, \ldots, m_{jk}\}$.

The model invokes several independence assumptions conditional on the θ_i and β: (1) the y_i are independent of the x_i; (2) the responses of a student to different items are independent (i.e., the distribution of the y_i is the product of the probabilities in (14.1) and (14.2)); (3) responses from different students are independent; and (4) y_{ij} are independent of $\theta_{ij'}$ conditional on θ_{ij}, $j \neq j'$, that is, item responses depend only on the proficiency to which they are assigned. With these independence assumptions, the density of y_i, $i = 1, \ldots, n$ conditional on θ_i can be represented as

$$\prod_{i=1}^{n} \left\{ \prod_{j=1}^{p} f_j \left(y_{ij} \mid \theta_{ij}, \beta_j \right) \right\}, \tag{14.3}$$

and each $f_j \left(y_{ij} \mid \theta_{ij}, \beta_j \right)$ is in turn the product of the response probabilities for the cognitive items given in (14.1) and (14.2). Items not presented to a student do not contribute to the likelihood function because they are missing completely at random (MCAR) by design.

The β_j are included in the likelihood term, $f_j \left(y_{ij} \mid \theta_{ij}, \beta_j \right)$, to explicitly denote the dependence of (14.3) on each β_j. To identify the logistic IRT model, the mean and variance of θ in the overall population are constrained to be zero and one. Mislevy and Bock (1982), and Muraki (1992) give details of these models.

Distribution of the latent proficiency conditional on the background variables

The θ_i vectors are assumed to be normally distributed conditional on the x_i. The mean of this conditional distribution is given by the multivariate multiple linear regression, $\Gamma' x_i$, where $\Gamma = \left[\gamma_1 \mid, \ldots, \mid \gamma_p \right]$ and γ_j, $j = 1, \ldots, p$ are unknown regression parameter vectors of length q. The common (unknown) p dimensional conditional variance–covariance matrix is Σ with elements Σ_{jk}, (Mislevy, Johnson, and Muraki, 1992). The distribution of θ_i can be viewed as a normal prior density conditional on x_i and the parameters Γ and Σ, before observing the cognitive data y_i: $\phi \left(\theta_i; \Gamma' x_i, \Sigma \right)$.

Estimation using multiple imputation

Applying the independence assumptions, the regression model and the item response model fully specify the distribution of observed data. The likelihood function for the parameters (β, Γ, Σ) is the distribution of the data (y_i, x_i), $i = 1, \ldots, n$, given (β, Γ, Σ):

$$\text{lik} \left(\beta, \Gamma, \Sigma \right) = \prod_{i=1}^{n} \int \phi \left(\theta_i; \Gamma' x_i, \Sigma \right) \prod_{j=1}^{p} f_j \left(y_{ij} \mid \theta_{ij}, \beta_j \right) d\theta_i. \tag{14.4}$$

The integrand in (14.4) is proportional to the posterior distribution of θ_i with β, Γ, and Σ regarded as known:

$$f\left(\theta_i; x_i, y_i, \beta, \Gamma, \Sigma\right) \propto \phi\left(\theta_i; \Gamma'x_i, \Sigma\right) \prod_{j=1}^{p} f_j\left(y_{ij} \mid \theta_{ij}, \beta_j\right). \qquad (14.5)$$

The θ_i are regarded as "missing" data and imputed. The missing item responses are not directly imputed as part of the multiple imputation procedures. Thomas and Gan (1997) provide details on the numerical methods used to generate multiple imputations from (14.5) and the distribution of (β, Γ, Σ) given (x, y). Following Rubin and Schenker (1986), five multiply imputed data sets are created. The standard methods in Rubin (1987b) are applied to combine the multiple estimates. The same methods are used for primary reporting and for secondary analyses, which are based on the same multiply imputed data sets with confidential data excluded.

Simplifying normal error approximation

The contribution of the IRT component to the likelihood function in (14.5), $\Pi_{j=1}^{p} f_j$ $\left(y_{ij} \mid \theta_{ij}, \beta_j\right)$, approaches normality as the number of items increases (Chang and Stout, 1993). Replacing the IRT component by a corresponding product of normal densities reduces the problem to one with analytic solutions, which have been extensively studied. For each proficiency, θ_{ij}, $j = 1, \cdots, p$, reuse the symbol y_{ij} to represent the maximum likelihood estimate (MLE) of θ_{ij} based on $f_j\left(y_{ij} \mid \theta_{ij}, \beta_j\right)$. The redefined response y_{ij} is a scalar, in contrast to the original y_{ij}, which is a vector of item responses. Let τ_{ij} represent the variance based on the observed information from $f_j\left(y_{ij} \mid \theta_{ij}, \beta_j\right)$. The normal approximation is then

$$f_j\left(y_{ij} \mid \theta_{ij}, \beta_j\right) \propto \phi\left(\theta_{ij}; y_{ij}, \tau_{ij}\right). \qquad (14.6)$$

The τ_{ij} can be viewed as measurement error variances for the estimation of the θ_{ij}, and the estimation problem is an example of generalized least squares (Johnson, 1984). The τ_{ij} are determined by the item parameters and the item sampling design. When there are no items measuring the θ_{ij}, the y_{ij} are set equal to an overall mean value, and the τ_{ij} are (effectively) infinite so there is no contribution to the likelihood function in (14.5) for the jth proficiency from the students. When the number of items measuring θ_{ij} is small, the modal approximation can be unstable. Mislevy (1992) and Thomas (1993) propose alternative normal approximations that perform better in this situation.

14.3 Split and balanced designs for estimating population parameters

Split and balanced designs

A balanced design is one in which each student is assigned the same number of items measuring each proficiency, although the number of items measuring different proficiencies can differ. An example of a balanced design assigns 10 items measuring geometry and five items measuring algebra to each student. More items may be assigned to a proficiency if it is deemed more important. The simplified balanced designs considered here assume that the same number items are assigned for each proficiency.

A split design assigns differing numbers of items to different students. A typical example of a split design assigns some students 5 algebra items and 5 geometry items, some students 10 algebra items only, and some students 10 geometry items. Such designs may be used for statistical reasons, but they occur more commonly when test designers create extended collections of related items involving a complex task measuring a single proficiency. These items take most of the allotted testing time so it is not feasible to include items measuring other proficiencies. Discussion is limited here to bivariate traits, although there may be as many as five traits in practice and similar design issues apply.

Zeger and Thomas (1997) consider the following class of split designs for bivariate traits that have a balanced design as a limiting case (when $m = 0$):

1. A fixed number of items, n_I, are administered to each student, and they are divided between items measuring either proficiency.

2. A subsample of $n_I - m$ students are assigned $1/2$ the items to measure each proficiency. An additional subsample of $m/2$ students are assigned items measuring only the first proficiency, and the remaining $m/2$ students are assigned items measuring the second proficiency.

Simplifying assumptions

The simplifying normal approximation to the IRT model is applied to the class of designs along with the following assumptions:

1. The individual student estimates of the proficiencies, (y_{i1}, y_{i2}), are bivariate normal as in (14.6) with a common measurement error variance, τ.

2. The variances of the proficiencies in the two subdomains are equal ($\Sigma_{11} = \Sigma_{22}$).

3. The first two assumptions imply that if n_I items measuring one of the proficiencies are utilized, the variance of the estimate of the proficiency is τ/n_I. If $1/2$ of the items are selected to measure each proficiency, the variance for each proficiency is $2\tau/n_I$.

Several considerations relating to these assumptions are as follows:

1. A very difficult item does not differentiate weak students from mediocre students, but is very effective in demonstrating the performance of the best students, for example. In practice, measurement error variances are heterogeneous and may depend on the proficiency level.

2. It is implicitly assumed that there is no unplanned item missingness. Such missing does occur in NAEP further altering the idealized designs (and more importantly, may not be missing at random (MAR)).

3. The variances for different proficiencies are set to one in the overall population as part of the identification of the IRT model. The variances conditional on covariates are not exactly equal, but have not varied widely in past surveys.

The design and accompanying assumptions about the variances produce mathematical simplification by creating symmetry between the proficiencies. The results in Section 14.4 on ML estimation indicate that the optimality properties extend to other designs with multiple differing split types.

Optimal designs

The optimal design (minimizing the variance of the MLE of the mean estimators) in this class has $m = 0$, that is, all students receive the items measuring both proficiencies. Despite several simplifications to the actual NAEP setting, the simplified setting produced accurate predictions of the efficiencies of several NAEP designs, both approximately balanced and split designs. The robustness of the actual NAEP results to the moderate deviations from exact balance and normality is another example of the robustness of regression estimators to normality and moderate heterogeneity, which results in slightly nonoptimal weighting. The loss in efficiency due to the use of split designs was in the range of 15 to 20%.

14.4   Maximum likelihood estimation

With the simplifying normal approximation, the optimal balanced designs yield standard multivariate ML estimation. The conditional distribution in (14.6) implies the distribution of y_{ij}, $j = 1, 2$ is

$$y_{ij} \sim \phi\left(\boldsymbol{\gamma}'\boldsymbol{x}_i, \Sigma_{jj} + 2\tau/n_I\right),$$

and cov $(y_{i1}, y_{i2}) = \Sigma_{12}$. The complex NAEP estimation procedures are thus reduced (approximately) to standard multivariate multiple regression. One immediate consequence is that despite the missing data for each proficiency, which is well predicted by observed data from other proficiencies, the estimation is essentially univariate. Despite no borrowing of information between the correlated proficiencies, the balanced design/estimation is optimal.

Zeger and Thomas (1997) also show that if there is balanced missing data for a proficiency, its estimation does not depend on data from other proficiencies regardless of the item sampling design for the other proficiencies.

It is instructive to compare the balanced NAEP setting to the classic formulas in Little and Rubin (1987) for bivariate normal data with one variable subject to all-or-none missing data (assumed to be MAR). This corresponds to the setting in which any observed y_{ij} equal θ_{ij}. With the first proficiency completely observed for each subject, a subsample of the second proficiency fully observed, but an additional subsample with no data on the second proficiency, the MLE from equation (6.9) of Little and Rubin is:

$$\hat{\mu}_2 = \bar{y}_2 + \hat{\beta}_{21|1} \left(\hat{\mu}_1 - \bar{y}_1 \right), \tag{14.7}$$

where (μ_1, μ_2) refer to population means without covariates, $\hat{\beta}_{21|1}$ is the regression of y_2 on y_1, $(\bar{y}_1, \bar{y}_2)$ are sample means restricted to the students with fully observed proficiencies, and $\hat{\mu}_1$ is the mean of y_1 in the complete sample. The estimator in (14.7) is a poststratification adjustment of the subsample with y_2 observations to the full sample. Little and Rubin show that there can be large gains in precision if the correlation between the two proficiencies is high, approaching complete information for correlations near one. The adjustment can also reduce bias when the y_2 are MAR, but not MCAR.

The corresponding estimator for the second proficiency in a balanced design is

$$\hat{\mu}_2 = \bar{y}_2 + \frac{\hat{\beta}_{21|1}}{n} \sum_{i=1}^{n} \left(\hat{\mu}_1 - y_{i1} \right), \tag{14.8}$$

where $\hat{\beta}_{21|1}$ now estimates the attenuated regression coefficient, $\Sigma_{12}/(\Sigma_{11} + \tau/n_I)$. The poststratification adjustment is exactly zero in this simplified version of the NAEP balanced setting because the students with observed and missing measurements coincide. The lack of adjustment occurs despite the substantial improvement in the estimators of the individual proficiencies, θ_{i2}. Corresponding to (6.12) of Little and Rubin, (14.8) can be rewritten

$$\hat{\mu}_2 = \frac{1}{n} \sum_{i=1}^{n} \left(y_{i2} - \hat{\beta}_{21|1} \left(y_{i1} - \hat{\mu}_1 \right) \right)$$

$$\equiv \frac{1}{n} \sum_{i=1}^{n} \hat{\theta}_{i2}, \tag{14.9}$$

where $\hat{\theta}_{i2}$ is the regression adjusted estimate of θ_{i2}.

The next section shows that a similar situation exists when using the extensive collection of secondary variables to improve the estimates for subpopulation parameters defined by the primary covariates. The item sampling design also determines the need for poststratification with respect to the background covariates.

14.5   The role of secondary covariates

The primary reporting of NAEP is focused on subpopulations defined by a few "primary" demographic variables and school type (i.e., public, private). The multiply imputed data sets upon which it is based, however, are created using numerous secondary covariates. The primary reporting is evaluated here by considering a single binary primary covariate, x_1, and a single binary secondary covariate, x_2 (for details, see Thomas, 2002). The simplified setting based on the normal approximating model with analytic MLEs is considered first, and then the consequences for imputation-based estimation are discussed. Denote the mean of the first proficiency in the subpopulation with $x_1 = 1$ by $\mu_{1(1)}$.

First, consider the situation with a balanced design. Ignoring the secondary covariate, the estimator of $\mu_{1(1)}$ is just $\bar{y}_{1(1)}$, the mean of the measurements of the first proficiency among subjects with $x_1 = 1$. Under a saturated model for x_1 and x_2, the MLEs for the means of the first proficiency among subjects with $x_1 = 1$ and ($x_2 = 0$ or $x_2 = 1$) are the corresponding sample means, $\bar{y}_{1(10)}$ and $\bar{y}_{1(11)}$. Setting $\hat{\lambda}$ equal to the observed proportion of students with $x_2 = 1$ among students with $x_1 = 1$, the MLE of $\mu_{1(1)}$ based on x_1 and x_2 is:

$$\hat{\mu}_{1(1)} = \hat{\lambda}\bar{y}_{1(11)} + \left(1 - \hat{\lambda}\right)\bar{y}_{1(10)}. \tag{14.10}$$

It is easy to check that $\hat{\mu}_{1(1)} = \bar{y}_{1(1)}$, so the secondary covariate does not contribute to the estimation of the means of the primary subpopulations.

With a split design, the analytic formulas are more complex, even in the simplified normal model setting. To further reduce complexity, the contribution of the items measuring the second proficiency are ignored, and a split design is evaluated in which all items measure the first proficiency, or no items measure the first proficiency. This further simplification yields an MLE similar to the one with the balanced design:

$$\tilde{\mu}_{1(1)} = \hat{\lambda}\bar{y}_{\text{obs}_{1(11)}} + \left(1 - \hat{\lambda}\right)\bar{y}_{\text{obs}_{1(10)}}, \tag{14.11}$$

where $\bar{y}_{\text{obs}_{1(10)}}$ and $\bar{y}_{\text{obs}_{1(11)}}$ are the means of the first proficiency among the subsamples of students with $x_1 = 1$ and ($x_2 = 0$ or $x_2 = 1$) who receive items measuring the first proficiency. Unlike the MLE for the balanced design, $\tilde{\mu}_1$ does not equal $\bar{y}_{1(1)}$ because the weighting of the two subpopulations is based on the full sample proportion, $\hat{\lambda}$, which is not restricted to the subsample of students with measurements of the first proficiency. The $\tilde{\mu}_{1(1)}$ estimator is a poststratification estimator, averaging the sample means of students assigned cognitive items within the secondary subpopulations using the complete sample proportions of the covariates.

The effect of the design on the role of the secondary covariates has been confirmed in the much more complex NAEP models utilizing multiple imputation to evaluate them. Examination of imputations from actual NAEP models showed that the imputations created using secondary covariates differ from imputations created without the secondary covariates because the proficiency estimates for individual students are more accurate. The posterior means were more dispersed when secondary covariates were included because there was less shrinkage to a common mean, while the corresponding posterior variances were smaller producing the same combined variability in the imputed values. As with the contributions from individual students to the MLE (e.g., equation 14.9), the changes in the posterior means under the (approximately) balanced designs, when averaged over the primary reporting subpopulations, are nearly unchanged from the average of the posterior means for students computed without regard to the secondary covariates.

The fact that approximately balanced designs for missing data nearly eliminate the role of secondary covariates from the primary estimation does not imply that their inclusion is unimportant. The relationship between the secondary covariates and the proficiencies is also important, and the subject of secondary analyses based on public-release imputed data. The controversies (Fay, 1992, 1996) regarding the inclusion/exclusion of variables from the imputers' model were known to the NAEP researchers from the earliest implementation of multiple imputation in NAEP. Methodological and empirical studies led to the adaption of principal components as an approach to reduce the high dimensionality and multicollinearity induced by the inclusion of all secondary covariates in the imputation model. To avoid attenuation of correlations between the proficiencies and the secondary covariates, principal components capturing 90% of the variation in the covariates were recommended for inclusion in the model in Section 14.1 (Mazzeo, J., Johnson, E., Bowker, D., and Fong, Y. 1992). The inclusion of secondary covariates may also improve the MAR approximation for unplanned missing item responses.

14.6   Conclusions

Rubin (1996) succinctly describes the prevailing advice about the inclusion of numerous predictors when forming multiple imputations:

"Thus, the danger with an imputer's model is generally leaving out predictors rather than including too many, and the advice has always been to include as many variables as possible when doing multiple imputation."

The results in Sections 14.3 to 14.5 do not contradict this advice, rather they show that when missing data are planned, an appropriate balanced design can reduce the role of the highly complex multivariate imputation model while yielding highly efficient estimation. Although it has not been rigorously demonstrated, the reduced role of the secondary covariates in primary analyses with balanced missing data designs suggests increased robustness of the most important analyses to the difficult-to-specify multivariate regression model including the numerous secondary covariates, while also insuring the approximate validity of secondary analyses.

15

Propensity score estimation with missing data

Ralph B. D'Agostino Jr.[1]

15.1 Introduction

This chapter covers a breadth of topics that will be discussed in detail elsewhere throughout the first three parts of this book (missing-data modeling, causal inference and observational studies, and statistical computation). The work in this chapter will not necessarily introduce new methods in any of these areas but shows how these separate techniques and methods can be combined to address a common problem in applied research. We shall briefly introduce concepts of missing data, propensity scores, and statistical computation and then focus on the use of these tools in application.

In many applications, research to determine the effectiveness of a particular treatment cannot be carried out using a controlled clinical trial. In settings such as these, observational studies must be used to make inference concerning the effectiveness of a particular nonrandomized treatment. The treated and nontreated (i.e., control) groups in these observational studies may have substantial differences in observed covariates, and these differences can lead to biased estimates of treatment effects unless properly handled. Propensity score methods are popular tools used for balancing the distribution of the covariates in the two groups to reduce this bias.

[1]Department of Public Health Sciences, Wake Forest University School of Medicine, Winston-Salem, N.C. The author would like to thank the editors and reviewers for their helpful comments and corrections. He would also like to thank his family for their love and support in this work. This work was supported in part by National Cancer Institute Grant 1 R01 CA79934.

Applied Bayesian Modeling and Causal Inference from Incomplete-Data Perspectives.
Edited by A. Gelman and X-L. Meng © 2004 John Wiley & Sons, Ltd ISBN: 0-470-09043-X

This method has been shown to confer a greater reduction in bias than standard adjustment methods, such as ANCOVA (analysis of covariance), in many circumstances (D'Agostino Jr., 1998; Rubin and Thomas, 2000). In order to estimate the propensity score, defined as the conditional probability of being treated given the observed covariates, we must model the distribution of the treatment indicator given the observed covariates. An additional complication that often occurs is that missing data may be present among the covariates and in some cases the pattern of the missing covariates may be prognostically important. When this occurs, the propensity score needs to be modeled conditional on both the observed values of the covariates and the patterns of missing data.

Background on propensity scores

Since introduced, propensity scores have been used in observational studies in many fields to adjust for imbalances on pretreatment covariates, X, between a treated group, indicated by $Z = 1$, and a control group indicated by $Z = 0$ (D'Agostino Jr., 1998; Rosenbaum and Rubin, 1983a; Rubin 1997). Propensity scores are a one-dimensional summary of multidimensional covariates, X, such that when the propensity scores are balanced across the treatment and control groups, the distribution of all the covariates, X, are balanced in expectation across the two groups. Typically, matched sampling (e.g., Heckman et al., 1996; Lytle et al., 1999; Rosenbaum and Rubin, 1985a; Takizawa et al., 1999; Willoughby et al., 1990) or subclassification (e.g., Barker et al., 1998; Conners et al., 1996; Nakamura et al., 1999; Rosenbaum and Rubin 1984; U.S. GAO, 1994) on estimated propensity scores is used, often in combination with model-based adjustments (e.g., Curley et al., 1998; Lieberman et al., 1996; Rich, 1998; Rubin and Thomas, 2000; Smith et al., 1998).

The propensity score for an individual is the probability of being treated conditional on the individual's covariate values. To estimate propensity scores for individuals, one must model the distribution of Z given the observed covariates, X. There is a large technical literature on propensity score methods with complete data. (Rubin and Thomas, 1992, 1996, 2000; Rubin, 1978a). In practice, however, typically some covariate values will be missing, and so it is not clear how the propensity score should be estimated. Often, the missingness itself may be predictive about which treatment is received in the sense that the treatment assignment mechanism is ignorable (Rubin, 1986) given the observed values of X and the observed pattern of missing covariates but not ignorable given only the former.

There are several possible ways to estimate propensity scores in the presence of missing covariates. We follow the approach of D'Agostino and Rubin (2000) and model the joint distribution of (Z, X, R), where R is the missing covariate indicator ($R = 1$ for observed, $R = 0$ for missing). The particular approach we will illustrate in our application is based on a general location model (Olkin and Tate, 1961) accounting for the missing data (Schafer, 1997). This modeling implies a conditional distribution for Z given (X_{obs}, R), that is, a generalized propensity score: probabilities of $Z = 1$ versus $Z = 0$ for each unit as a function of its observed covariate values X_{obs}, and its missing data pattern R. Because X is missing when $R = 0$, a saturated model for (X, R) cannot be fit, even with the general location model. We impose loglinear constraints on the categorical variables, which include the missing value indicators for covariates whose missingness is related to treatment assignment. In the special case of no missing data and only continuous covariates, the approach reduces to estimating propensity scores by discriminant analysis, which is practically very close to logistic regression (Rubin and Thomas, 1992). Our methods use as basic computational tools the EM (Dempster, Laird, and Rubin, 1977) and ECM (Meng and Rubin, 1993) algorithms applied to the general location model. We will briefly illustrate this method by estimating propensity scores for an applied example and use these estimated propensity scores to select matched samples that have similar distributions of observed covariates and missing value indicators. We also provide suggestions for diagnostic procedures to assess the success of the matching in creating balanced distributions of these observed covariates and missing value indicators. We illustrate these procedures in the context of a matched-sampling study of the effects of postterm pregnancy.

The problem we are describing is different from most missing data problems in which the goal is parameter estimation. We are not interested in obtaining one set of estimated parameters for a logistic regression or discriminant analysis, or a posterior distribution for these parameters, or even drawing inferences about these parameters. Rather, parameters particular to each pattern of missing data serve only in intermediate calculations to obtain estimated propensity scores for each subject. Moreover, the propensity scores themselves serve only as devices to balance the observed distribution of covariates and patterns of missing covariates across the treated and control groups. Consequently, the success of the propensity score estimation is assessed by this resultant balance rather than by the fit of the models used to create the estimated propensity scores. This goal is not special to the case with missing values in covariates, but rather has been the goal with propensity score estimation from the start.

15.2 Notation

Estimation of propensity scores

With complete data, Rosenbaum and Rubin (1983) introduced the propensity score for subject $i (i = 1, \ldots, N)$ as the conditional probability of receiving a particular treatment $(Z_i = 1)$ versus control $(Z_i = 0)$ given a vector of observed covariates, x_i:

$$e(x_i) = pr(Z_i = 1 \mid X_i = x_i), \tag{15.1}$$

where it assumed that, given the X's, the Z_i are independent:

$$pr(Z_1 = z_1, \ldots, Z_N = z_N \mid X_1 = x_1, \ldots, X_N = x_N)$$

$$= \prod_{i=1}^{N} e(x_i)^{z_i} \{1 - e(x_i)\}^{1-z_i}. \qquad (15.2)$$

Rosenbaum and Rubin (1983) showed that for a specific value of the propensity score, the difference between the treatment and control means for all units with that value of the propensity score is an unbiased estimate of the average treatment effect at that propensity score, if the treatment assignment is strongly ignorable given the covariates. Thus, matching, subclassification, or regression (covariance) adjustment on the propensity score tends to produce unbiased estimates of the treatment effects when treatment assignment is strongly ignorable, which occurs when the treatment assignment, Z, and the potential outcomes, Y, are conditionally independent given the covariates X: $Pr(Z|X, Y) = Pr(Z|X)$.

Propensity scores with incomplete data

Let the response indicator be R_{ij}, $(j = 1, \ldots, T)$, which is one when the value of the jth covariate for the ith subject is observed and zero when it is missing; R_{ij} is fully observed by definition. Also, let $X = (X_{\text{obs}}, X_{\text{mis}})$, where $X_{\text{obs}} = \{X_{ij}|R_{ij} = 1\}$ denotes the observed parts and $X_{\text{mis}} = \{X_{ij}|R_{ij} = 0\}$ denotes the missing components of X.

The generalized propensity score for subject i, which conditions on all of the observed covariate information, is

$$e_i^* = e_i^*(X_{\text{obs},i}, R_i) = pr(Z_i = 1 \mid X_{\text{obs},i}, R_i), \qquad (15.3)$$

Rosenbaum and Rubin (1985) showed that with missing covariate data and strongly ignorable treatment assignment given X_{obs} and R, the generalized propensity score, e_i^* in (3), plays the same role as the usual propensity score, e_i in (1) with no missing covariate data. Treatment assignment is strongly ignorable given (X_{obs}, R) if $Pr(Z|X, Y, R) = Pr(Z|X_{\text{obs}}, R)$. If in addition, the missing data mechanism is such that $Pr(R|X, Z) = Pr(R|X_{\text{obs}})$, then $Pr(Z|X, Y, R) = Pr(Z|X_{\text{obs}})$, and R itself can be ignored in the modeling. It is important to emphasize that, just as with propensity score matching with no missing data, the success of a propensity score estimation method is to be assessed by the quality of the balance in the (X_{obs}, R) distributions between control and treated groups that has been achieved by matching on it. Consequently, the usual concerns with the fit of a particular model (i.e., the general location model) are not relevant if such balance is achieved.

General location model with complete data

The distribution of (X, Z) is defined by the marginal distribution of the categorical variables, Z and the categorical covariates, U, and the conditional distribution of continuous covariates, say V, given (U, Z) (thus, $X = (U, V)$). (U_{ij}, Z_i) locates the ith subject in one of the m cells of the table formed by (U, Z).

We assume that (U, Z) are iid multinomial random variables, and conditional on U_i, Z_i, we assume that V_i is K-variate normal with mean that depends on the cell but with a common covariance. This is the general location model (Olkin and Tate, 1961) with parameters $\Pi =$, cell probabilities from the multinomial distribution, $\Gamma =$, the matrix of cell means, and $\Omega =$, the positive definite covariance matrix common to all cells; $\theta = (\Pi, \Gamma, \Omega)$.

Krzanowski (1980, 1982), Little and Schlucter (1985), and Little and Rubin (1987) describe restricted general location models having fewer parameters. One way to reduce the number of parameters to be estimated is to constrain Π by a loglinear model (Goodman, 1968; Bishop, Fienberg, and Holland, 1975); for example, three-way and higher-order interactions are set to zero. ML estimates of the parameters for these models have closed form solutions for many configurations, but if they do not have a closed form, they can be found by using an iterative procedure such as iterative proportional fitting (IPF; Bishop, Fienberg, and Holland, 1975).

A second way to reduce the number of parameters to be estimated in the general location model is to impose ANOVA-like restrictions on the means, Γ, using a known design matrix A to define Γ in terms of a lower-dimensional matrix of unknown regression coefficients β; if A is the identity matrix, then there are no restrictions. With standard models and complete data, the parameter estimates for β and Ω can be found using standard regression techniques (Anderson, 1958, Chapter 8).

Although the restrictions described previously reduce the number of parameters to be estimated in the model, we could also generalize the model to increase the number of parameters to be estimated. For instance, the assumption of a common covariance Ω across all cells of the contingency table can be relaxed to allow for possibly different covariance matrices to be estimated in different cells. This, however, can require substantial sample sizes in each cell. A more useful extension may be to estimate separate covariance matrices only for the treated and control groups. Other extensions involve proportional covariance matrices and more general ellipsoidal distributions (e.g., t-distributions as in Liu and Rubin 1995, 1998).

Fitting the general location model with missing data

The basic method for finding estimates for the parameters of the general location model when there are ignorably missing data is outlined in Little and Rubin (1987, Chapter 10) and Schafer (1997) and is based on the EM algorithm (Dempster, Laird, and Rubin, 1977) and the ECM algorithm (Meng and Rubin, 1993), which is used when loglinear restrictions have been placed on the general location model such that IPF is needed with complete data. Of particular importance for our situation, where we want to include explicitly the response indicator R in the modeling, is that R is a fully observed collection of categorical (in fact binary) covariates. For notational convenience, let $U^* = (U, R)$, with corresponding changes to the other notation. Because X_{ij} is missing when $R_{ij} = 0$, some restrictions are needed to obtain unique maximum likelihood estimates of parameters for the joint distribution of (Z, X, R), which we need to obtain unique maximum likelihood estimates (MLEs) of the conditional distribution of Z given X_{obs} and R.

At each iteration of the EM algorithm, the E-step computes the expected values of the complete-data sufficient statistics given the observed data and the current estimates of the parameters, $\theta^{*(t)} = (\Gamma^{*(t)}, \Omega^{*(t)}, \Pi^{*(t)})$, where t indexes iterations. The M-step computes the ML estimates for the parameters using the estimated values of the sufficient statistics. These become the current estimates of the parameters to be used in the next E-step calculations. The E and M steps are repeated until convergence. When there are loglinear constraints on the categorical covariates, ECM is used instead of EM; the M-step of EM is replaced by CM steps, which perform one cycle of IPF. The complete-data sufficient statistics for this model are the raw sums of squares and cross products of the V's ($\Sigma V_i^T V_i$), the sums of the V's in each cell (cell totals), and the cell frequencies from the table defined by (Z, U^*).

Once EM or ECM has converged, we have ML estimates, $\hat{\theta}^*$, for the parameters $\theta^* = (\Pi^*, \Gamma^*, \Omega^*)$ for the joint distribution of (V, Z, U^*), which we use to calculate an estimated propensity score for each subject, $\hat{e}_i^*$, as in (3) with $X_{obs,i} = (V_{obs,i}, U_{obs,i})$.

$$\hat{e}_i^* = pr(Z_i = 1 | V_{obs,i}, U^*_{obs,i}, R_i, \hat{\theta}^*). \tag{15.4}$$

To find the estimated propensity score (4) from $\hat{\theta}^*$ and the observed data, we simply run one E-step using the converged MLE $\hat{\theta}^*$, *but now treating Z_i as missing.*

15.3 Applied example: March of Dimes data

Description of the data

To illustrate these methods, we take data from a March of Dimes observational study examining the effects of postterm birth versus term birth on neuropsychiatric, social, and academic achievements of 5- to 10-year-old children. The investigators were interested in selecting and interviewing a sample of 5- to 10-year-old term and postterm children from a large database of birth records collected. Since the database of birth records consisted of more than 4,000 potential children, it was financially infeasible to try and recruit all potential children. Therefore, the initial issue they faced was how to select the sample to facilitate inference for the effect of being postterm. It was decided that the best approach to address this question was to identify potential matches for each postterm child from the pool of term children and then recruit a subset of the matched pairs into the study. A complication was that for some children, some covariates had missing values.

From this applied data set, a sample of 4,500 potential children were used in this illustration. Of these 4,500 children, 4,155 (92.3%) were term babies and 345

(7.7%) were postterm babies. In the propensity score model we fit, there were 25 covariates that were felt to be scientifically significant for predicting postterm birth and prognostically important for predicting outcomes, and thus, if left uncontrolled, could confound estimated treatment effects. Among the covariates included in the propensity score model, there were a few (e.g., infant's weight or medical induction during labor) that may be considered improper in the sense that their values were not determined prior to the "treatment assignment" to being term or postterm. That is, in the hypothetical experiment underlying the observational study, before week 42 a decision could have been made to induce labor for the postterm babies and the effect of not doing so is the effect we seek. Formally, any covariate measured after 41 weeks is thus an improper covariate because it could be affected by treatment.

For example, infant's weight had the largest initial imbalance, but can be considered to be an outcome of being postterm, and not a proper covariate. Despite this, as with the other improper covariates, the investigating physicians felt strongly that this variable needed to be controlled. Because this improper covariate can be thought of as a proxy for unmeasured proper pretreatment covariates that predict fetal disorders, the physicians and investigators felt they needed to explicitly control these variables as if they were proper covariates, if useful inferences were to be drawn about policy-relevant advice concerning postterm pregnancies. Still, we acknowledge that the inclusion of such improper covariates may actually adjust away part of the true treatment effect. However, this limitation occurs regardless of which method for control is employed (i.e., matching or covariate modeling). In addition, the focus of this example is to illustrate the estimation and use of propensity scores with missing covariate data, which could have been applied using only proper covariates.

Tables 15.1 and 15.2 present descriptive statistics for the covariates and fitted propensity score, separately for the term and postterm groups. It is important to emphasize that these statistics are descriptive and not inferential in the sense that they do not purport to estimate relevant population parameters, but rather simply describe the two samples and their differences. Table 15.1 presents, for each continuous covariate and the propensity score, the mean and standard deviation using available cases; also presented are standardized percentage differences prior to and following matching, defined as the mean difference between postterm and term groups as a percentage of the standard deviation: $\left(\dfrac{100(\bar{x}_p - \bar{x}_t)}{\sqrt{(s_p^2 + s_t^2)/2}} \right)$, where $\bar{x}_p$ and $\bar{x}_t$ are the sample means in the postterm and term groups respectively, and s_p^2 and s_t^2 are the corresponding sample variances, again based on available cases.

Table 15.1 presents the proportion of women in each category in the term and postterm groups; also presented are the corresponding results for the missing data indicators (last 10 rows) for the 10 covariates with any missing values (either continuous or categorical). The third column displays these proportions for the term group chosen as matches for the postterm group. The fourth and fifth columns display the absolute differences in percent between the term and postterm

Covariate	Mean (SD) Term	Postterm Mean (SD)	Standardized Difference in %[b] Initial	Final
Binary predictors (modeled as continuous)				
Antepartum complications	0.72 (.45)	0.72 (.45)	1	2
Previous obstetrical history	0.47 (.50)	0.40 (.49)	−14	1
Vaginal bleeding	0.12 (.33)	0.11 (.31)	−4	1
Second stage indicator[a]	0.81 (.39)	0.77 (.42)	−10	3
Ordinal predictors				
Delivery mode	1.26 (.51)	1.30 (.51)	8	2
Labor complications	0.58 (.63)	0.66 (.59)	14	10
Class[a]	2.37 (.77)	2.31 (.77)	−8	3
Diabetes[c]	0.15 (1.05)	0.11 (.82)	−4	3
Fetal distress	0.04 (.64)	0.15 (1.2)	11	3
Induction	0.17 (.88)	0.41 (1.2)	23	11
Pelvic adequacy (clinic)[a]	0.19 (.68)	0.19 (.67)	0	2
Pelvic adequacy (X-ray)[a]	1.71 (.94)	1.69 (.79)	−3	12
Placental problems	0.11 (.1.04)	0.09 (.93)	−2	2
Previous perinatal mortality[d]	0.22 (1.49)	0.15 (1.13)	−6	7
Urinary tract disorders	0.11 (.51)	0.13 (.53)	4	2
Continuous predictors				
Child's age (months from 1980–range 0–48)	23.4 (13.0)	23.9 (11.4)	4	7
Infant's weight (grams)[a]	3338 (461)	3626 (533)	58	11
Length of first stage (min)[a]	784 (571)	910 (665)	20	1
Length of second stage (min)[a]	53.8 (65)	59.5 (66)	9	3
Time since membranes ruptured (min)[a]	454 (791)	414 (651)	−6	3
Mother's age (years)	28.8 (5)	28.2 (5)	−12	5
Parity	0.77 (1.0)	0.66 (1.1)	−10	3
Total length of labor (min)[a]	841 (589)	968 (688)	20	1
Propensity score	.072 (.081)	.168 (.180)	69	2
Logit of propensity score	−2.95 (1.02)	−1.95 (1.29)	86	2

[a] Covariate suffers from some missing data.

[b] The standardized difference in % is mean difference as a percentage of the average standard deviation: $\frac{100(\bar{x}_p - \bar{x}_t)}{\sqrt{(s_p^2 + s_t^2)/2}}$, where for each covariate $\bar{x}_p$ and $\bar{x}_t$ are the sample means in the postterm and term groups, respectively, and s_p^2 and s_t^2 are the corresponding sample variances.

[c] Diabetes—0 = None, 1 = diabetes insipidus or glucosuria, 5 = abnormal glucose tolerance test, and 10 = diabetes mellitus.

[d] Previous perinatal mortality—0 = no previous child deaths, 5 = previous late death (in first year of life), 10 = previous stillbirth or neonatal death, and 20 = previous stillborn and previous neonatal death (or any combination of 2 or more perinatal mortalities).

Table 15.1 Descriptive statistics for the variables that were included in the propensity score models as continuous variables. These statistics are shown for the term and postterm groups before and after matching and include the means, standard deviations, and the standardized differences in percent for each variable estimated using available cases only.

Covariate		Term	Postterm	Term Matches	Differences in % Initial	Final
Race	White	.70	.72	.72	2	0
	Nonwhite	.30	.28	.28	2	0
Gender	Male	.49	.51	.49	2	2
	Female	.51	.49	.51	2	2
	Vertex	.77	.72	.73	5	1
Delivery mode	Cesarean	.21	.27	.25	6	2
	Other	.02	.01	.02	1	1
	No labor (cesarean)	.08	.06	.06	2	0
Labor	No complications	.26	.21	.21	5	0
Complications	Some complications	.66	.73	.75	7	2

Missing value indicators (proportion observed)

	Term	Postterm	Term Matches	Differences in % Initial	Final
Pelvic adequacy (X-ray)	.05	.10	.09	5	1
Length of 2nd stage of labor	.78	.74	.72	4	2
Race	.95	.95	.95	0	0
2nd stage of labor indicator	.99	1.00	1.00	1	0
Class	.99	.99	1.00	0	1
Pelvic adequacy (clinic)	.85	.90	.88	5	2
Infant's weight	.99	1.00	1.00	1	0
Length first stage of labor	.89	.91	.91	2	0
Time membranes ruptured	.97	.97	.97	0	0
Length of labor	.89	.91	.91	2	0

Table 15.2 Descriptive statistics for categorical variables and missing value indicators. These statistics are shown for the term and postterm groups before and after matching and include the observed proportions in each category and the differences between term and postterm groups for each variable.

groups for each of the categorical covariates and missing data indicators before and after matching.

The initial term versus postterm group differences summarized in Tables 15.1 and 15.2 indicate the possible extent of biased comparisons of outcomes due to different distributions of observed covariates and patterns of missing data in the initial term and postterm groups. That is, ideally all such descriptive statistics should suggest the same distribution in the term and postterm groups, as they would be in expectation if the treatment indicator (term vs postterm) had been randomly assigned. As can be seen by examining these tables, there exists considerable initial bias between the term and postterm groups. For instance, nine of the continuous covariates have initial standardized differences larger than 10%. In addition, there are substantial differences between the groups based on the estimated propensity score. The missingness rates appear similar except that there seems to be a trend for some indicators of potential complications to be more observed in the postterm group (e.g., pelvic adequacy, both x-ray and clinical), suggesting a greater need for such medical tests among the postterm subjects. In addition, the missing data indicator for the length of the second stage of labor shows that more individuals in the term group had this variable observed than in the postterm group (78 vs 74%).

Specific propensity score model fit

The generalized propensity score model we fit used all 23 continuous covariates in Table 15.1 and the two categorical covariates (race and child's gender) in Table 15.2. In addition, among the 10 variables that contained missing values, the missingness on two were differentially distributed in the treatment groups and believed to be prognostically important: the results of a pelvic x-ray and the length of the second stage of labor. In addition, for the pelvic x-ray variable there was a large statistically significant difference in the rate of missingness between groups based on a chi-square test. Therefore, the propensity score model fit included the missing value indicators for these two variables as additional categorical covariates in the model. This propensity score model without any constraints would have 1,043 parameters, therefore we placed some constraints on the model. Loglinear constraints were placed on the cell probabilities so that the three-way and higher interactions were set to zero, and thus we estimated five main effects (one for the treatment indicator, two for the categorical covariates, gender, and race, and two for the two missing value indicators) and 10 two-way interactions. The design matrix relating the means of the continuous variables to the categorical variables includes an intercept, main effects for each of the categorical variables, and terms for each of the two-way interactions of the categorical variables. This constrained model has 659 total parameters including 15 for the contingency table, 368 regression coefficients, and 276 variance and covariances.

Matching using estimated propensity scores

We estimated the propensity score for the model using the ECM algorithm as illustrated above. Then we used the nearest available matching on these estimated propensity scores to choose matches for the postterm subjects. We randomly ordered the term and postterm subjects, and then selected the term subject with propensity score closest to the first postterm subject. Both subjects were then removed from the pools of subjects. We repeated this procedure for each postterm subject, which resulted in selecting a total of 345 term subjects from the 4,150 available ones. There are numerous other approaches that we could have used to select the matches, but chose this straightforward approach using propensity scores in order to focus on how well matching based on the propensity score model succeeds in balancing the distribution of observed covariates and missing-data indicators between the term and postterm groups.

Resultant distributions of propensity scores and their logits

We first compare the distributions of the propensity scores in the postterm group with those in the initial term group ($n = 4,150$), those in a randomly selected term group ($n = 345$), and the matched term group ($n = 345$), by examining some of the descriptive statistics from each group. We find that the median propensity score for the matched term group (0.109) was nearly equal to the median in the postterm group (0.11), whereas both these values were larger than even the 75th percentile propensity score in the unmatched and randomly selected term groups (0.089 and 0.090 respectively), indicating that the vast majority of term group babies had lower propensity scores than those in the postterm group. Thus, if a random sample of term babies had been selected for analyses the majority of those selected would not have resembled the postterm group. In addition, we find that the spread of propensity scores for the unmatched term group ($n = 4,150$) spans nearly the whole range of potential propensity scores (i.e., from 0 to nearly 1.0) This second feature allows us to find matches from the term group with propensity scores close to the propensity scores of the subjects in the postterm group.

The distribution of the estimated propensity scores can also be compared by examining columns 4 and 5 of Table 15.1. Column 4 contains the initial standardized differences in percent for the propensity score and its logit and column 5 contains the same statistic after matching using the model. The initial standardized differences in % was quite large (69%) as well as the initial variance ratio (5.00) (estimated as the ratio of the sample variance of the propensity score in the postterm group divided by the sample variance of the propensity score in the term group), which suggests that this model is successful in separating the term and postterm groups. When we examine the standardized differences after matching, we see that this difference was reduced to less than 2% and the variance ratio was reduced to 1.19.

Resultant covariate balance after matching

To assess further the relative success of the propensity score model for creating balanced matched samples, we compare balance on observed covariates and missing data indicators in the matched samples created by the model. It is important for practice to realize that, as in Rosenbaum and Rubin (1984, 1985), these assessments can be made before any resources have been committed to collecting outcome data on the matched controls. Also, it is important to realize that, since these comparisons involve only observed covariates and their missing-data indicators and not outcome variables, there is no chance of biasing results in favor of one treatment condition versus the other through the selection of matched controls.

Columns 4 and 5 from Table 15.1 compare the standardized differences in percent, after matching, for the continuous covariates. The matching performed well in reducing the bias of the background covariates with moderate to large initial standardized differences. For instance, the initial standardized difference for the length of the first stage of labor variable is 20%, and this was reduced to

1% after matching. Even the initial standardized difference for infant's weight is substantially reduced by the matching (from 58 to 11%)

In Table 15.2, which compares the available-case cell proportions for the categorical covariates and missing value indicators between the term and postterm groups before and after matching, we find that the initial imbalance in delivery mode and labor complications was moderate, with 26% of the postterm babies being cesarean birth versus 20% of the term pregnancies and 73% of the postterm pregnancies having some complications versus 66% of the term pregnancies. These differences were both significantly reduced after matching.

We acknowledge that there are many other plausible propensity score models that could be constructed using the 25 covariates and their missing-data indicators, and that among these there are likely to exist models that produce better balance than our propensity score model. Still, this model would provide the investigators with what they wanted, that is, propensity scores that could be used to select matches for the postterm babies from the available pool of term babies, where the bias that was observed between the term and postterm groups on many covariates and their missingness prior to matching was substantially reduced (and often essentially removed) by the matching.

15.4   Conclusion and future directions

This chapter has focused on using a model-based (the general location model) approach for estimating propensity scores in the presence of missing data. There are alternative methods for handling the missing data that are currently being developed. These include using a pattern mixture model (Rosenbaum and Rubin, 1984; Little, 1993) or multiple imputation methodology (Rubin, 1987b) for propensity score estimation.

We have presented an approach for estimating propensity scores in the presence of missing data using the EM and ECM algorithms as computing tools. The framework allows the investigator to put structure on the relationships among the covariates in the model, including missing value indicators for specific effects. In addition, we have illustrated our approach using the March of Dimes data. Simulation studies are underway to examine the effects of specifying different missing-data mechanisms on the data. In addition, user-friendly software is being developed to make implementation of these methods more easily available to investigators.

16

Sensitivity to nonignorability in frequentist inference

Guoguang Ma and Daniel F. Heitjan[1]

16.1 Missing data in clinical trials

The occurrence of missing data can harm both the power of a trial, through loss of sample size, and its validity, through selection bias. This is particularly vexing in psychiatric trials, where the collection of primary outcome data (depression scores, quality of life, etc.) depends on the willingness of subjects to submit to a series of possibly burdensome examinations. The potential loss of power is easy to overcome by simply increasing accrual targets, which is a typical feature of practical trial design. An explicit, quantitative assessment of the potential for bias, although justified, is not yet standard practice.

16.2 Ignorability and bias

In his seminal paper, "Inference and missing data", Rubin (1976a) established a general theory of ignorability for frequentist inferences from data subject to missingness. The key condition, now called *missing completely at random* (MCAR), is that the conditional probability of the observed missing data indicator, given the complete data, is the same for all possible values of the complete data. When

[1]Clinical Biostatistics, Merck & Co., Inc., Blue Bell, Pa., and Department of Biostatistics and Epidemiology, University of Pennsylvania, Philadelphia, Pa. The authors thank the referees and editors for a number of helpful suggestions. The USPHS supported this research under grant HL 68074.

Applied Bayesian Modeling and Causal Inference from Incomplete-Data Perspectives.
Edited by A. Gelman and X-L. Meng © 2004 John Wiley & Sons, Ltd ISBN: 0-470-09043-X

MCAR holds, the stochastic nature of the missingness mechanism is ignorable in frequentist inferences. In practice, this means that if you set out to sample 10 observations but only managed to get 5, for the purposes of frequentist inferences you can act as though your intention had been to collect only 5 in the first place (e.g., see Heitjan and Basu, 1996; Little and Rubin, 2002). Heitjan (1997) provides further justification for this view by showing that MCAR is equivalent to what he called *observed ancillarity* of the missingness indicator. Thus, when the data are MCAR, not only *can* one do conditional frequentist inference given the missingness indicator, but in a sense one *should*.

Unfortunately, we seldom know with certainty that missing data are MCAR, and consequently we seldom feel confident that our inferences are free of this type of bias. Thus, the methodology of nonignorable modeling (Schluchter 1992; Diggle and Kenward, 1994) and global analysis of sensitivity to nonignorability (Vach and Blettner, 1995; Rotnitzky, Robins, and Scharfstein, 1998; Scharfstein, Rotnitzky, and Robins, 1999) has been an area of rapid development in recent years. A related enterprise is the development of indices that measure sensitivity locally, in the neighborhood of an ignorable model (Copas and Li, 1997; Copas and Eguchi, 2001; Verbeke et al., 2001; Troxel, Ma, and Heitjan, 2004). These methods promise straightforward, practical, and robust assessments of sensitivity.

In this chapter, we illustrate the concept of local sensitivity by applying it in the planning of a clinical trial. The intention is to estimate mean depression score in psychiatric patients who are prone to drop out according to a mechanism that may not be MCAR. The methodology differs from the local sensitivity approaches described above in that it is explicitly frequentist and is applicable prior to data collection.

We begin by describing the model.

16.3   A nonignorable selection model

Consider a random vector $Y = (Y_1, \ldots, Y_n)$ whose components Y_i are independent and distributed according to densities $f_\theta^{Y_i}(y_i)$, $i = 1, \ldots, n$, governed by a common parameter of interest θ. Let G_i be a variable that equals 1 if Y_i is observed and 0 if Y_i is missing. Assume that the probability of being observed depends on y_i through a link function h and a parameter $\gamma = (\gamma_0, \gamma_1)$, as follows:

$$\Pr_{\gamma_0, \gamma_1}[G_i = 1 | Y_i = y_i] = h(\gamma_0 + \gamma_1 y_i).$$

If $\gamma_1 = 0$, every possible data set is MCAR, G is ancillary for θ, and it follows that one should base frequentist inferences on the conditional distribution of the data Y given $G = g$—that is, as if G were fixed at its observed value (Cox and Hinkley, 1974, pp. 31–33; Heitjan, 1997).

For practice, a key question is the extent to which the distribution of Y given $G = g$ under ignorability ($\gamma_1 = 0$) gives an adequate approximation to the correct distribution under nonignorability ($\gamma_1 \neq 0$). For a variety of reasons, we believe that it is sufficient in most cases to explore sensitivity in the neighborhood of $\gamma_1 = 0$. Thus, we devote the remainder of this chapter to exploring local approximations to various summaries of this distribution.

16.4   Sensitivity of the mean and variance

Suppose that we intend to sample n independent observations, $y = (y_1, \ldots, y_n)$, from a normal density with mean μ and variance τ. We end up observing only n_o units, and we are concerned that the missingness mechanism is nonignorable. If the target of estimation is the mean, then we seek to determine the extent to which the mean of the population of units that would be observed (the expectation of the observed data) differs from the mean of the entire population.

A straightforward calculation shows that, to first order,

$$E[Y_i | G_i = 1] \approx \mu + \gamma_1 \frac{h'(\gamma_0)}{h(\gamma_0)} \tau \tag{16.1}$$

for all observed units, where $h'(\cdot)$ is the first derivative of h. Thus if $\overline{Y}_o$ is the sample mean of the observed units, that is, $\overline{Y}_o = \sum G_i Y_i / \sum G_i$, then its expectation is also given by the right side of (16.1). We can further approximate the mean to second order as

$$E[Y_i | G_i = 1] \approx \mu + \gamma_1 \frac{h'(\gamma_0)}{h(\gamma_0)} \tau + \gamma_1^2 \left[\frac{h(\gamma_0)h''(\gamma_0) - [h'(\gamma_0)]^2}{h^2(\gamma_0)} \right] \mu \tau, \tag{16.2}$$

where $h''(\cdot)$ is the second derivative of h. The first-order sensitivity of the variance is zero:

$$\text{var}[Y_i | G_i = 1] \approx \tau, \tag{16.3}$$

and henceforth we assume that the variance is known.

In fact, with a little extra effort we can generalize these results to any moment. Let $M_{\gamma_1}(t)$ be the moment-generating function of a random variable Y, given that it is observed, when the nonignorability parameter is fixed at γ_1. The k-th derivative of this function is

$$M_{\gamma_1}^{(k)}(t) = \frac{\int u^k \exp(tu) f_\theta^Y(u) h(\gamma_0 + \gamma_1 u)\, du}{\int f_\theta^Y(u) h(\gamma_0 + \gamma_1 u)\, du}.$$

A straightforward differentiation shows that

$$\left. \frac{\partial M_{\gamma_1}^{(k)}(t)}{\partial \gamma_1} \right|_{\gamma_1 = 0} = \frac{h'(\gamma_0)}{h(\gamma_0)} \left[M_0^{(k+1)}(t) - M_0^{(k)}(t) E(Y) \right],$$

and

$$\frac{\partial^2 M_{\gamma_1}^{(k)}(t)}{\partial \gamma_1^2}\bigg|_{\gamma_1=0} = \frac{h''(\gamma_0)}{h(\gamma_0)}\left[M_0^{(k+2)}(t) - M_0^{(k)}(t)E(Y^2)\right]$$

$$-2\left[\frac{h'(\gamma_0)}{h(\gamma_0)}\right]^2 \left[M_0^{(k+1)}(t) - M_0^{(k)}(t)E(Y)\right]E(Y).$$

From this we readily obtain (16.1), (16.2), and (16.3). The moment formulas apply to any random variable whose distribution is sufficiently regular to permit passing the derivative with respect to γ_1 under the integral sign.

16.5   Sensitivity of the power

Suppose that we wish to test $\mu = \mu_0$ in a normal model with known variance, and that we expect to observe n_o of a possible n units. Let $\mu(\gamma_1)$ be the hypothetical mean of the observed data when the nonignorability parameter takes value γ_1. Define $P_{\gamma_1}(\alpha, \mu, g)$ to be the probability, conditional on $G = g$, of rejecting the null hypothesis $\mu = \mu_0$ in a one-sided test of level α when the mean in the underlying population is $\mu(\gamma_1) > \mu_0$,

$$P_{\gamma_1}(\alpha, \mu, g) = \Phi\left[\frac{\mu(\gamma_1) - \mu_0}{\sqrt{\tau/n_o}} + z_\alpha\right],$$

where Φ is the standard normal integral. This expression involves a level of approximation because the distribution of Y, given that it is observed, may not be exactly normal for $\gamma_1 \neq 0$. The departure for small γ_1 should be modest, however, and when n_o is large, the central limit theorem guarantees that $\overline{Y}_o$ is normal even if the parent distribution is not. Consequently, these results can be taken to hold in a large-sample sense for any random variable with sufficient finite moments.

We approximate the probability, to second order, by the Taylor series

$$P_{\gamma_1}(\alpha, \mu, g) \approx P_{\gamma_1}(\alpha, \mu, g)\big|_{\gamma_1=0} + \gamma_1 \frac{\partial P_{\gamma_1}(\alpha, \mu, g)}{\partial \gamma_1}\bigg|_{\gamma_1=0}$$

$$+\frac{\gamma_1^2}{2}\frac{\partial^2 P_{\gamma_1}(\alpha, \mu, g)}{\partial \gamma_1^2}\bigg|_{\gamma_1=0}. \tag{16.4}$$

By (16.2), $\mu(\gamma_1)$ is approximated by

$$\mu(\gamma_1) \approx \mu(0) + \gamma_1 \frac{h'(\gamma_0)}{h(\gamma_0)}\tau + \gamma_1^2\left[\frac{h(\gamma_0)h''(\gamma_0) - [h'(\gamma_0)]^2}{h^2(\gamma_0)}\right]\mu(0)\tau.$$

Letting $a = [\mu(0) - \mu_0]/\sqrt{\tau/n_o} + z_\alpha$, the derivatives of the probability with respect to γ_1 at $\gamma_1 = 0$ are

$$\frac{\partial P_{\gamma_1}(\alpha, \mu, g)}{\partial \gamma_1}\bigg|_{\gamma_1=0} = \phi(a)\frac{h'(\gamma_0)}{h(\gamma_0)}\sqrt{n_o\tau} \tag{16.5}$$

and

$$\left.\frac{\partial^2 P_{\gamma_1}(\alpha, \mu, g)}{\partial \gamma_1^2}\right|_{\gamma_1=0} = -a\,\phi(a)\left[\frac{h'(\gamma_0)}{h(\gamma_0)}\right]^2 n_0\tau$$

$$+ 2\phi(a)\left[\frac{h(\gamma_0)h''(\gamma_0) - [h'(\gamma_0)]^2}{h^2(\gamma_0)}\right]\mu(0)\sqrt{n_0\tau},$$

(16.6)

where ϕ is the standard normal density.

For a two-sided test, the power equals

$$P_{\gamma_1}(\alpha, \mu, g) = \Phi\left[\frac{\mu(\gamma_1) - \mu_0}{\sqrt{\tau/n_0}} + z_{\alpha/2}\right] + \Phi\left[\frac{\mu_0 - \mu(\gamma_1)}{\sqrt{\tau/n_0}} + z_{\alpha/2}\right]. \quad (16.7)$$

Letting

$$c_1 = \frac{\mu(0) - \mu_0}{\sqrt{\tau/n_0}} + z_{\alpha/2}, \quad c_2 = \frac{\mu_0 - \mu(0)}{\sqrt{\tau/n_0}} + z_{\alpha/2},$$

the derivatives evaluated at $\gamma_1 = 0$ take the following forms:

$$\left.\frac{\partial P_{\gamma_1}(\alpha, \mu, g)}{\partial \gamma_1}\right|_{\gamma_1=0} = \left[\phi(c_1) - \phi(c_2)\right]\frac{h'(\gamma_0)}{h(\gamma_0)}\sqrt{n_0\tau}$$

and

$$\left.\frac{\partial^2 P_{\gamma_1}(\alpha, \mu, g)}{\partial \gamma_1^2}\right|_{\gamma_1=0} = -\left[c_1\,\phi(c_1) + c_2\,\phi(c_2)\right]\left[\frac{h'(\gamma_0)}{h(\gamma_0)}\right]^2 n_0\tau + 2\left[\phi(c_1) - \phi(c_2)\right]$$

$$\times \left[\frac{h(\gamma_0)h''(\gamma_0) - [h'(\gamma_0)]^2}{h^2(\gamma_0)}\right]\mu(0)\sqrt{n_0\tau}.$$

If the smaller term in (16.7) can be ignored for a fixed hypothetical μ, the derivatives of the power for a two-sided test are approximately given by (16.5) and (16.6) with z_α replaced by $z_{\alpha/2}$.

We conducted a simulation experiment to illustrate the adequacy of these approximations. We generated 2,000 data sets of size $n = 10$ from a normal distribution with mean 2 and variance 4, and fixed $\gamma_0 = 0$. For a range of values of γ_1 (-0.4 to 0.4 by 0.05), we simulated a vector of the missingness indicator g given the true y, according to $\Pr[G_i = 1|Y_i = y_i] = h(\gamma_0 + \gamma_1 y_i)$, for $i = 1, \ldots, 10$, with the logistic link $h(u) = \exp(u)/[1 + \exp(u)]$. Although any number of points could be missing, we focused on the most common patterns, which for these parameters are those with 4 to 7 points observed. For each pattern, we counted the number of data sets with $(\bar{Y}_0 - \mu_0)\sqrt{n_0/\tau} > z_{1-\alpha}$ for a one-sided test, and $|\bar{Y}_0 - \mu_0|\sqrt{n_0/\tau} > z_{1-\alpha/2}$ for a two-sided test, where $\mu_0 = 0$. Finally, we calculated the empirical and approximate probabilities, which are plotted in Figure 16.1 for a one-sided test. Results are similar for a two-sided test.

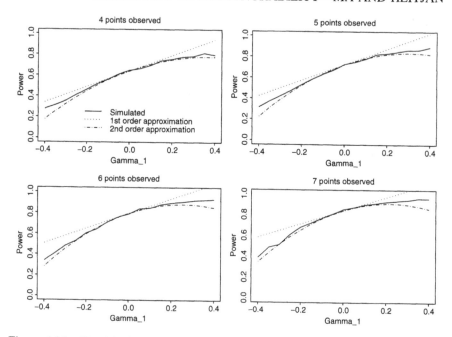

Figure 16.1   Empirical and approximate power of a one-sided test as a function of the nonignorability parameter.

Figure 16.1 shows that the first-order curve fits well and the second-order curve fits better. As γ_1 increases from zero, larger observations are more likely to be observed, and therefore the power to detect $\mu > \mu_0$ increases monotonically. Although the power increases with n_0, the sensitivity is similar in all four graphs.

16.6   Sensitivity of the coverage probability

Another important application is to the calculation of confidence interval coverage probabilities. Assuming normality with the variance known to be τ, the upper confidence bound for the mean is $\overline{Y}_0 + z_{1-\alpha}\sqrt{\tau/n_0}$. The coverage probability is then

$$CP_{\gamma_1}(\alpha, \mu, g) = \Phi\left[z_{1-\alpha} + \frac{\mu(\gamma_1) - \mu(0)}{\sqrt{\tau/n_0}}\right],$$

where Φ and $\mu(\gamma_1)$ are as in Section 16.5, and again the equation holds for large n_0. We approximate the coverage probability as

$$CP_{\gamma_1}(\alpha, \mu, g) \approx 1 - \alpha + \gamma_1 \frac{\partial CP_{\gamma_1}(\alpha, \mu, g)}{\partial \gamma_1}\bigg|_{\gamma_1=0} + \frac{\gamma_1^2}{2} \frac{\partial^2 CP_{\gamma_1}(\alpha, \mu, g)}{\partial \gamma_1^2}\bigg|_{\gamma_1=0}$$

$$(16.8)$$

where

$$\frac{\partial CP_{\gamma_1}(\alpha, \mu, g)}{\partial \gamma_1}\bigg|_{\gamma_1=0} = \phi(z_{1-\alpha})\frac{h'(\gamma_0)}{h(\gamma_0)}\sqrt{n_0\tau},$$

and

$$\frac{\partial^2 CP_{\gamma_1}(\alpha, \mu, g)}{\partial \gamma_1^2}\bigg|_{\gamma_1=0} = -z_{1-\alpha}\phi(z_{1-\alpha})\left[\frac{h'(\gamma_0)}{h(\gamma_0)}\right]^2 n_0\tau$$

$$+ 2\phi(z_{1-\alpha})\left[\frac{h(\gamma_0)h''(\gamma_0) - [h'(\gamma_0)]^2}{h^2(\gamma_0)}\right]\mu(0)\sqrt{n_0\tau}.$$

The first derivative of the coverage probability does not depend on $\mu(0)$, while the second derivative does.

For a two-sided interval, the coverage probability is

$$CP_{\gamma_1}(\alpha, \mu, g) = \Phi\left[\frac{\mu(0) - \mu(\gamma_1)}{\sqrt{\tau/n_0}} + z_{1-\alpha/2}\right] + \Phi\left[\frac{\mu(\gamma_1) - \mu(0)}{\sqrt{\tau/n_0}} + z_{1-\alpha/2}\right] - 1.$$

The first-order sensitivity of the two-sided coverage probability is identically zero, because any bias at all, in either direction, erodes the probability with equal magnitude, and consequently the coverage probability reaches its maximum value at $\gamma_1 = 0$. Other derivatives are

$$\frac{\partial^2 CP_{\gamma_1}(\alpha, \mu, g)}{\partial \gamma_1^2}\bigg|_{\gamma_1=0} = -2z_{1-\alpha/2}\phi(z_{1-\alpha/2})n_0\tau\left[\frac{h'(\gamma_0)}{h(\gamma_0)}\right]^2$$

and

$$\frac{\partial^3 CP_{\gamma_1}(\alpha, \mu, g)}{\partial \gamma_1^3}\bigg|_{\gamma_1=0} = -12z_{1-\alpha/2}\phi(z_{1-\alpha/2})n_0\tau\frac{h'(\gamma_0)}{h(\gamma_0)}$$

$$\times\left[\frac{h(\gamma_0)h''(\gamma_0) - [h'(\gamma_0)]^2}{h^2(\gamma_0)}\right]\mu(0).$$

Therefore, the coverage probability of a two-sided confidence interval approximately equals

$$CP_{\gamma_1}(\alpha, \mu, g) \approx 1 - \alpha + \frac{\gamma_1^2}{2}\frac{\partial^2 CP_{\gamma_1}(\alpha, \mu, g)}{\partial \gamma_1^2}\bigg|_{\gamma_1=0}$$

$$+ \frac{\gamma_1^3}{6}\frac{\partial^3 CP_{\gamma_1}(\alpha, \mu, g)}{\partial \gamma_1^3}\bigg|_{\gamma_1=0}. \tag{16.9}$$

The simulation process is similar to that of Section 16.5: First, we generated 2,000 datasets of size 10 from the normal distribution with $\mu = 2$ and $\tau = 4$. Next, we generated vectors of g values conditional on the simulated y vectors for given $\gamma_0 = 0$ and γ_1. Finally, for each n_o, we tabulated empirical coverage probabilities for comparison with approximations (16.8) and (16.9).

Figures 16.2 and 16.3 plot the coverage probabilities of one-sided and two-sided 95% confidence intervals, for $n_o = 5$ and 6, against γ_1 (results are similar for 4 and 7 points observed). For the one-sided interval, again the first-order approximation

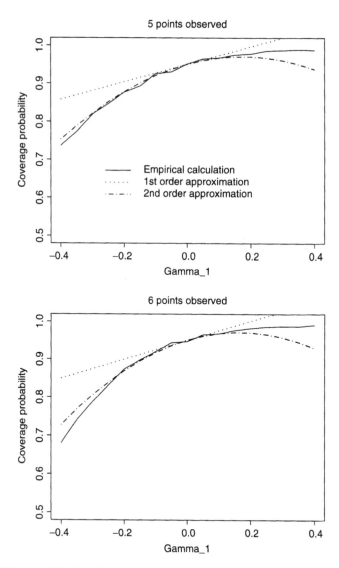

Figure 16.2   Empirical and approximate coverage probability of a one-sided interval as a function of the nonignorability parameter.

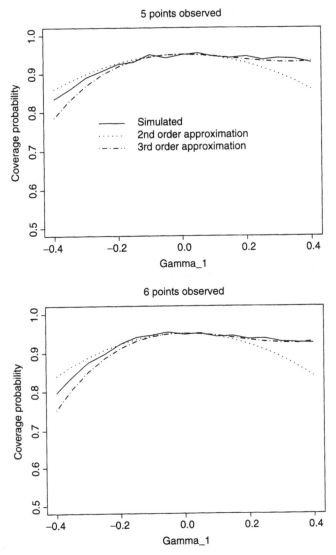

Figure 16.3   Empirical and approximate coverage probability of a two-sided interval as a function of the nonignorability parameter.

is good but the second-order approximation is better, and probably adequate for most purposes. When γ_1 increases positively, more large values are observed, and consequently, the coverage probability exceeds 95%; conversely, coverage probabilities can be much less than 95% for negative γ_1. For a two-sided interval (Figure 16.3), the third-order approximation is generally sufficient. As indicated above, the coverage probability reaches its nominal, and maximum, value at $\gamma_1 = 0$.

16.7   An example

The data are from a 12-week randomized trial of the antidepressant desipramine in outpatients with a diagnosis of cocaine dependence and a depressive disorder. The study hypothesized that desipramine would be superior to placebo in reducing the symptoms of depression and therefore also cocaine use. To demonstrate our method, we discuss only the data from the desipramine arm and the outcome measure HAMD-21. Of 55 patients randomized to desipramine, only 25 were left by week 12.

Suppose that we wish to conduct a further study to test $H_0: \mu = 11$ versus $H_a: \mu < 11$ when, as in the desipramine trial, $n_o = 25$ and $\tau = 35$. If $\gamma_1 = 0$ and $\mu(0) = 8$, we have 81% power to detect the difference of 3 points at level 0.05 s. Figure 16.4 plots the power of this test as a function of γ_1, using an expression analogous to (16.4). The graph shows that a value of $\gamma_1 = .04$, meaning that a 4-unit increase in HAMD-21 (e.g., a change from the lowest to the highest level on the anxiety subscale) is associated with an increase in the odds of being observed of $\exp(4 \times .04) = 1.17$, would be sufficient to reduce the power to roughly 65%. Thus, the power is sensitive to moderate departures from ignorability.

We approximate the coverage probability of a 95% two-sided interval by (16.9), which we plot against γ_1 in Figure 16.5. The graph shows that the coverage

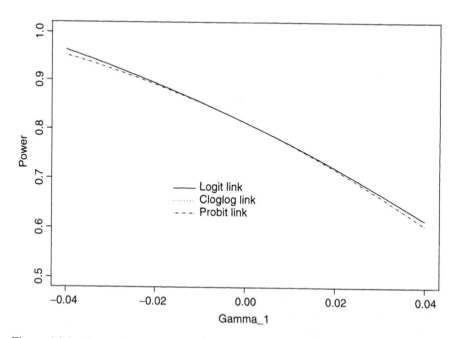

Figure 16.4   Approximate power of a one-sided test as a function of the nonignorability parameter using different link functions.

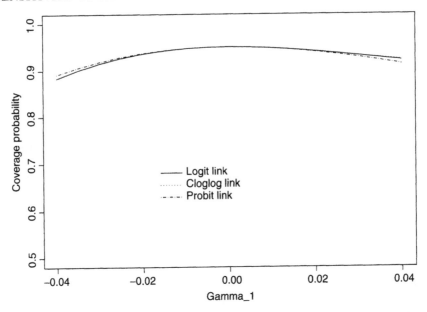

Figure 16.5 Approximate coverage probability of a two-sided interval as a function of the nonignorability parameter using different link functions.

probability of this interval is less sensitive to nonignorability than the power. The relatively modest values of $\gamma_1 = -.04$ can, however, reduce the coverage probability to less than 90%.

The plots also show the effect on sensitivity of using a complementary log–log or probit link in place of the logistic. The curves are directly comparable, in that we have transformed the binary regression slope parameters for the probit and complementary log–log selection models to have the same meaning, in terms of odds ratios, as the logistic γ_1. It is evident that the general shapes of the curves and measures of sensitivity are much the same for all the three models.

16.8   Discussion

In this article, we have demonstrated a simple method for computing the approximate sensitivity to nonignorability of the operating characteristics of frequentist data summaries. Simulations show that the approximations are valid across a relevant range of nonignorability parameters.

We anticipate using the method as a quick check on the validity of frequentist procedures when data are missing. For example, in clinical trials it is common to inflate enrollment goals beyond the nominal target sample size to adjust for anticipated dropout. Thus, if a power calculation shows that one needs 40 subjects, but one expects to lose 20% to attrition, the study would plan to enroll 50. Our findings suggest that still more patients are needed to counteract nonignorability, whose bias can erode power even further. More to the point, if the size of a test (its power under the null) is sensitive under the anticipated missingness mechanism, it may be prudent to avoid testing altogether.

Although our model allows any link function, the logistic may be the most convenient because its γ_1 parameter is a log odds ratio. Evidence to date (Xie, 2003) suggests that the choice of link makes little difference, as long as the dependence of the missingness probability on y is monotone. Thus, our methods may not be helpful with variables like self-reported income, where both high and low values are more likely to be missing. They should work well with health-related quality of life, where evidence suggests that the missingness probability is inversely related to y.

One can easily extend our method to other contexts such as the normal linear model with regression coefficient β and predictors X. Simply note that the expectation of $\hat{\beta}_0$, the estimated regression coefficient from the observed subjects, is

$$E(\hat{\beta}_0|G = g) = (X_0'X_0)^{-1}X_0'E(Y_0|G = g)$$

where X_0 is the submatrix of X for observed subjects and $E(Y_0|G = g)$ is a vector of (16.2) for subjects with $g_i = 1$. According to (16.3), the conditional variance of $\hat{\beta}_0$ given $G = g$ is

$$\text{var}(\hat{\beta}_0|G = g) = (X_0'X_0)^{-1}X_0'\text{var}(Y_0|G = g)[(X_0'X_0)^{-1}X_0']'$$
$$\approx \tau(X_0'X_0)^{-1},$$

which again needs no adjustment, to first order. For more complicated models such as nonlinear regression, survival curve estimation, and survival regression, the analysis steps may be more complex. The idea should be straightforward to carry through, however, numerically if not analytically.

This chapter has emphasized approaches to evaluating sensitivity prior to data collection. Once the data are available, one may wish to base inferences on sample moments corrected for nonignorability according to formulas such as (16.1), (16.2), and (16.3). Such analyses are potentially complicated by the need to estimate γ_0 for various tentative values of γ_1. In data analysis, we generally espouse a Bayesian/likelihood sensitivity analysis as proposed by Troxel, Ma, and Heitjan (2004).

Part III

Statistical modeling and computation

17

Statistical modeling and computation

D. Michael Titterington[1]

To provide even a brief overview of an area as wide ranging and active as "statistical modeling and computation" is a daunting task, and in view of the limitations of the space available I can only scratch the surface.

At first sight, it might seem more appropriate to consider the two topics separately, because the more cerebral activity of modeling seems rather different from the nuts-and-bolts issues in computation. However, they are of course very closely linked, in that, in general, new ideas in modeling are only immediately worthwhile if they are computationally feasible. In general, therefore, the two lines of research have pulled each other along, encouraged by the carrot of the need to analyze, in realistic fashion, ever more complex data structures thrown up by pressing real-life problems. Modeling ideas then develop that stretch the current capabilities of computational hardware and techniques, and improvement on the computational side challenges the modeler to make the most of the new resources. Sometimes, of course, methodological ideas are proposed that may not be immediately implementable, but the inventor predicts, with justifiable confidence, that computer power will soon be such that the methods will be routine; I recall noting such a prediction in one of Rubin's own early expositions of the ideas of multiple imputation (Rubin, 1978b)!

As I have said, I can only give a brief sketch of the total picture here, and I shall concentrate on the topics of regression and latent-variable models, partly because, although they are rather specific, they do subsume a wide range of

[1]Department of Statistics, University of Glasgow, Scotland.

Applied Bayesian Modeling and Causal Inference from Incomplete-Data Perspectives.
Edited by A. Gelman and X.-L. Meng © 2004 John Wiley & Sons, Ltd ISBN: 0-470-09043-X

particular manifestations, and partly because they include scenarios of personal interest to me. Another theme will appear that I feel is also of current importance, namely the existence of strongly related activity in the computer science research community.

17.1   Regression models

A vast array of ideas is covered by the general heading of regression models, in which one or more response variables are related to one or more covariates. Here the simplest scenario is provided by linear models with uncorrelated Gaussian errors, but recent and current work goes way beyond that, and even way beyond the rich class of generalized linear models: parametric developments include nonlinear regression models, models including random effects, and models that accommodate measurement error on the covariates; semiparametric models allow more flexibility for part of the regression function, often the intercept term, while retaining a parametric formulation for the component of the model involving many of the covariates; and approaches to nonparametric regression aim simply to acknowledge local smoothness without the straightjacket of a global parametric component. One particular area of much current activity concerns longitudinal data (see for example Diggle et al., 2002), to which all these types of approach have been applied, and which has popularized the generic ideas of generalized estimating equations and so-called "sandwich" variance estimators of the uncertainty associated with estimators of parametric elements of the model.

Purely nonparametric approaches to regression have now matured, based on splines and other types of basis function. Further seminal advances may be hard to come by unless some major new step is taken, but the realistic application of these approaches to problems with more than a very small number of covariates still represents a major challenge. It is perhaps fortunate that many important applications involve only one covariate, such as time, or only two, perhaps representing spatial coordinates, or just three, as would be appropriate for spatiotemporal modeling of climatic behavior or three-dimensional image modeling.

As promised, I shall refer more than once to developments in the computer science or artificial intelligence literature, and here it is appropriate to comment on the class of models known as *artificial neural networks* and, in particular, those called *multilayer perceptrons* or *feed-forward networks*. As indicated for example in Cheng and Titterington (1994), these have a direct interpretation as complicated nonlinear classifiers or regression models, depending on the nature of the response variable or variables. They have certainly been applied to complicated scenarios with vast numbers of covariates, for example, in optical character recognition in which there is a covariate for each pixel in a fine-resolution pixelated image. Strictly speaking, these models are parametric, but they are so rich that their effect is essentially nonparametric.

17.2   Latent-variable problems

The incorporation of latent variables represents one way of enriching an otherwise simple model with a view to rendering it more realistic and widely applicable. Mixture models provide a good example of this, with the latent variable corresponding to the mixture-component indicator of an observation. Of course, in some applications, such a latent variable is actually a real but missing variable, such as the disease category of an undiagnosed patient, and then the mixture distribution models the marginal distribution of certain observed quantities, but there seems to be an ever-increasing use of mixture models as flexible structures in which the latent variable does not necessarily have a physical meaning. Of course, there may be some *post hoc* attempt to interpret the meaning of the latent variable, as in factor analysis, which corresponds to the case of continuous latent variables.

If we restrict ourselves for the time being to a multinomial latent variable, as is the case with mixture models, it is undeniable that interest is still remarkably high, especially if one includes certain extensions. With mixture data, the latent variables or states for different observations are assumed independent, but one might consider imposing some dependence among then. For example, if the latent states are assumed to follow a Markov chain then we have the class of hidden Markov chains, more usually called *hidden Markov models*. For some time now, these models have found application in areas such as ecology and speech modeling, and more recently they have been widely used in the context of DNA sequencing. If we generalize further, move into two dimensions and assume that the hidden states follow a Markov random field then we produce the type of model that became popular in statistical image analysis (Geman and Geman, 1984; Besag, 1986) and which kick-started the revolution in Gibbs sampling and Bayesian computation in general; see later. Here again the computer science community has shown considerable interest, with further variations on mixture models, such as the so-called hierarchical mixtures of experts models (Jordan and Jacobs, 1994) and mixtures of factor analyzers (Ghahramani and Beal, 2000; Fokoué and Titterington, 2003), the latter of which contains both discrete and continuous latent variables.

17.3   Computation: non-Bayesian

As mentioned in the introduction, the practical viability of models and methods relies on computational feasibility, and, if a non-Bayesian approach is adopted, the most crucial issue in modern statistics is the practicality of maximum likelihood estimation. For many decades, the numerical difficulties of implementing problems of maximum likelihood estimation in Gaussian mixture problems restricted parameter estimation to *ad hoc* methods and Karl Pearson's version of the method of moments. In general, in the context of latent-variable problems, the fact that the contributors to the likelihood are themselves marginal densities without a neat closed-form expression renders the likelihood unwieldy and not amenable to explicit maximization. This led to the need for numerical methods, such as Newton–Raphson, gradient ascent, and of course the EM algorithm and its many refinements and derivatives.

The impact of the initial EM paper by Dempster, Laird, and Rubin (1977) has been astonishing and continuing. Its level of citation seems still to be very high and its pervasiveness is truly impressive. Again, I should like to emphasize its fame and influence on the computer science and machine learning literature, where many versions of the algorithm have been developed for the models, often based on latent variables, used there. Some nice foundational work has also appeared, such as the interpretation of the EM steps as a pair of maximizations; see for example Amari (1995) and Neal and Hinton (1999).

Various adaptations have been made to the EM algorithm to speed it up or to deal with difficulties of implementation. An initial attraction of the method was that the M-step is easy if complete-data maximum likelihood is easy, so that the iterative stages could be explicit provided the E-step is also available in closed form. This is not always possible. For example, given data from a hidden Markov random field, neither the E-step nor the M-step can be done explicitly. A number of ways of dealing with these problems, none of them totally satisfactory, have been formulated. For instance, the expectation required in the E-step, of the high-dimensional distribution of the latent states given the data, might be approximated by Monte Carlo, based on simulations, but generation of such simulations requires time-consuming Gibbs sampling. An alternative idea is to approximate the conditional distribution, which reflects complicated interdependence among the latent states, by something much simpler, such as an independence model. This leads into the ideas of variational approximations and the use of so-called mean-field approximations, which first evolved in the statistical-physics literature. These modifications seem to work quite well empirically (see for instance Zhang, 1992), but their theoretical properties remain unresolved and of interest, to me at least. Variational approximations have also been used to obtain computationally feasible lower bounds for the values of unwieldy likelihoods; for example, Jordan et al., (1999) apply this approach to certain graphical models with hidden variables, such as hidden Markov models and Boltzmann machines. Again, some theoretical issues are unresolved in general, such as the consistency or otherwise of the estimator defined by the maximizer of the lower-bound surface.

17.4   Computation: Bayesian

By direct analogy with the problems incurred by maximum likelihood estimation, Bayesian analysis of the complex models that are required for explaining data sets of interest in the modern world is nontrivial. Straightforward conjugate analysis of exponential-family models is not sufficiently rich. The EM algorithm certainly helped here also, in often facilitating the computation of posterior modes, but the Bayesian computational revolution began in earnest with Geman and Geman

(1984), and Gelfand and Smith (1990), as the full potential of Gibbs sampling and other forms of Markov chain Monte Carlo (MCMC) methods started to be revealed; of course, the methodology was not totally new, having many antecedents in the statistical-physics literature and in Hastings (1970). Gibbs sampling is arguably the simplest and most pervasive of the techniques, but other methods include versions of Metropolis–Hastings algorithms, importance sampling, and so on. Important issues concern the convergence of the algorithms, and strategies for generating multiple samples from the posterior distribution of interest; see for example, the work of Gelman and Rubin (1992). Helpful recent thumbnail sketches of these topics are given by Cappé and Robert (2000) and Gelfand (2000). The former paper identifies perfect sampling and adaptive or sequential sampling as important current areas. With perfect sampling, one can guarantee convergence of the sampling algorithm at a certain stage, but there are doubts about the scope of problems to which it can confidently be applied. Sequential approaches include the group of techniques known as *particle filtering*; see for example Doucet et al. (2001).

These simulation-based methods have been complemented by some deterministic techniques. In the statistical literature, the most well-known idea has been the method of Laplace approximations (see for example Tierney and Kadane, 1986) that has been particularly valuable in the evaluation of integrals of interest to Bayesians. On a different tack, the computer science literature has publicized a class of so-called variational Bayes approximations. These have been applied in particular to latent-variable problems such as mixtures, as approximations to the joint posterior distribution of the model parameters and the latent variable indicators. One general approach is to choose a variational posterior of a prescribed simplified form that is as close as possible to the true posterior according to Kullback–Leibler divergence. Typically, the level of "simplification" involves assuming that the indicators and the model parameters are independent, *a posteriori*. In many cases, this leads to workable algorithms with good empirical results. However, theoretical issues are still being resolved; in principle, MCMC should reflect the true posterior distribution, once convergence has occurred, whereas that cannot really happen with the variational approach. Some key references are Corduneanu and Bishop (2001) and Ueda and Ghahramani (2003), with more citations provided in Titterington (2004).

17.5   Prospects for the future

The evolution of statistical modeling and computation will surely continue in tandem as it has done over the last 100 years or so, with models becoming more flexible in order to provide meaningful answers to questions about increasingly complicated and voluminous data sets. The wealth of application areas will guarantee this; a more-or-less randomly chosen issue of JASA Applications Section, from September 2003, dealt with a range of topics, from a multistage model for the life cycle of grasshoppers to sampling mechanisms of long-period comets, including on the way the application of wavelets to colon carcinogenesis, the assessment of higher education, hierarchical Bayes modeling of consumer purchasing, and papers about climate change. The nature of the modeling process will no doubt change, perhaps becoming more hierarchical or more nonparametric, and the distinction, if such exists, between mainstream statistics and machine learning will become even more blurred (Hastie et al., 2001); I have several times noticed that, according to the announcements of conferences in machine learning, "Bayesian statistics", "graphical models", and even "statistics", are in fact already thought of as being topics within the umbrella of machine learning! Notwithstanding this trend, long-standing ideas may well have new leases of life, as has recently been the case with multiple testing on a massive scale, in contexts such as functional imaging (Genovese et al., 2002) and simultaneous inference for the differential expression of thousands of genes in microarray analysis (Storey and Tibshirani, 2003).

18

Treatment effects in before-after data

Andrew Gelman[1]

18.1 Default statistical models of treatment effects

The default analyses for experiments and observational studies assume constant treatment effects. The usual modeling or Bayesian approach with ignorable treatment assignment starts with a constant treatment effect; for example, $y_i = \beta_0 + \beta_1 T_i + \beta_2 x_{2i} + \beta_3 x_{3i} + \cdots + \epsilon_i$, where T_i is the treatment variable (most simply, an indicator that equals 1 for treated units and 0 for controls). In Fisher's classical test, the null hypothesis is that treatment effects are zero for all units. More generally, this approach can be inverted to obtain confidence intervals for a constant treatment effect. Neyman (1923) allowed the possibility for varying effects (see Rubin, 1990) but only as a goal toward estimating or testing hypotheses about average treatment effects.

Before-after designs have been much discussed in the statistical literature (see Brogan and Kutner, 1980; Laird, 1983; Crager, 1987; Stanek, 1988; Stein, 1989; Singer and Andrade, 1997; Yang and Tsiatis, 2001). It is recognized that treatment effects can vary with pretreatment covariates ($x_2, x_3, \ldots$ in the above model), and that these interactions can be substantively important (see Dehejia, 2004). We argue here that interaction between treatment and covariates is a general phenomenon

[1]Department of Statistics and Department of Political Science, Columbia University, New York. We thank Gary King, Iain Pardoe, Don Rubin, Hal Stern, and Alan Zaslavsky for helpful conversations and the National Science Foundation for financial support.

that can be seen as deriving from an underlying variance components model. We posit fundamental variation among experimental (or observational) units that is not fully captured in pretreatment predictors and manifests itself in experimental or observational outcomes.

18.2 Before-after correlation is typically larger for controls than for treated units

Our point is not merely that treatment effects vary—in practice, everything varies—but that they vary in systematic, predictable ways. We begin by reviewing a ubiquitous pattern in experiments and observational studies with before-after data: the correlation between "before" and "after" measurements is commonly higher for controls than in the treatment group.

An observational study of legislative redistricting

Figure 18.1 gives an example from our research on the effects of redistricting on the partisan bias of electoral systems (Gelman and King, 1994). The symbols in the graph represent state legislatures in election years (e.g., California in 1974), with the estimated "partisan bias" (a measure of the fairness of the electoral system) of the legislature in that year plotted versus the estimated partisan bias in the previous election. The small dots in the graph represent "control" cases in which there was no redistricting, and the larger symbols correspond to "treated" cases, or redistrictings. The treatment has three levels—corresponding to redistrictings controlled by Democrats, Republicans, or both parties—but here we consider all treatments together. Elections come every two years and redistricting typically happens every 10 years, so most of the data points are controls. The correlation between before and after measurements is much larger for controls than treated cases. (The regression lines for the three levels of treatment are constrained to be parallel and equally spaced because there were not enough data points to accurately estimate separate slopes or separate effects for the two parties.)

From the usual standpoint of estimating treatment effects, the interaction between treatment and x (estimated partisan bias in previous election) in Figure 18.1 is dramatic—and, in fact, we had not thought to include an interaction in our model until it jumped out at us from the graph. Stepping back a bit, however, the different slopes for the two groups should be no surprise at all. In the control cases with no redistricting, the state legislature changes very little, and so the partisan bias will probably change very little from the previous election. In contrast, when the legislative districts are redrawn, larger and more unpredictable changes occur.

In fact, in this example, the interaction effect of redistricting—that it tends to reduce partisan bias—is larger than the original object of this study, which was the partisan advantage of redistricting (the slight difference between the lines

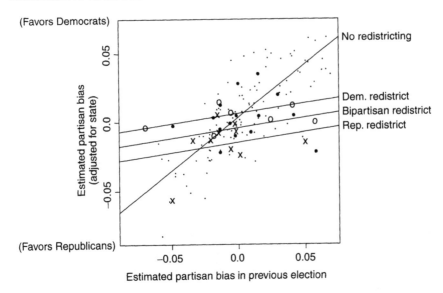

Figure 18.1   Effect of redistricting on partisan bias. Each symbol represents a state and election year, with dots indicating control cases (years with no redistricting) and the other symbols corresponding to different types of redistricting. As indicated by the fitted regression lines, the "before" value is much more predictive of the "after" value for the control cases than for the treated (redistricting) cases. In contrast to the minor differences between Democratic, bipartisan, and Republican redistricting, the dominant effect of the treatment is to bring the expected value of partisan bias toward 0, and this effect would not be discovered with a model that assumed parallel regression lines for treated and control cases. From Gelman and King (1994).

for Democratic, bipartisan, and Republican treatment lines in Figure 18.1). It was crucial to model the variation in the treatment effects to see this effect.

An experiment with pretest and posttest data

Figure 18.2 summarizes before-after correlations from an educational experiment performed on a set of elementary-school classes.[2] In each of the four grades, the classes were randomized into treated and control groups, with pretest and posttests taken for each class. Figure 18.2 shows the correlation between before and after measurements, computed separately among the control and treated classes. At each grade level, the correlation is higher for the controls.

---

[2] The treatment in this experiment was exposure to a new educational television show called "The Electric Company." The experiment was conducted around 1970 and used as an example in Don Rubin's course at Harvard University in 1985.

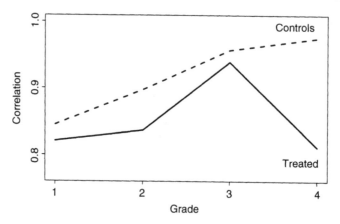

Figure 18.2 Correlation of pretest and posttest scores for an educational experiment, for control and treated classrooms in each of the four grades. Correlations are higher in the control groups, which is consistent with models of varying treatment effects.

As in our previous example, the pattern of correlations makes sense: the pretest is a particularly effective predictor of posttest scores for the control classes, where no intervention has been imposed (except for a year of schooling). In the treatment group, it is reasonable to expect the intervention to have different effects in different classrooms, thus attenuating the correlation of before and after measurements.

Congressional elections with incumbents and open seats

We give one more example of before-after correlations, in an observational study of the effect of incumbency in elections in the US House of Representatives.[3] The units in this example are Congressional districts, the before and after measurements are the Democratic Party's share of the vote in two successive elections, and the "treatment" is incumbency. For simplicity, we separately analyze in each year the seats held by Democrats and by Republicans.

In the context of our discussion here, the "control" districts are those where the incumbents are running for reelection, and the "treated" districts are the open seats, where the incumbent party is running a new candidate. We use this labeling because the races with incumbents represent less change from the previous election, whereas running a new candidate can be viewed as an intervention. The effect of incumbency in a given district is then the negative of the treatment effect as defined here.

Figure 18.3 shows the correlations between the Democratic vote shares in each pair of two successive elections, computed separately for controls (incumbents

---

[3] See Gelman and King (1990) and Gelman and Huang (2004) for details.

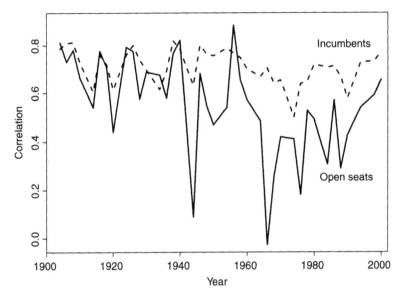

Figure 18.3   Correlations of party vote share in each pair of successive Congressional elections in the past century, computed separately for the incumbents running for reelection (the "control group") and open seats (the "treatment group"). Correlations are consistently higher in the control group, which makes sense since there is less change between before and after in these districts. In the early part of the century, when correlations in the two groups were about the same, the effect of incumbency was very small.

running) and treated districts (open seats).[4] As in our previous examples, the before-after correlation is much higher in the control group. Again, this picture is consistent with the idea that there is little change among the controls, whereas a varying treatment effect reduces the predictive importance of past data.

A careful look at Figure 18.3 reveals that the before-after correlations within the two groups did not diverge until the second half of the century. A separate analysis (not shown here) estimates the average advantage of incumbency in Congressional elections to be near zero for the first half of the century, then increasing dramatically through the 1950s and 1960s to its current high level. Thus, as the treatment effect increased, its variation also increased. (The jaggedness of the solid line in Figure 18.3 can largely be explained as sampling variability given the small number of open seats, especially in recent decades.)

---

[4]We exclude uncontested elections and years ending in "2," when district lines are redrawn. Within each group (incumbents running and open seats), we compute correlations separately for the Democratic- and Republican-held seats: Figure 18.3 presents the averages of the within-party correlations for each pair of election years.

18.3   A class of models for varying treatment effects

When only "after" data are available in an experiment, it is not possible to see the consequences of varying treatment effects, and the classical t interval gives appropriate superpopulation inference for average treatment effects (see Gelman, Carlin, Stern, and Rubin, 2003; Section 7.5). In contrast, treatment effects that vary as a function of "before" data can be modeled and estimated in a number of ways.

Plots such as Figure 18.1 suggest regression models with treatment effects interacted with pretreatment covariates. We would like to think more generally of treatments that can have varying effects, both additive and subtractive. For example, suppose we label the "before" and "after" measurements for unit j as y_{jt}, $t = 0, 1$, and fit the two-error-term model,

$$\text{before: } y_{j0} = (X\beta)_{j0} + \alpha_j + \gamma_{j0} + \epsilon_{j0}$$

$$\text{after: } y_{j1} = T_j\theta + (X\beta)_{j1} + \alpha_j + \gamma_{j1} + \epsilon_{j1}, \tag{18.1}$$

where T represents the indicator for treatment (which in this setup occurs between the "before" and "after" measurements) and θ is the average treatment effect— the usual object of inference in an observational study. The matrix X represents other linear predictors in the regression model (e.g., demographic variables for a model of individuals, or district-level characteristics for a model of election outcomes), and the unit-level term α_j represents persistent variation among units not explained by the predictors. The error terms $\epsilon_{j0}, \epsilon_{j1}$ are the usual independent observation-level errors.

The terms γ_{j0}, γ_{j1} take model (18.1) beyond the usual longitudinal or panel-data hierarchical regression framework, and our key innovation is in linking this variance component with the treatment, so that it is affected differently by the treatment and controls. Various models are possible here, all of which allow treatment effects to vary by unit and have the by-product that before-after correlation is higher for controls than treated units. We list some possibilities here.

Replacement treatment error.   Suppose that under the control condition, γ_j is unchanged (that is $\gamma_{j1} \equiv \gamma_{j0}$), but under the treatment, γ_{j0} and γ_{j1} are independent draws from the same probability distribution. In this model, the treatment has the effect of replacing a random error component. This could make sense if the control corresponded to staying with a particular regimen and the treatment corresponded to switching to a new approach. For example, in the redistricting example in Figure 18.1, the treatment replaces an old districting plan with a new one.

Additive treatment error.   Suppose that $\gamma_{j0} \equiv 0$ for all units, and $\gamma_{j1} = 0$ for controls, but is drawn from a distribution for treated units. In this model, the treatment adds a source of variability that was not present before. This could happen if the treatment is a new, active intervention (for example, the educational TV program in Figure 18.2).

Subtractive treatment error. For a different model, suppose that γ_{j0} comes from some probability distribution, and under the control condition, $\gamma_{j1} \equiv \gamma_{j0}$), but under the treatment, $\gamma_{j1} \equiv 0$. In this model, the treatment subtracts a source of variability. This could apply to a setting in which an active intervention has already been applied to the "before" measurements, and the control and treatment conditions correspond to staying with or dropping the intervention. For example, in the incumbency example in Figure 18.3, the "treatment" corresponds to an open seat—the disappearance of an incumbent (see Gelman and Huang, 2004).

More formally, using the potential-outcome notation of Rubin (1974), the error terms γ_{j1} could be written as γ_{Tj1}, where $T = 0$ or 1 corresponds to the control and treatment conditions. In any case, these models, or more general distributions on these error terms, capture the idea that the treatment *changes* the affected units as well as having some average additive effect. Similar models are used in animal breeding to model genetic variation and treatment effects (see Lynch and Walsh, 1988; Sargent and Hodges, 1997) present related ideas for hierarchical models of complex regression interactions. We would also like to formulate a class of models in which treatments with larger main effects naturally have larger variation, as this is another property that often seems to hold in practice.

18.4   Discussion

It has been argued that statistical models should be adapted individually to applied problems (see, for example, Chapter 27 in this volume). However, in practice, default procedures and models are used in a wide variety of settings. This is not merely for convenience (or because certain models are easier to access in statistical software packages such as SPSS) but because default models often work. Methods such as t-intervals, the analysis of variance, and least-squares regression have been effective in all sorts of problems (see, for example, Snedecor and Cochran, 1989), and much of the methodological research of the past few decades has resulted in extensions of these and other approaches. Our current toolbox of default methods includes t models for robust regression and multivariate imputation (generalized from the normal; see Liu, 1995), wavelet decompositions (generalized from Fourier analysis; see Chapter 31 in this volume), generalized linear models (McCullagh and Nelder, 1989), splines and locally weighted regressions (Wahba, 1978; Cleveland, 1979), and model averaging for regressions and density estimates (Hoeting, Madigan, Raftery, and Volinsky, 1999; Richardson and Green, 1997). All these methods have been demonstrated for specific examples but are intended to be flexible generalizations of previous default approaches.

In this chapter, we have tried to motivate an expansion of the default model of experiments and observational studies to allow for treatment effects to vary among units. This variation can sometimes be expressed as interactions with pretreatment measurements but more generally can be understood as effects on unobserved unit-level variance components of the sort that are used in instrumental variables and principal stratification (see Chapters 8 and 9 in this volume). Our models are still under development and we hope they will reach "default" stage sometime in the not so distant future, as a small part of a general applied framework for causal inference deriving ultimately from the potential-outcome perspective of Rubin (1974).

19

Multimodality in mixture models and factor models

Eric Loken[1]

As a psychology graduate student, I worked with Kagan who viewed infant temperament as a set of qualitatively different types, rather than as continuous dimensions along which infants differ (Kagan, 1994). I had been using cluster analysis to explore latent group structure when Don Rubin introduced me to more formal analyses he and Hal Stern had completed on a subset of the same infant data (Rubin and Stern, 1994). For my dissertation, I followed closely the work of Stern et al. (1994), using latent class analysis—a simple type of mixture model—to explore Kagan's data for evidence of temperament types. Inevitably, working with Don led me to adopt a Bayesian approach to mixture modeling, and of the many challenges I encountered, one of the most interesting was the label-switching problem. In a mixture model, each case is assumed to have a missing indicator denoting the latent class to which it belongs. But this means there are multiple equivalent modes, as the likelihood and posterior distribution are invariant to arbitrary reorderings of the class membership labels. The problem is relatively easy to recognize in maximum likelihood (ML) estimation as the "different" solutions are of equal likelihood and just represent permutations of the class indicators. In a Bayesian setting, however, label switching might occur during simulation of the posterior distribution, and posterior summaries will be biased and have inflated variance if the problem is not addressed.

The first section of this chapter shows how the label-switching problem can affect posterior inferences. In the context of a latent class example, I present a

---

[1]Department of Human Development and Family Studies, Pennsylvania State University, University Park, Pa.

Applied Bayesian Modeling and Causal Inference from Incomplete-Data Perspectives.
Edited by A. Gelman and X-L. Meng © 2004 John Wiley & Sons, Ltd ISBN: 0-470-09043-X

simple technique to limit label switching and improve posterior inference. The second section extends the discussion to models with continuous latent variables. There, in the context of an orthogonal factor model, I discuss a similar concern about multimodality, and show how certain choices of identification constraints can lead to problems impacting both ML and Bayesian inference.

19.1 Multimodality in mixture models

The label-switching problem in mixture models has long been recognized. Common solutions have typically centered on applying constraints during simulation or postprocessing the posterior draws (Celeux, Hurn, and Robert, 2000; McLachlan and Peel, 2000; Richardson and Green, 1997; Stephens, 2000), with the goal of keeping the simulation confined to one modal region in the posterior distribution. A novel method that appears to have promise is to preclassify one or more observations in order to dampen or eliminate the nuisance modes and leave the mode of interest largely unaffected. The soundness of the idea is still under exploration. I report here on an example using latent class analysis (see also Chung, Loken, and Schafer (in press) for an example with a mixture of exponentials; an argument regarding preclassifying to identify a model was made in a different context by Taylor (1995) and Zhuang et al. (2000)).

A latent class model is a simple mixture model with categorical indicators and the key assumption that the indicators are conditionally independent given class membership. Suppose a researcher has binary data (low/high) on four measures of infant reactivity. Specifically, the researcher might observe the infant's motor activity, crying, smiles, and vocalizations (let m, c, s, v represent the observed variables). If there are two latent classes, the probability of observing a specific data pattern $[m, c, s, v]$ is

$$\pi_{mcsv} = \sum_{i=1}^{2} \pi_{z=i} \pi_{m|z=i} \pi_{c|z=i} \pi_{s|z=i} \pi_{v|z=i},$$

where π_z represents the probability a case is in class z, and $\pi_{m|z=i}$ represents the conditional probability of the motor score given an infant belongs to latent class i. Four binary indicators yield 16 possible observed data patterns. If N_r represents the number of observations with pattern r, the likelihood function, given model parameters θ, is

$$p(y|\theta) = \prod_{r=1}^{16} \left[\sum_{i=1}^{2} \pi_{z=i} \pi_{m|z=i} \pi_{c|z=i} \pi_{s|z=i} \pi_{v|z=i} \right]^{N_r}$$

Table 19.1 gives population parameters for two latent classes and four binary indicators. In one class, the infants have high motor and cry reactivity, low smiling, and low vocalizations; in the other class, the opposite pattern occurs. One sample of 100 observations was drawn for this example.

	Class 1	Class 2
π_z	0.6	0.4
$\pi_{m=1|z}$	0.7	0.3
$\pi_{c=1|z}$	0.7	0.3
$\pi_{s=1|z}$	0.3	0.7
$\pi_{v=1|z}$	0.3	0.7

Table 19.1 Population parameters for two latent classes with four manifest binary indicators.

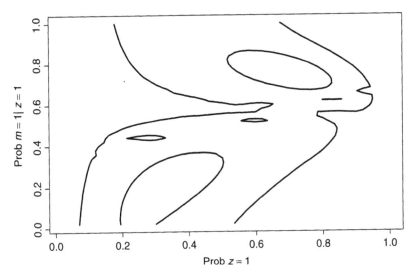

Figure 19.1 Profile likelihood with $\pi_{m|z=1}$ and π_z fixed. Inner contour is for $\log L > -267$, outer contour is for $\log L > -270$ ($\log L_{\max} = -266.4$).

Before proceeding to a Bayesian analysis, it is instructive to use some graphical techniques to represent the contours of the likelihood. Figure 19.1 shows the profile likelihood for these data, calculated by maximizing over the free parameters after fixing $\pi_{m=1|z=1}$ and $\pi_{z=1}$ at specific values. The figure has rough edges because the likelihood is maximized on a 100×100 grid and because only three estimates were made at each location (requiring 30,000 runs of EM). The most notable feature is the two modal regions corresponding to estimates in the neighborhood of the two equivalent maximum likelihood estimates (MLEs). Note also that the likelihood is actually quite flat, indicating a good deal of uncertainty in the estimates.

From a Bayesian perspective, the posterior distribution, $p(\theta|y)$, is proportional to the product of the prior distribution and the likelihood. As the class label and the observed indicators are binary variables, the natural conjugate prior for all parameters is the Beta distribution. Therefore, we set $p(\pi_{m|z}) = Beta(\alpha_{m|z}, \beta_{m|z})$

(and similar for the other three indicators), and $p(\pi_z) = Beta(\alpha_z, \beta_z)$. The full posterior distribution is

$$p(\theta|y) = p(\pi_z)p(\pi_{m|z})p(\pi_{c|z})p(\pi_{s|z})p(\pi_{v|z})$$

$$\times \prod_{r=1}^{16} \left[\sum_{i=1}^{2} \pi_{z=i}\pi_{m|z=i}\pi_{c|z=i}\pi_{s|z=i}\pi_{v|z=i} \right]^{N_r}$$

The posterior distribution can be simulated following Stern et al. (1994, 1995). The Markov chain Monte Carlo (MCMC) algorithm iterates between the conditional distribution of the class indicators given the current parameter estimates, and the distribution of the parameters given the complete data (i.e., including the latent class indicators). However, it is possible (and in this example highly likely) that the simulation will move back and forth between the two modal regions shown in Figure 19.1. Figure 19.2 shows that mode switching is clearly evident in a chain of 40,000 draws from the posterior distribution for $\pi_{m|z=1}$.

Posterior means derived by averaging over chains such as the one shown in Figure 19.2 will be considerably biased to 0.5. Table 19.2 shows that the posterior means are clearly biased in the expected direction relative to the population parameters shown in Table 19.1.

Solutions to the label-switching problem often involve applying constraints or somehow regrouping the posterior draws (Richardson and Green, 1997). However, we can see from the first figure that care must be taken in applying an appropriate constraint. For instance, setting $\pi_{z=1} > \pi_{z=2}$ (i.e., setting class 1 to be the majority

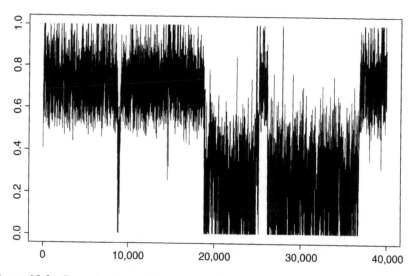

Figure 19.2   Posterior draws for $\pi_{m|z=1}$. Abrupt shifts represent mode switching, where the labels $z = 1$ and $z = 2$ change meaning.

	Class 1	Class 2
π_z	0.52 (.24)	0.48 (.24)
$\pi_{m=1\|z}$	0.52 (.29)	0.44 (.29)
$\pi_{c=1\|z}$	0.52 (.20)	0.45 (.20)
$\pi_{s=1\|z}$	0.48 (.18)	0.52 (.19)
$\pi_{v=1\|z}$	0.51 (.30)	0.58 (.30)

Table 19.2 Posterior means (standard deviations) calculated over MCMC chain. Owing to label switching, the means are biased to 0.5 and the variances are large.

class) would not be an effective constraint because it would fix the MCMC algorithm to the right side of the figure, and still allow label switching for the motor parameter. In this case, the constraint $\pi_{m=1|z=1} > \pi_{m=1|z=2}$ would probably be better, defining class 1 as the class in which the infants are most likely to have low motor scores.

A different approach might be to assign one of the observations to a specific class. If the observation is judiciously chosen, the effect should be to leave one of the modes of interest virtually untouched, and to "dampen" the nuisance mode. How can we choose an observation to preclassify? In the simulated data, there were 14 observations with data pattern [1,1,2,2] (i.e., low motor and cry, and high smiling and vocalizing). If we assume that one of these observations is known to be in class 1, the modified likelihood looks strikingly different. Figure 19.3 shows the profile likelihood calculated in the same manner as in Figure 19.1, except with

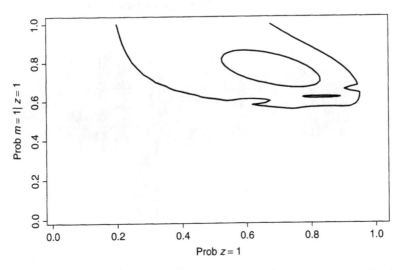

Figure 19.3 Profile likelihood with $\pi_{m|z=1}$ and π_z fixed. Inner contour line is for $\log L > -267$, outer contour is for $\log L > -270$. ($\log L_{max} = -266.4$).

the modification that the latent group membership for one case is set to $z = 1$ with probability 1. The modification has the desired result of significantly dampening the second mode while leaving the first mode substantively similar.

The preclassification trick is easy to implement in posterior simulation; all we really need to do is "move" the one case from the observed data into the prior. We accomplish this by adding 1 to the appropriate α and β hyperparameters of the prior distribution and removing the observation from the observed data, ensuring that the case is always classified to the same latent class at each imputation step of the MCMC algorithm.

Figure 19.4 shows posterior draws generated under the modified model. There is still some minor mode switching, but because the density in that region has been greatly reduced, the algorithm quickly jumps back to the desired mode. The posterior summaries are now much closer to the true values as the draws from one mode dominate the chain (see Table 19.3).

The technique of preclassifying one observation to restrict the MCMC for a two-class mixture model to one dominant mode appears promising, but requires more study. It is an appealing approach because it is extremely easy to implement and it does not require continuous monitoring and rejecting draws during simulation of the posterior distribution (as is necessary with a constraint). The technique also involves less subjective information than might be thought at first. Although classifying one of the observations does imply a stronger prior, the modification can actually be viewed as following from the definition of the model. According to the model, each case belongs to one of the latent classes, so preclassifying one case simply defines class 1 as the class to which that observation belongs. In fact, in a latent class model, the assumption is even weaker because it only stipulates that *one of* the cases with a specific observed data pattern belongs in class 1.

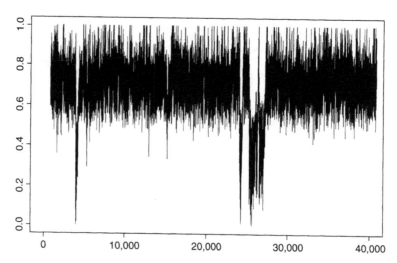

Figure 19.4 Posterior draws for $\pi_{m|z=1}$ under modified posterior.

	Class 1	Class 2
π_z	0.67 (.16)	0.33 (.16)
$\pi_{a=1|z}$	0.71 (.13)	0.25 (.21)
$\pi_{b=1|z}$	0.62 (.09)	0.34 (.17)
$\pi_{c=1|z}$	0.37 (.09)	0.62 (.16)
$\pi_{d=1|z}$	0.29 (.15)	0.79 (.21)

Table 19.3 Posterior means (standard deviations) after preclassifying one case. The posterior means are closer to the population means, and the variance is significantly reduced.

Nevertheless, it is clear that some observations are better than others to preclassify. The classified observation should have high posterior probability for one of the two classes. Preclassifying an observation with 0.50 posterior probability of being in either class does nothing to dampen either mode. Clearly the success of the technique is not exactly blind to the data.

19.2 Multimodal posterior distributions in continuous latent variable models

Having become interested in the problem of nuisance modes in the posterior distributions for mixture models, I also started to wonder if there were similar issues in models with latent continuous variables. One day Rubin dug into his file cabinet to give me a copy of Rubin and Thayer (1982), the paper in which he outlines an EM algorithm to estimate exploratory and confirmatory factor models. The empirical example was a four factor confirmatory model analyzing a set of nine academic tests (Joreskog, 1969). Following Joreskog's notation, the model is

$$ \mathbf{y} = \mathbf{\Lambda}\mathbf{x} + \mathbf{z}, $$

where for this example $\mathbf{y}$ is a $9 \times n$ matrix of observed scores; $\mathbf{x}$ is a $4 \times n$ matrix of factor scores; and $\mathbf{z}$ is $9 \times n$ matrix of disturbance terms. In this example, $\mathbf{\Lambda}$ is the 9×4 matrix of factor loadings connecting $\mathbf{y}$ and $\mathbf{x}$. The four factors are uncorrelated, centered at 0, and have unit variance. We also assume that $E(\mathbf{z}\mathbf{z}^t) = \mathbf{\Psi}$ is a diagonal matrix of uniquenesses. The factor loadings and uniquenesses combine to reproduce the variance–covariance matrix of $\mathbf{y}$ as follows:

$$ \mathbf{\Sigma} = \mathbf{\Lambda}\mathbf{\Lambda}^{\mathrm{T}} + \mathbf{\Psi}. $$

Assuming a multivariate normal distribution for y, the log-likelihood function is

$$ p(y|\mathbf{\Lambda}, \mathbf{\Psi}) = -\frac{1}{2}n \left[\log |\mathbf{\Sigma}| + tr(\mathbf{S}\mathbf{\Sigma}^{-1}) \right]. $$

As initially suggested in Dempster, Laird, and Rubin (1977), Rubin and Thayer argued that EM is an effective tool for understanding the likelihood in factor analysis. Applying it to the Joreskog data, they ran the model from three different sets of starting values and found three different modes. They argued that the estimation procedures implemented in SEM (structural equation modeling) packages such as LISREL may miss important features of the likelihood, and they pointed out that parameter inferences based on symmetric, normal approximations to a unimodal likelihood may not be accurate.

Bentler and Tanaka (1983) quickly responded that only one of the three modes identified by Rubin and Thayer (1982) met the technical criteria for a local maximum, and asserted that multimodality in factor models is actually quite rare. It is true that Rubin and Thayer may have been hasty in their initial paper (the preprint that Don gave me had a few inked-in corrections). However, they did raise some substantive issues with regard to ML estimation of factor models. Even if the other "modes" they identified were not strictly local maxima, they still represented high probability regions of the likelihood several standard deviations removed from the global maximum, an unexpected result if the standard quadratic approximations to the likelihood were appropriate.

Multimodality analogous to the problem of equivalent modes in mixture models also arises in factor models. It is well known that the factor loadings are only uniquely estimated up to $\boldsymbol{\Lambda}\boldsymbol{\Lambda}^{\mathrm{T}}$: most major structural equations packages require that at least $k(k-1)/2$ loadings are fixed to identify a k factor solution.

Consider the MLEs for the Rubin and Thayer example given in Table 19.4. Bentler and Tanaka (1983) fixed λ_{12} to identify the first two factors. Of course, this is only a local identification, as there are still four equivalent solutions corresponding to reversals of the polarity of factors 1 and 2.

A simple way to get a visual representation of the likelihood in this case is to rotate the first two factors *maintaining the fixed constraint*. If there were no fixed

Variable	f1	f2	f3	f4
1	0.71	0*	0.15	0*
2	0.74	0.04	0.22	0*
3	0.25	0.86	0.33	0*
4	0.24	0.80	0.08	0*
5	0.64	0.08	0*	0.37
6	0.72	0.07	0*	0.15
7	0.57	0.19	0*	0.35
8	0.39	0.73	0*	0.02
9	0.36	0.74	0*	-0.12

Table 19.4 Factor loadings (Λ) at MLE for Rubin and Thayer example. Asterisks indicate loadings that were fixed to zero.

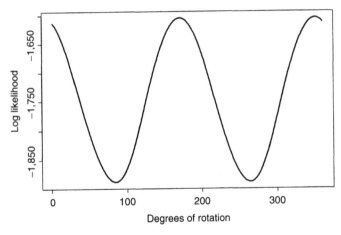

Figure 19.5   Tracing values of log-likelihood under orthogonal rotation of the loadings on factors 1 and 2 from Table 19.4. At each point in the rotation, the constraint $\lambda_{12} = 0$ is reapplied. The second mode at 180 degrees of rotation corresponds to reversing the polarity of the two factors.

loadings, the resulting plot of the log-likelihood versus the θ of rotation would be a flat line. Under the constraint, the plot traces a curve as in Figure 19.5 with a second peak at 180 degrees of rotation. (The curve in Figure 19.5 is *not* a profile likelihood. I have held the uniquenesses at the same value as in Table 19.4, and so the depth of the valleys here is exaggerated. The purpose is just to illustrate the shape and the second equivalent mode.)

All of the above is very well known, but a couple of important points are worth emphasizing. First, a researcher carrying out a Bayesian analysis of this factor model would have to be attentive to the multiple modal regions. From the ML perspective, the four possible modes due to reversals of the first two factors are immediately seen to have the same substantive interpretation. However, posterior simulation must not be allowed to sample from more than one of the symmetric modes, or the summaries will be biased to zero and the variances will be inflated. Most researchers who apply Bayesian methods for latent variable models do not distinguish between local and global identification of the models.

The constraints sufficient for identification in ML estimation are generally not sufficient for Bayesian estimation. But at least in the zero loading case, the equivalent modes are widely separated and under careful consideration reasonable constraints can be applied to restrict the simulation to one mode (see for example Geweke and Zhou, 1996). Something more interesting, however, occurs when a loading is fixed to a nonzero constant in order to identify the first two factors. Originally, Joreskog (1969) thought that as long as $k(k-1)/2$ loadings were fixed, a k factor model would be locally identified. However, Jennrich (1978) showed that when nonzero loadings were fixed, there were still equivalent modes in the likelihood, but they were not as transparent as in the zero loading case. Joreskog

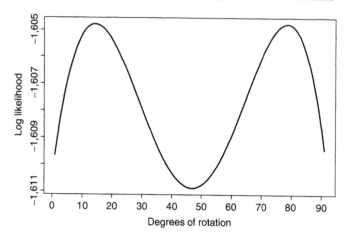

Figure 19.6   Tracing the log-likelihood with the constraint $\lambda_{12} = 0.6$. A symmetric mode is now only a few degrees of rotation removed from the mode of interest.

and Sorbom (1979) acknowledged this point but concluded that even models with nonzero constraints were usually locally identified.

The simple rotational technique above can be used to gain some understanding as to the shape of the likelihood when a nonzero loading is fixed. Figure 19.6 shows the likelihood plotted for an orthogonal rotation of a few degrees of the first two factors, now maintaining a constraint $\lambda_{12} = 0.6$. Note the two modes separated by only 60 degrees of rotation. In fact, if we fix λ_{12} at larger values approaching 0.71 (the MLE for λ_{11} when λ_{12} is fixed to zero), the modes converge.

The double mode shown in Figure 6 is a concern for both ML and Bayesian estimation and inference. From the ML perspective, it is clear that under certain identification constraints, a quadratic approximation to the mode is inappropriate, as a second high density region can lie adjacent to it. From a Bayesian perspective, such a double mode will incur mode switching, and it is doubtful whether any sensible constraints could be applied to restrict the algorithm to one mode.

Although not exactly the type of multimodality addressed by Rubin and Thayer (1982), the above examples reinforce their well-founded concerns about inference in factor models. The likelihood and posterior distributions for these models have some peculiar properties, and at the very least, researchers employing a Bayesian approach must recognize a multimodality problem in factor models analogous to the label-switching problem in mixture models.

19.3   Summary

In this chapter, we have used some simple graphical techniques to explore and understand symmetries in the likelihood and posterior distributions of latent variable models. Simple profile likelihood contour plots show the equivalent modes in a mixture model that correspond to label switching. The same plots also show the impact of the trick of "preclassifying" one observation to help identify the model and improve Bayesian inference.

Similarly, a simple representation of the likelihood under orthogonal rotations of a two-factor model gave a clear picture of the equivalent modes. In the case where the model is identified by fixing a zero loading, the rotation plot reveals an expected second mode at 180 degrees. However, in the case where a nonzero loading is fixed, the rotation plot shows that the symmetric mode may actually be quite close to the mode of interest, presenting problems for both ML and Bayesian inference.

20

Modeling the covariance and correlation matrix of repeated measures

W. John Boscardin and Xiao Zhang[1]

20.1   Introduction

Advances in computational methods have brought about a recent resurgence of interest in modeling the covariance and correlation matrix of repeated measures data. Examples include: Daniels and Kass (2001) who shrink toward known structures; Barnard, McCulloch, and Meng (2000) who develop one element at a time jumping rules for MCMC sampling of correlation matrices; Chib and Greenberg (1998) who place a truncated multivariate normal prior distribution on the vector of correlations; Shi, Weiss, and Taylor (1996) who develop a semiparametric approach to covariance matrix estimation for longitudinal data; and Daniels and Pourahmadi (2002), and Pourahmadi and Daniels (2002) who introduce covariate models for the parameters of the covariance matrix.

In this chapter, we describe an approach to setting up a model for either the covariance or correlation matrix of repeated measures and discuss MCMC computational methods for this approach. This methodology unifies concepts described in Boscardin and Weiss (2004) and Zhang, Boscardin, and Belin (2004). In the covariance matrix case, the model bridges two important approaches: (i) the inverse-Wishart prior

[1]Department of Biostatistics, University of California, Los Angeles. This work was partially supported by NIH grants MH60213, NS30308, and AI28697.

Applied Bayesian Modeling and Causal Inference from Incomplete-Data Perspectives.
Edited by A. Gelman and X-L. Meng   © 2004 John Wiley & Sons, Ltd   ISBN: 0-470-09043-X

distribution with given scale matrix (e.g., Schafer, 1997) and (ii) parametric models such as those estimated in the SAS Proc Mixed (Wolfinger, Stroup, Milliken, and Littell, 1990) and the R/S-plus package lme (Pinheiro and Bates, 2000).

More specifically, we consider the problem of performing inference about the covariance or correlation matrix of zero-mean multivariate Gaussian data,

$$y_i | \Sigma \sim N_p(0, \Sigma), \tag{20.1}$$

where $i = 1, \ldots, n$ indexes subjects, each of whom is measured p times (e. g., at p common time points, or on p multiple measures). Extensions to the nonzero-mean case or the regression setting (so that the mean vector is $X_i \beta$) and to unbalanced or missing data settings are straightforward and will be presented in Section 20.5. We will discuss the case in which all or part of Σ is a correlation matrix. This is particularly useful if the y_i vector (or part of it) contains latent variables as in the repeated measures multivariate probit model.

For concreteness, the setting we have in mind might involve repeated measures data on a moderate number of subjects (e. g., $n = 50$). Each subject may be potentially measured on each of 40 to 50 variables (e. g., quantities A, B, and C at each of 15 timepoints), but only 10 to 20 variables are recorded per subject on average. Thus, we would be interested in estimating a covariance matrix of dimension 40 to 50 using 50 data vectors that are typically two-thirds missing. Inference would be extremely vague if no assumptions were made about Σ, yet we would like to avoid a completely parametric model. We now indicate how to achieve a compromise between these extremes.

20.2   Modeling the covariance matrix

We propose a hierarchical prior distribution for Σ that is centered around a parametric family. Briefly, the idea is to first choose a reasonable parametric family $\Omega(\theta)$ that might capture the important features of Σ. For example, $\Omega(\cdot)$ might be an autoregressive matrix of order one, in which case θ is a two-dimensional vector of the lag-one correlation ρ and the variance parameter σ^2. Next, we assume Σ has a prior distribution on the space of covariance matrices, which we denote CovMat (e.g., inverse-Wishart or Wishart), centered around this parametric structure with hyperparameters v, for example, degrees of freedom for the Wishart distributions. Thus,

$$\Sigma | v, \theta \sim \text{CovMat}_v(\Omega(\theta)). \tag{20.2}$$

The model is completed with prior distributions on v and θ. In most settings, v will be a degrees of freedom parameter; for computational simplicity, we allow it to have support on the positive real numbers instead of the natural numbers, and thus a gamma distribution would be a natural choice for a prior distribution. In the examples of Sections 20.7 and 20.8, v is fixed and might be thought of as a tuning parameter for the amount of smoothing performed. Similar models have appeared in the pre-MCMC literature (Chen, 1979; Dickey, Lindley, and Press, 1985), but computational limitations only allowed for very specific cases to be used.

If Σ is an unrestricted covariance matrix, then we achieve conditional conjugacy by using the inverse-Wishart distribution:

$$\Sigma | v, \theta \sim \text{Inv-Wishart}_v ((v\Omega(\theta))^{-1}). \tag{20.3}$$

Here, the degrees of freedom parameter, v, regulates the fidelity to the parametric model. A large value of v corresponds to little possible deviation from the parametric model; conversely, a small value of v allows Σ to be quite different from $\Omega(\theta)$. Model (20.3) bridges several important special cases. First, parametric models for Σ correspond to the case $v \to \infty$. Second, the commonly used inverse-Wishart prior with known scale matrix (e.g., Schafer, 1997) corresponds to a point mass prior for θ. Finally, letting $v \to -1$ and $\Omega(\theta) \to 0$ corresponds to a commonly used noninformative prior on Σ.

As part of the inference procedure described next, we obtain an estimate $\Omega(\hat{\theta})$, where $\hat{\theta}$ is some estimate of θ, for example, the posterior mean. This might be thought of as the best parametric approximation to Σ. We also obtain direct inference about and an estimate of Σ itself, for example, the posterior mean of Σ. These quantities can be compared to see where departures from the parametric model are most noticeable.

Computation for the hierarchical covariance model

Let Y be an n by p matrix whose ith row is the data vector for the ith case, y_i^T. Our goal is to generate simulations from the distribution $p(\Sigma, \theta, v|Y)$. The posterior density is proportional to the likelihood in equation (20.1) times the prior density given in equation (20.2). In the inverse-Wishart case of equation (20.3), we have

$$p(\Sigma, \theta, v|Y) \propto \left(2^{vp/2} \Gamma_p \left(\frac{v}{2} \right) \right)^{-1} |\Sigma|^{-n/2} \exp\left(-\frac{1}{2} \text{tr}(\Sigma^{-1} Y^T Y) \right) \tag{20.4}$$

$$\times |v\Omega(\theta)|^{v/2} |\Sigma|^{-(v+p+1)/2} \exp\left(-\frac{1}{2} \text{tr}(\Sigma^{-1} v\Omega(\theta)) \right) p(\theta)p(v),$$

where $\Gamma_p(v/2) \equiv \prod_{j=1}^{p} \Gamma((v+1-j)/2)$. As this is not a tractable density function, we use a Metropolis-within-Gibbs approach to generate simulations that converge in distribution to simulations from the joint posterior distribution.

At the mth step of the algorithm, we have values $(\Sigma^{(m)}, \theta^{(m)}, v^{(m)})$:

1. $[\Sigma, \theta|v^{(m)}, Y]$. As described in Boscardin and Weiss (2004), we are able to draw directly from the joint conditional distribution of Σ and θ. We

accomplish this by integrating Σ out from equation (20.4), so that θ can be drawn conditional on only Y and ν. This helps to reduce autocorrelation in the sampler (Liu, 1994). The details of the two substeps are

(a) $[\Sigma|\theta^{(m)}, \nu^{(m)}, Y]$. Generate $\Sigma^{(m+1)}$ from Inv-Wishart$_{\nu+n}((\nu\Omega(\theta) + Y^T Y)^{-1})$.

(b) $[\theta|\nu^{(m)}, Y]$. If the sampler is currently at $\theta^{(m)}$, propose a candidate θ^* according to a jumping kernel $q(\cdot|\theta^{(m)})$ (e.g., multivariate normal with variance and correlation tuned in burn-in period). Set $\theta^{(m+1)} = \theta^*$ with probability min$(1, \alpha_\theta)$, where

$$\alpha_\theta = \frac{p(\theta^*)q(\theta^{(m)}|\theta^*)\,|\Omega(\theta^*)|^{\nu/2}\,|\nu\Omega(\theta^*) + Y^T Y|^{-(\nu+n)/2}}{p(\theta^{(m)})q(\theta^*|\theta^{(m)})\,|\Omega(\theta^{(m)})|^{\nu/2}\,|\nu\Omega(\theta^{(m)}) + Y^T Y|^{-(\nu+n)/2}},$$

(20.5)

otherwise set $\theta^{(m+1)} = \theta^{(m)}$.

2. $[\nu|\Sigma^{(m+1)}, \theta^{(m+1)}, Y]$. For the examples in Sections 20.7 and 20.8, we fix ν to a set value, so that it is a tuning parameter for the amount of smoothing. Should we wish the data to determine the amount of smoothing, we can easily include ν in the MCMC simulation. A convenient prior distribution is $\nu \sim$ Gamma(α_ν, β_ν). No conjugacy is obtained, so we would perform a Metropolis–Hastings step here.

20.3   Modeling the correlation matrix

In many settings, including the multivariate probit model described in Section 20.6, the vector y_i in equation (20.1) may actually represent a scale-free, latent vector, in which case we are only interested in estimating its correlation matrix R. We might consider attempting to mimic the approach of equation (20.2) and put a prior distribution, which we will denote CorMat, on R. Unfortunately, distributions that are supported only on the space of proper correlation matrices are at a real premium. One of the few available ones, in the sense that we can both simulate directly from it and compute its density function, is what we term the Wishart correlation (WC) distribution (Gupta and Nagar, 2000). The WC density function is obtained by integrating out the variance components from the density function for a Wishart with a nonsingular diagonal scale matrix. This leads to $p(R) = WC_\nu(R) = \Gamma(\nu/2)^p / \Gamma_p(\nu/2)|R|^{(\nu-p-1)/2}$. Nondiagonal scale matrices make this integration intractable. An inverse-Wishart correlation (IWC) density can be similarly obtained by integrating out the diagonal elements of an inverse-Wishart density function with diagonal scale matrix.

The WC and IWC distributions lack flexibility for two reasons: (i) they are centered around the identity matrix, which means the posterior density is shrunk toward a matrix with zero correlations, and (ii) there is no way to allow a generalization to the $\Omega(\theta)$-type family of centers. As an additional complication, we do not get conjugacy for the Σ step of an MCMC algorithm by using either the WC or IWC distribution.

To circumvent these issues, we use parameter expansion (Liu, Rubin, and Wu, 1998) to create models for the correlation matrix. The central idea is to augment the correlation matrix R, with a diagonal matrix of variances D to create a covariance matrix $\Sigma = D^{1/2}RD^{1/2}$. We then place a Wishart prior distribution on Σ with v degrees of freedom and scale matrix $\Omega(\theta)/v$, so that Σ has prior mean $\Omega(\theta)$. The change of parameterization from Σ to (R, D) gives a Jacobian term $J_{\Sigma \to R, D} = (\prod_{j=1}^{p} D_{jj})^{(p-1)/2}$. We term the prior density on (R, D) the PX-Wishart prior and denote $p(R, D) = PX_v(R, D|\Omega(\theta)/v) = J_{\Sigma \to R, D} \text{Wishart}_v(\Sigma|\Omega(\theta)/v)$. The conjugacy that is gained in the covariance matrix situation from using the inverse-Wishart distribution is unfortunately lost in general here because of this Jacobian term; one exception is described in Liu (2001), where a Jeffreys prior distribution for R leads to an inverse-Wishart full conditional distribution for Σ. We will thus use the Wishart distribution in what follows for simplicity.

Computation for the hierarchical correlation model

The joint posterior for R, D, θ, and v can be found as:

$$p(R, D, \theta, v|Y) \propto \left(2^{vp/2}\Gamma_p\left(\frac{v}{2}\right)\right)^{-1} |R|^{-n/2} \exp\left(-\frac{1}{2}\text{tr}(R^{-1}Y^T Y)\right)$$

$$\times \left(\prod_{j=1}^{p} D_{jj}\right)^{(p-1)/2} |\Omega(\theta)/v|^{-v/2}|D^{1/2}RD^{1/2}|^{(v-p-1)/2}$$

$$\times \exp\left(-\frac{1}{2}\text{tr}(D^{1/2}RD^{1/2}(\Omega(\theta)/v)^{-1})\right) p(\theta)p(v). \quad (20.6)$$

Notice that inference for D does not depend on the data.

This posterior density is even more intractable than in the covariance matrix case, and we therefore use an MCMC algorithm to perform inference. At the mth step of the algorithm, we have values $(R^{(m)}, D^{(m)}, \theta^{(m)}, v^{(m)})$:

1. $[R, D|\theta^{(m)}, v^{(m)}, Y]$. As mentioned at the end of the previous section, we now need a Metropolis–Hastings step for (R, D). We extend the same idea that we used for putting a model on R to simulating a candidate for this step: generate $\Sigma^* = (D^*)^{1/2}R^*(D^*)^{1/2}$ from $\text{Wishart}_{v_0}(\Sigma^{(m)}/v_0)$. Set $(R^{(m+1)}, D^{(m+1)}) = (R^*, D^*)$ with probability α_{RD}, where

$$\alpha_{RD} = \frac{q((R^{(m)}, D^{(m)})|(R^*, D^*))p(R^*, D^*, \theta^{(m)}, v^{(m)}|Y)}{q((R^*, D^*)|(R^{(m)}, D^{(m)}))p(R^{(m)}, D^{(m)}, \theta^{(m)}, v^{(m)}|Y)}, \quad (20.7)$$

$q((R_1, D_1)|(R_0, D_0)) = J_{\Sigma_1 \to R_1, D_1} \text{Wishart}_{v_0}(\Sigma_1|\Sigma_0/v_0)$ is the jumping kernel, and $p(R, D, \theta, v|Y)$, which is proportional to $p(R, D|\theta, v, Y)$, is the posterior density from equation (20.6). The covariance matrix $\Sigma^{(m+1)}$ determines $R^{(m+1)}$ and $D^{(m+1)}$, so we condition on $\Sigma^{(m+1)}$ in the next two steps for simplicity.

2. $[\theta \mid \Sigma^{(m+1)}, \nu^{(m)}, Y]$. It is no longer possible to analytically integrate out Σ. We propose a candidate θ^* according to a jumping kernel $q(\cdot \mid \theta^{(m)})$, and accept θ^* with the appropriate probability.

3. $[\nu \mid \Sigma^{(m+1)}, \theta^{(m+1)}, Y]$ can be performed similarly to the ν step for the covariance sampler.

20.4   Modeling a mixed covariance-correlation matrix

The model of Section 20.3 can be generalized to the case in which a portion of y_i has meaningful variance components, and the remainder is scale-free. We define v_i to be the portion of y_i with meaningful scale and z_i the scaleless portion. In defining a split of the p by 1 vector y_i into two subvectors, say v_i and z_i, we will use a slight abuse of notation and write $y_i = (u_i, z_i)$, even though the correct notation is $y_i = (u_i^T, z_i^T)^T$. Let Λ denote the combination covariance-correlation matrix of the $y_i = (v_i, z_i)$ vectors

$$\Lambda = \begin{pmatrix} \Sigma_{vv} & \Sigma_{vz} \\ \Sigma_{vz}^T & R_{zz} \end{pmatrix} = \begin{pmatrix} D_{vv}^{1/2} R_{vv} D_{vv}^{1/2} & D_{vv}^{1/2} R_{vz} \\ R_{vz}^T D_{vv}^{1/2} & R_{zz} \end{pmatrix}. \tag{20.8}$$

We expand the parameterization of Λ to include a diagonal scale matrix D_{zz} for z_i. The expanded matrix is called Σ and we can write $\Sigma = D^{1/2} R D^{1/2}$, where

$$D = \begin{pmatrix} D_{vv} & 0 \\ 0 & D_{zz} \end{pmatrix}, \text{ and } R = \begin{pmatrix} R_{vv} & R_{vz} \\ R_{vz}^T & R_{zz} \end{pmatrix}. \tag{20.9}$$

We define a prior density on (R, D) as the product of the Jacobian for transforming Σ to (R, D) times a Wishart density for Σ with ν degrees of freedom and scale matrix $\Omega(\theta)/\nu$. Computation proceeds as in Section 20.3.

20.5   Nonzero means and unbalanced data

The models we have presented can be easily extended to the case of nonzero means. These are often expressed as k parameter regression models, where the p-dimensional mean vector of y_i is modeled as $X_i \beta$, with X_i a p by k design matrix for the ith subject. Also, in many applied settings, each subject has measurements at only a subset of size p_i of p possible recording times. We regard the data vectors

y_i as incomplete observations of a p vector and handle this through a framework for ignorable missingness (Little and Rubin, 2002).

To extend the covariance matrix MCMC algorithm to handle a regression model for the mean and ignorably missing data, we proceed as follows. Partition y_i as (y_i^{obs}, y_i^{mis}). Given the missing data y_i^{mis} and the regression coefficients β, we subtract off $X_i \beta$ from the complete vector y_i. This gives zero-mean data so that the MCMC algorithm goes through exactly as above for Σ, θ, and v. The β and y_i^{mis} steps are straightforward Bayesian regression computations as described in Schafer (1997), for example.

20.6 Multivariate probit model

Sections 20.3 and 20.4 discuss the case where all or part of the data vector is scale-free. The usual setting for this is that the scale-free continuous vector is a latent vector for a multivariate ordinal probit model (Chib and Greenberg, 1998; Liu, 2001). We assume that the data vector y_i consists of a continuous portion v_i of length p_1 and an ordinal portion c_i of length p_2 (with $p_1 + p_2 = p$). The element c_{ij} takes values on the discrete set $0, 1, \ldots, J_j - 1$, and $c_{ij} = l$ if and only if the latent variable z_{ij} is in the range $(\gamma_{j,l-1}, \gamma_{j,l}]$. We set $\gamma_{j,0} = -\infty$, $\gamma_{j,J_j-1} = +\infty$ for notational simplicity, and $\gamma_{j,1} = 0$ for identifiability of the cutpoints. The vector c_i will then follow a multivariate probit model if we assume $z_i \sim N_{p_2}(X_i^{(2)}\beta, R)$, where $X_i^{(2)}$ is the p_2 by k design matrix for the linear model and R is a p_2 by p_2 correlation matrix.

To model v_i and c_i simultaneously, we assume that given the latent vector z_i, the vector $y_i = (v_i, z_i) \sim N_p(X_i\beta, \Lambda)$, where X_i is vertically partitioned as $X_i^{(1)}$ (the p_1 by k design matrix for v_i) and $X_i^{(2)}$ (as defined in the previous paragraph), and Λ is the covariance-correlation matrix of the vector (v_i, z_i) as defined in equation (20.8). The model is completed by placing a prior distribution on the ragged array $\{\gamma_{j,l}, j = 1, \ldots, p_2, l = 2, \ldots, J_j - 1\}$; a common noninformative choice is $p(\gamma_{j,l}) \propto 1$.

Ignorably missing data for portions of the v_i or z_i vectors are handled exactly as in Section 20.5; sample missing components given the observed components, β, and Λ from the appropriate conditional multivariate normal distribution. Given complete data, y_i, and the regression parameters, β, subtract off $X_i\beta$ from y_i to reduce the computation for Λ to the case described in Section 20.4.

There are two additional steps for the MCMC procedure, which we write assuming zero-mean complete data for simplicity:

- $[z_{ij}|z_{i,-j}, v_i, \Lambda, c_{ij} = l, \gamma_{j,l-1}, \gamma_{j,l}]$. z_{ij} given the other elements of z_i and v_i is sampled from an interval truncated univariate normal with conditional mean and variance derived from the standard results for conditioning on a part of a multivariate normal vector (since the full vector (v_i, z_i) is multivariate normal with mean zero and covariance-correlation matrix Λ). The interval for truncation is $(\gamma_{j,l-1}, \gamma_{j,l}]$ so that z_{ij} will be compatible with c_{ij}.

- $[\gamma_{j,l}|\gamma_{j,l-1}, \gamma_{j,l+1}, z_{1,j}, \ldots, z_{n,j}, c_{1,j}, \ldots, c_{n,j}]$. Draw $\gamma_{j,l} \sim U(a, b)$ where $a = \max\{\max_i\{z_{ij} : c_{ij} = l\}, \gamma_{j,l-1}\}$ and $b = \min\{\min_i\{z_{ij} : c_{ij} = l+1\}, \gamma_{j,l+1}\}$. In other words, we sample a cutpoint from the widest range possible given the constraints that it must be compatible with the data and must lie in between the neighboring cutpoints.

20.7   Example: covariance modeling

We now apply the covariance matrix version of our repeated measures regression model with missingness to data from the UCLA Brain Injury Research Center (BIRC) (Glenn et al., 2003). The UCLA BIRC studies moderate to severe head trauma patients, focusing on short-term postinjury brain metabolism and long-term neuropsychological outcome. Metabolic data are collected longitudinally on a variety of markers including arterial-venous (AV) differences of oxygen, glucose, and lactate, and cerebral blood flow rates (CBF). AV differences are measured as differences in concentrations in a particular substance (measured in milligrams per milliliter) between arterial uptake and venous release in the brain. CBF is measured in units of milliliters per 100 grams per minute. These metabolic measurements are potentially taken twice per day during the time that patients are in the study. Most patients are studied for at least seven or eight days postinjury. Informative dropout is an issue that will not be discussed here; instead, we focus on the first seven days postinjury for which dropout is not as much of a problem. During this timeframe, many of the twice-daily measurements are missing for reasons that might be considered ignorable, for example, qualified personnel are not always available to perform the measurements twice per day. In practice, most patients end up with approximately one measurement per day on a particular parameter. We examine here the relationship of CBF and AV difference in oxygen (AVDO$_2$), and whether considering AV difference in glucose (AVDglc) contributes toward inference on this relationship.

To investigate this, we use data through postinjury hour 180 on CBF, AVDO$_2$, and AVDglc. CBF was transformed by square root and AVDglc by natural logarithm to make the normality assumptions more plausible. We abbreviate the transformed quantities as C (square root of CBF), O (AVDO$_2$), and G (logarithm of AVDglc). A total of $n = 56$ patients had at least one measurement for one of these three parameters in this time frame. We gridded the postinjury hours into 15 twelve-hour epochs, so that each patient could potentially contribute a vector of length $p = 45$, that is, fifteen timepoints times three measures. In reality, patients had anywhere from 1 to 27 total measurements on the three parameters, recorded at between 1 and 9 of the 15 possible timepoints. The total number of measures was 292 for C, 214 for O, and 215 for G.

We analyze the data using the model in Section 20.2. The complete data for the ith patient are ordered as $y_i = (C_{i,1}, O_{i,1}, G_{i,1}, \ldots, C_{i,15}, O_{i,15}, G_{i,15})^T$. The X_i matrix is constant across subjects, and indicates a common mean for each of C, O, and G. Thus, X_i has 45 rows and 3 columns, and consists of 15 three by three identity matrices stacked atop one another. The 45 by 45 covariance matrix Σ was assumed to be centered around a multivariate compound symmetry matrix generalized to allow AR-1 correlation attenuation between time points (Galecki, 1994). To give a bit more detail, $\Omega(\theta)$ is a 45 by 45 matrix partitioned as a 15 by 15 array of 3 by 3 submatrices. The (i, j)th submatrix of $\Omega(\theta)$ is $M + \rho^{|i-j|}V$, where the 3 by 3 matrix M represents the between subject variability of the mean values for blood flow, oxygen, and glucose, the 3 by 3 matrix V represents the within subject variability of these measures, and ρ measures how the within subject variability attenuates across timepoints. So $\theta \equiv (M, V, \rho)$ has 13 parameters instead of the 1,035 in a general 45 by 45 covariance matrix. In the special case of $\rho = 0$, we get the multivariate compound symmetry model with $M + V$ on the diagonal blocks of $\Omega(\theta)$ and M on the off-diagonal blocks. For this computation, we fixed the value of v to be 40. The six variance components in θ were given independent inverse-gamma prior distributions with $\alpha = 4$ and $\beta = 2$, the six intrasubject and intersubject correlation components were given independent beta prior distributions rescaled to run from -1 to 1 with $\alpha = 3$ and $\beta = 3$, and finally, the AR correlation was given a similar beta prior but with $\alpha = 5$ and $\beta = 3$. We generated 50,000 simulations from the MCMC algorithm with results saved at every 50th iteration to produce a manageable sample size from the posterior. Various diagnostics suggested that the sampler had converged.

Figure 20.1(a) shows the prior and posterior marginal distributions of the three elements in θ that relate to the correlation of C and O: the leftmost panel shows the correlation in M, that is, the correlation between subject means; the middle panel gives the (intrasubject) correlation in the V matrix between C and O; and the rightmost panel displays the AR-1 parameter describing how the intrasubject correlation decays with time difference. Notice that there is quite a bit of information in the data on all three of these parameters, and it is clear that the parametric underpinning to Σ has strong evidence for negative correlations of C and O, both intersubject and intrasubject. It is also evident that the AR-1 attenuation is strongly supported by the data.

To examine the individual elements of Σ, Figure 20.2 gives the posterior median and a 90% posterior interval for all 225 correlations between a C measurement and an O measurement, $Corr(C_{t_1}, O_{t_2})$ for t_1 and t_2 running from 1 to 15. These correlations are grouped by the t_1 index, so that the 15 bars above the portion of the figure corresponding to row number 1, give correlations of the first C measurement with O measurements 1 through 15. Many of the individual elements of Σ have high posterior probability of being below zero.

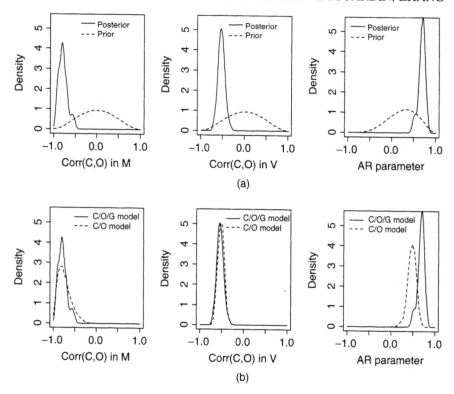

Figure 20.1 (a) Correlations between square root of blood flow (C) and arterial-venous oxygen differences (O) for the parametric center of Σ. Left panel gives correlation of subject level means. Middle panel gives intrasubject correlation. Right panel gives AR-1 parameter describing the attenuation of intrasubject correlation with time gap. (b) Comparison of estimation precision for the same three parameters between the three and two repeated measure model.

To investigate the contribution of the glucose values to information about correlations of blood flow and oxygen, we fit a reduced model that used only the C and O data, so that Σ is now 30 by 30. Figure 20.1(b) compares posterior inference for θ: the left panel compares inference about the between subject means correlation between C and O (from the M matrix), the middle panel the intrasubject correlation of C and O (from the V matrix), and the right panel the AR-1 parameter. Within each panel, the solid density estimate is for the three-measure model (C, O, and G), and the dashed estimate for the two-measure model (C and O only). The figure shows that inference about the between subject means correlation and the AR-1 parameter is more precise if we use the glucose data, but that there is not very much information about the intrasubject correlation of C and O in the glucose data.

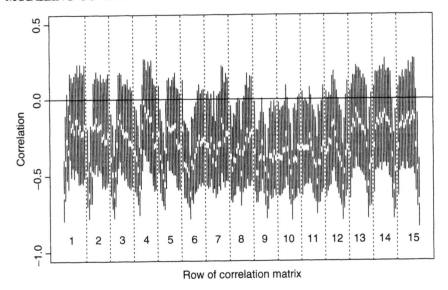

Figure 20.2 Posterior medians and 90% posterior intervals for individual corre-lation elements between C and O. Grouping is by timepoint for C. Within a group, the 15 bars run from timepoint 1 through 15 for O.

20.8   Example: mixed data

We now give an example of how to apply the methodology of Section 20.4 to a mixture of continuous and ordinal repeated measures data, again from the UCLA BIRC, on $n = 51$ subjects. For each subject, we use two ordinal measures collected near the time of injury: (i) *Pupils* is 0, 1, or 2 indicating the number of abnormal pupils; and (ii) *Improve* is 0, 1, or 2 indicating whether the subject's Glasgow Coma Scale score improved, stayed the same, or worsened in the early hours postinjury. We also include four continuous variables collected during long-term follow-up visits: (i) *NRS6* and *NRS12* give the total score at 6 and 12 months postinjury on the Neurobehavioral Rating Scale with higher scores indicating poorer condition; and (ii) *Peg6* and *Peg12* are related to the time required at 6 and 12 months postinjury to complete a test involving putting pegs in a pegboard. The two *Peg* variables were created by taking the average of the logarithm of left-hand time and logarithm of right-hand time. This transformation was chosen to give approximate normality. We chose a compound symmetry structure for $\Omega(\theta)$ with θ fixed at an intraclass correlation value of 0.5. The degrees of freedom parameter was set to $\nu = 8$. We ran the MCMC sampler for 100,000 iterations. The Metropolis–Hastings step used a jumping degrees of freedom of $\nu_0 = 200$ and achieved an acceptance rate of approximately 20%. All diagnostics indicated excellent convergence of the algorithm.

	NRS6	NRS12	Peg6	Peg12	Pupils	Improve
NRS6	1	.77 (.06)	.60 (.09)	.45 (.12)	.32 (.14)	.14 (.14)
NRS12	.99	1	.42 (.11)	.38 (.12)	.19 (.15)	.18 (.14)
Peg6	.99	.99	1	.81 (.06)	.27 (.15)	.27 (.13)
Peg12	.99	.99	.99	1	.27 (.15)	.15 (.14)
Pupils	.99	.90	.96	.95	1	−.03 (.17)
Improve	.84	.91	.98	.86	.43	1

Table 20.1 Above diagonal: posterior means and standard deviations of the correlation matrix for the mixed data. Below diagonal: posterior probability of positive correlation.

	NRS12	Peg12
Pupils	.19 (.15)	.27 (.16)
Improve	.19 (.14)	.17 (.15)

Table 20.2 Posterior means and standard deviations of partial correlations between long-term outcomes and each of the two early diagnostics while controlling for the other early diagnostic.

The upper triangular portion of Table 20.1 gives the posterior means and standard deviations of the correlation matrix R, and the lower triangular portion gives the posterior probability of a positive correlation. Here we see that the strongest categorical-continuous correlations are between *Pupils* and *NRS6* and both *Peg* times and between *Improve* and *Peg6*; all four of these have at least a 95% posterior probability of being positive.

To investigate the correlation structure more carefully, we examined the posterior distribution of the partial correlations in R. The partial correlation of quantities 1 and 2 controlling for quantity 3 is calculated for each posterior simulation as $r_{12.3} = (r_{12} - r_{13}r_{23})/\sqrt{(1 - r_{13}^2)(1 - r_{23}^2)}$. Table 20.2 presents the posterior means and standard deviations of the partial correlations for the two long-term outcomes, *NRS12* and *Peg12*, with each of the two early diagnostic measures *Pupils* and *Improve* controlling for the other early diagnostic. The strongest partial correlation is between *Pupils* and *Peg12* controlling for *Improve*, suggesting that patients with normal pupils are faster at peg placement one year postinjury, even after controlling for initial GCS status.

21

Robit regression: a simple robust alternative to logistic and probit regression

Chuanhai Liu[1]

21.1   Introduction

The logistic and probit regression models are commonly used in practice to analyze binary response data, but many authors (see, Pregibon (1982) and the references therein) have shown that their maximum likelihood estimators are not robust. This chapter considers *robit regression*, which replaces the normal distribution in probit regression with a *t*-distribution with known or unknown degrees of freedom. The use of the *t*-distribution for robust estimation in the different contexts where the response variables are typically modeled with the normal distribution has been addressed by many authors (e.g., Rubin, 1983; Lange, Little, and Taylor 1989; Liu and Rubin, 1995). As an alternative to logistic regression, the corresponding *t*-distribution has been previously suggested in the literature by Mudholkar and George (1978), and Albert and Chib (1993). Mudholkar and George (1978) discovered that a *t*-distribution with 9 degrees of freedom has the same kurtosis as the logistic regression. Albert and Chib (1993) suggested the use of a *t*-distribution with 8 degrees of freedom and provided the detailed implementation of the Gibbs sampler for Bayesian estimation.

[1] Statistics and Data Mining Research, Bell Laboratories, Murray Hill, N.J. The author thanks Dr. Diane Lambert for her numerous insightful and constructive comments.

Applied Bayesian Modeling and Causal Inference from Incomplete-Data Perspectives.
Edited by A. Gelman and X-L. Meng   © 2004 John Wiley & Sons, Ltd   ISBN: 0-470-09043-X

It is shown that (i) the maximum likelihood estimators are robust if the number of degrees of freedom is known; (ii) the robit regression model with about seven degrees of freedom provides an excellent approximation to the logistic regression model; and (iii) the robit regression model with a large number of degrees of freedom approximates the probit regression model. Thus, in a certain sense, the robit regression model provides a rich class of models, including logistic and probit regression models as special cases, for analysis of binary response data.

This chapter also provides efficient EM-type algorithms (Dempster, Laird, and Rubin, 1977; Liu, Rubin, and Wu, 1998) for finding the maximum likelihood estimates of the regression coefficients in the robit model. These algorithms provide information that can be used to identify outliers that have too much influence on the maximum likelihood estimates of the regression coefficient under the logistic and probit models.

The rest of the chapter is arranged as follows. Section 21.2 describes the robit model and its relationship with the probit and logistic models. Section 21.3 shows that the robust maximum likelihood estimators of the regression coefficients are robust. Section 21.4 formulates a complete-data model for robit regression that can be used for maximum likelihood estimation using EM-type algorithms and for identifying outliers under logistic and probit models. Section 21.5 provides detailed implementation of the EM, ECME, and PX-EM algorithm for maximum likelihood estimation of the robit model. Section 21.6 illustrates the methodology with an example. Finally, Section 21.7 concludes with a few remarks.

21.2   The robit model

The logistic and probit models

Suppose that the observed data consist of n independent observations $\{(x_i, y_i) : i = 1, \ldots, n\}$ with a p-dimensional covariate vector x_i and binary response y_i that is either 0 or 1. The logistic regression model is specified by

$$\text{logit } \Pr(y_i = 1|x_i, \beta) = \log\frac{\Pr(y_i = 1|x_i, \beta)}{1 - \Pr(y_i = 1|x_i, \beta)} = x_i'\beta \qquad (i = 1, \ldots, n).$$

$$(21.1)$$

The logistic regression model can also be derived by assuming that there are latent variables $z_i = x_i'\beta + e_i$, where e_i follows the *logistic* distribution function,

$$F_{\text{logistic}}(x) = \frac{\exp\{x\}}{1 + \exp\{x\}} \qquad (21.2)$$

and y_i is 1 if $z_i > 0$ and 0 otherwise. Then, the logistic regression model (21.1) is obtained as the marginal distribution of y_i. The maximum likelihood estimates of β can be obtained using the iterative reweighted least squares.

The probit model (e.g., Albert and Chib, 1993), for which

$$\Pr(y_i = 1|x_i, \beta) = 1 - \Pr(y_i = 0|x_i, \beta) = \Phi(x_i'\beta) \qquad (i = 1, \ldots, n),$$

is obtained by replacing the logistic distribution for the latent error terms e_i with the standard normal distribution, where $\phi(x)$ and $\Phi(x)$ are the density and distribution functions of the standard normal distribution respectively. The maximum likelihood estimates of β in the probit model can be obtained using the EM algorithm (Dempster, Laird, and Rubin, 1977) or the PX-EM algorithm (Liu, Rubin, and Wu, 1998).

The robit model: a simple extension of the probit model

To have a robust model, following Lange, Little, and Taylor (1989), who replaced the normal distribution in the linear regression model with a t-distribution to obtain robust estimators of linear regression coefficients, replace the normal distribution in the probit regression model with the t-distribution with ν degrees of freedom. For computational simplicity, which itself is important in the current state of the art in statistics as discussed by Liu (2000), Albert and Chib (1993) suggested the use of a t-distribution with 8 degrees of freedom and provided a detailed implementation of the Gibbs sampler for Bayesian estimation.

We call this model *robit* regression, and denote by robit (ν) the robit regression model with ν degrees of freedom. More formally, the robit regression model for the data $\{(x_i, y_i) : i = 1, \ldots, n\}$ is

$$\Pr(y_i = 1|x_i, \beta) = 1 - \Pr(y_i = 0|x_i, \beta) = F_\nu(x_i'\beta) \qquad (i = 1, \ldots, n),$$

where $F_\nu(x)$ denotes the cdf of the **t** random variable with center zero, scale parameter one, and ν degrees of freedom. $F_\nu(x)$ has the density function

$$f_\nu(x) \equiv \frac{\Gamma((\nu + 1)/2)}{(\pi\nu)^{1/2}\Gamma(\nu/2)(1 + x^2/\nu)^{(\nu+1)/2}} \qquad (x \in (-\infty, \infty)).$$

As $\nu \to \infty$, the robit(ν) model becomes the probit regression model.

The robit regression model with 7 degrees of freedom: an approximation to the logistic model

Empirically, the robit link with about 7 degrees of freedom approximates the logistic link, as Figure 21.1 suggests.[2] The quantiles below the 0.01 and 0.99 quantiles swing away from the reference line (dotted diagonal line), suggesting that the tail probabilities of the robit regression model are heavier than those of the logistic distribution. It is this tail property that distinguishes the robit and logistic links in terms of robust estimation. To balance robustness and approximation to the logistic model, one may like to use the t-distribution with even smaller number of degrees of freedom, such as 6 or 5.

---

[2]The scale parameter $\sigma = 1.5484$ in Figure 21.1 was chosen by numerically minimizing $\max_{x_i}\{|F_\nu(x_i/\sigma) - F_{\text{logistic}}(x_i)| : x_i = -10 + 0.002i, i = 1, \ldots, 1000\}$ over σ. For $\sigma = 1.5484$, the maximum distance is about 0.0006.

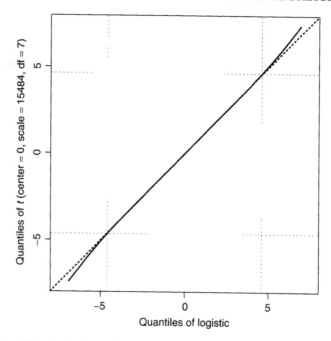

Figure 21.1   The Q–Q plot of the robit (7) model and the logistic model in the range corresponding to the probability range from 0.001 to 0.999. The horizontal and vertical dotted lines represent the 0.01 and 0.99 quantiles. The diagonal dotted line is the reference line indicating how well the two distributions match with each other.

21.3   Robustness of likelihood-based inference using logistic, probit, and robit regression models

Consider the effects of a potential observation (x, y) on the estimates of $p(y_i|x_i, \beta)$ for all i, or on the estimate of the regression coefficient vector β and consider the effective sample size s $(s > 0)$ of the potential observation. Without loss of generality, take $y = 1$. Let s $(s > 0)$ be the effective sample size. Denote by $\hat{\beta}_{+(x,y),s}$ the ML estimate of β with (y, x) included, that is,

$$\hat{\beta}_{+(x,y),s} = \arg\max_{\beta}\{\ell_{+(x,y),s}(\beta) \equiv \ell(\beta|Y_{\text{obs}}) + s\log(p(y|x, \beta))\},$$

where $\ell(\beta|Y_{\text{obs}})$ denotes the log-likelihood given the observed data. If the ML estimates $\hat{\beta}$ and $\hat{\beta}_{+(x,y),s}$ are unique and finite, the *potential* influence of (x, y) is defined as

$$I(x, y) \equiv \lim_{s\to+0} \frac{\hat{\beta}_{+(x,y),s} - \hat{\beta}}{s}. \tag{21.3}$$

If the Hessian matrix $H(\hat{\beta}) = \partial^2 \ell(\hat{\beta})/(\partial\beta\partial\beta')$ is negative definite, then

$$I(x, y) = -H^{-1}(\hat{\beta})\frac{\partial \log p(y|x, \hat{\beta})}{\partial \beta}.$$

Given the observed data, $H(\hat{\beta})$ is fixed and can be viewed as a scaling matrix for the factor $\partial \log p(y|x, \hat{\beta})/\partial\beta$. Given the observed data, $\hat{\beta}$ is also constant. To avoid the trivial cases, assume that all the components of $\hat{\beta}$ are nonzero so

$$\hat{\beta}'\frac{\partial \log p(y|x, \hat{\beta})}{\partial \beta}$$

is a convenient scalar factor. For the logistic regression model,

$$\hat{\beta}'\frac{\partial \log p(y|x, \hat{\beta})}{\partial \beta} = \frac{x'\hat{\beta}}{1 + \exp(x'\hat{\beta})},$$

implying that the influence can be unbounded. For the probit regression model,

$$\frac{\partial \log p(y|x, \hat{\beta})}{\partial \beta} = \frac{\phi(x'\hat{\beta})}{\Phi(x'\hat{\beta})}x'\hat{\beta}.$$

When $x'\hat{\beta} \to -\infty$, this factor is approximately $-(x'\hat{\beta})^2$. This quadratic function in x indicates that the influence of (y, x) is unbounded and is more extreme than the influence under the logistic regression model.

For the robit regression model,

$$\frac{\partial \log p(y|x, \hat{\beta})}{\partial \beta} = \frac{f_\nu(x'\hat{\beta})}{F_\nu(x'\hat{\beta})}x'\hat{\beta}.$$

This factor is bounded, and thereby the $I(x, y)$ is bounded because

$$\lim_{x'\hat{\beta} \to -\infty} \frac{f_\nu(x'\hat{\beta})}{F_\nu(x'\hat{\beta})}x'\hat{\beta} = \lim_{u \to -\infty}\frac{f_\nu(u)}{F_\nu(u)}u = -\lim_{u \to -\infty}\frac{(\nu + 1)u}{\nu + u^2}u = \nu + 1$$

and

$$\lim_{x'\hat{\beta} \to \infty} \frac{f_\nu(x'\hat{\beta})}{F_\nu(x'\hat{\beta})}x'\hat{\beta} = \lim_{\mu \to \infty}\frac{f_\nu(\mu)}{F_\nu(\mu)}\mu = 0.$$

21.4 Complete data for simple maximum likelihood estimation

Let y_i denote the univariate binary response of the i-th individual, and let x_i denote the p-dimensional vector of covariates for $i = 1, \ldots, n$. Let

$$\tau_i|\theta \sim \text{Gamma}(\nu/2, 2/\nu) \quad \text{and} \quad z_i|(\tau_i, \theta) \sim N(x_i'\beta, 1/\tau_i) \quad (i = 1, \ldots, n),$$

where $\theta = (\beta, v)$ with β being the p-dimensional vector of regression coefficients and v being the number of degrees of freedom. In the literature, τ_i is called weight, for example, in the context of iterative reweighted least squares. Then the robit regression model is completed by specifying

$$y_i = \begin{cases} 1, & \text{if } z_i > 0; \\ 0, & \text{if } z_i \leq 0. \end{cases} \tag{21.4}$$

This complete-data model belongs to the exponential family. The sufficient statistics for θ are

$$S_\tau = \sum_{i=1}^n \tau_i, \quad S_{\tau xx} = \sum_{i=1}^n \tau_i x_i x_i', \quad S_{\tau zz} = \sum_{i=1}^n \tau_i z_i^2,$$

$$S_{\tau xz} = \sum_{i=1}^n \tau_i x_i z_i, \quad \text{and } S_{\log \tau - \tau} = \sum_{i=1}^n (\log \tau_i - \tau_i); \tag{21.5}$$

and the complete-data maximum likelihood estimate of $\theta = (\beta, v)$ is given by $\hat{\beta} = S_{\tau xx}^{-1} S_{\tau xz}$ and

$$\hat{v} = \arg\max_v \left[-n \log \Gamma(v/2) + n(v/2)\log(v/2) + (v/2)S_{\log \tau - \tau} \right].$$

Let $\mu_i = x_i'\beta$, denote by t_v the t-deviate with location 0, scale parameter 1, degrees of freedom v, and denote by $f_v(.)$ the probability density of t_v, that is, $f_v(z) = c_v(1 + z^2/v)^{-(v+1)/2}$ with the normalizing constant $c_v = (\pi v)^{-1/2}\Gamma((v + 1)/2)\Gamma^{-1}(v/2)$. Then

$$\hat{\tau}_i \equiv \mathrm{E}(\tau_i | Y_{\text{obs}}, \theta) = \frac{y_i - (2y_i - 1)\Pr(t_{v+2} < -(1 + 2/v)^{1/2}\mu_i)}{y_i - (2y_i - 1)\Pr(t_v < -\mu_i)}, \tag{21.6}$$

$$\mathrm{E}(\tau_i(z_i - \mu_i) | Y_{\text{obs}}, \theta) = \hat{\tau}_i \frac{(2y_i - 1) f_{t_v}(\mu_i)}{y_i - (2y_i - 1)\Pr(t_{v+2} < -(1 + 2/v)^{1/2}\mu_i)},$$

$$\mathrm{E}(\tau_i(z_i - \mu_i)^2 | Y_{\text{obs}}, \theta) = v + 1 - v\hat{\tau}_i,$$

where $I(.)$ is the indicator function. With

$$\hat{z}_i \equiv \mu_i + \frac{(2y_i - 1) f_{t_v}(\mu_i)}{y_i - (2y_i - 1)\Pr(t_{v+2} < -(1 + 2/v)^{1/2}\mu_i)}, \tag{21.7}$$

it follows then

$$\mathrm{E}(\tau_i z_i | Y_{\text{obs}}, \theta) = \mathrm{E}(\tau_i(z_i - \mu_i) | Y_{\text{obs}}, \theta) + \mu_i \mathrm{E}(\tau_i | Y_{\text{obs}}, \theta) = \hat{\tau}_i \hat{z}_i,$$

and

$$\mathrm{E}(\tau_i z_i^2 | Y_{\text{obs}}, \theta) = v + 1 - v\hat{\tau}_i + \hat{\tau}_i \left[\mu_i^2 + 2\mu_i(\hat{z}_i - \mu_i) \right]. \tag{21.8}$$

When the conditional expectation of the sufficient statistics is calculated at the ML estimate of θ,

$$\hat{\beta} = \left(\sum_{i=1}^{n} \hat{\tau}_i x_i x_i'\right)^{-1} \left(\sum_{i=1}^{n} \hat{\tau}_i x_i \hat{z}_i\right),$$

which is the ML estimate of β in the linear regression $\hat{z}_i \sim N(x_i'\beta, 1/\hat{\tau}_i)$.

Letting $\nu \to \infty$ gives the complete-data probit regression model and the conditional expectations of the associated sufficient statistics:

$$\lim_{\nu \to \infty} \hat{\tau}_i = 1, \quad \lim_{\nu \to \infty} \hat{z}_i = \mu_i + \frac{(2y_i - 1)\phi(\mu_i)}{y_i - (2y_i - 1)\Phi(-\mu_i)},$$

and $\lim_{\nu \to \infty} E(z_i^2|Y_{\text{obs}}, \theta) = 1 + \mu_i \hat{z}_i$. The last equality is obtained using the fact that $\nu + 1 - \nu\hat{\tau}_i \to 1 - \mu_i z_i + \mu_i^2$ as $\nu \to \infty$.

21.5 Maximum likelihood estimation using EM-type algorithms

MLE of the regression coefficients β with known number of degrees of freedom ν using EM

With the complete data $\{(x_i, y_i, z_i, \tau_i) : i = 1, \ldots, n\}$ described in Section 21.4, the EM algorithm for finding the MLE of β with known ν is as follows. At iteration $t + 1$ with input $\beta^{(t)}$,

E-step of EM. Compute $\hat{\tau}_i$ and $\hat{z}_i$ for all $i = 1, \ldots, n$ in (21.6) and (21.7) with $\theta = (\beta^{(t)}, \nu)$, and then the expected sufficient statistics $\hat{S}_{\tau xx} = \sum_{i=1}^{n} \hat{\tau}_i x_i x_i'$ and $\hat{S}_{\tau xz} = \sum_{i=1}^{n} \hat{\tau}_i x_i \hat{z}_i$.

M-step of EM. Update β: $\beta^{(t+1)} = \hat{S}_{\tau xx}^{-1} \hat{S}_{\tau xz}$.

MLE of $\theta = (\beta, \nu)$ with unknown number of degrees of freedom ν using ECME

To use the EM algorithm to find the MLE of $\theta = (\beta, \nu)$ when the number of degrees of freedom ν is unknown, compute

$$E((\log\tau_i - \tau_i)|Y_{\text{obs}}, \theta) = \psi((\nu + 1)/2) - \log((\nu + 1)/2)$$
$$+ E\left(\log\frac{\nu + 1}{\nu + (z_i - \mu_i)^2}\bigg| Y_{\text{obs}}, \theta\right) - \hat{\tau}_i \quad (21.9)$$

for all $i = 1, \ldots, n$, where $\psi(\alpha) \equiv d\log(\Gamma(\alpha))/d\alpha = \Gamma'(\alpha)/\Gamma(\alpha)$ is the digamma function. Because there are no (obvious) numerical methods for computing the conditional expectation term in (21.9) and ECME typically converges dramatically faster than EM, we use ECME with two constrained maximization (CM) steps: one CM step maximizes the expected complete-data log-likelihood over β with ν fixed at its current estimate; and the other CM step maximizes the constrained actual likelihood over ν with β fixed at its current estimate, where the constrained log-likelihood function of ν given β is

$$\ell(\nu|\beta, Y_{\text{obs}}) = \sum_{i=1}^{n} \log\left(y_i(1 - \Pr(t_\nu < -\mu_i)) + (1 - y_i)\Pr(t_\nu < -\mu_i)\right). \quad (21.10)$$

The ECME algorithm for finding the MLE of $\theta = (\beta, \nu)$ is as follows. At iteration $t + 1$ with input $\theta^{(t)} = (\beta^{(t)}, \nu^{(t)})$,

E-step of ECME. The same as the E-step of EM: condition on the current parameter estimates, $\theta^{(t)} = (\beta^{(t)}, \nu^{(t)})$.

CM-step 1 of ECME. The same as the M-step of EM.

CM-step 2 of ECME. Search for the $\nu^{(t+1)}$ that maximizes $\ell(\nu|\beta^{(t+1)}, Y_{\text{obs}})$.

Then update ν using, for example, the half-interval method (Carnahan, Luther, and Wilks, 1969) to maximize $\ell(\nu|\beta, Y_{\text{obs}})$ in the likelihood function (21.10).

MLE of the robit model using PX-EM: a more efficient algorithm for computing $(\hat{\beta}, \hat{\nu})$

Liu, Rubin, and Wu (1998) show that the PX-EM algorithm, which makes use of the extra information captured in the imputed complete data, converges much faster than the EM algorithm for finding the MLE of the t-distribution and the probit regression model. Here PX-EM is used to find the MLE of the robit model, which involves both the t-distribution and the probit model. To make use of the extra information captured in the complete data, following Liu, Rubin, and Wu (1998), the complete-data model is extended as

$$(\tau_i/\alpha)|\theta^* \sim \text{Gamma}(\nu^*/2, \nu^*/2), \quad z_i|(\tau_i, \theta^*) \sim \text{N}(x_i'\beta^*, \sigma^2/\tau_i),$$

and

$$y_i = I(z_i \geq 0)$$

for $i = 1, \ldots, n$, where $\theta^* = (\beta^*, \nu^*, \alpha, \sigma)$ with $\alpha > 0$ and $\sigma > 0$. The observed-data model is preserved with the reduction function

$$\beta = (\alpha/\sigma)\beta^* \quad \text{and} \quad \nu = \nu^*. \quad (21.11)$$

The complete data sufficient statistics for the expanded parameters θ^* are given in (21.5). The complete-data MLE of θ^* is given by

$$\hat{\alpha} = n^{-1} \sum_{i=1}^{n} \tau_i, \quad \hat{\sigma}^2 = n^{-1}(S_{\tau zz} - S_{\tau xz}' S_{\tau xx}^{-1} S_{\tau xz}),$$

and $\hat{\beta}^*$ and $\hat{v}^*$ are the same as $\hat{\beta}$ and $\hat{v}$ respectively. Compared to the EM and ECME algorithms in Section 21.5, the corresponding PX-EM and PX-ECME algorithms require only simple extra computation, namely, the conditional expectations of S_τ and $S_{\tau zz}$. The PX-EM algorithm for finding the regression coefficients β with known number of degrees of freedom v is then a simple extension of the EM algorithm and is given as follows. At iteration $t + 1$ with input $\beta^{(t)}$,

E-step of PX-EM. The same as the E-step of EM, except for the extra calculation of the conditional expectations $\hat{S}_\tau = \sum_{i=1}^n \hat{\tau}_i$ and $\hat{S}_{\tau zz} = n(v + 1) - v \sum_{i=1}^n \hat{\tau}_i + \sum_{i=1}^n \hat{\tau}_i(2\mu_i\hat{z}_i - \mu_i^2)$.

M-step of PX-EM. Compute the estimates $\hat{\beta}^* = \hat{S}_{\tau xx}^{-1}\hat{S}_{\tau xy}$, $\hat{\alpha} = n^{-1}\hat{S}_\tau$, and $\hat{\sigma}^2 = n^{-1}(\hat{S}_{\tau zz} - \hat{S}_{\tau xz}'\hat{S}_{\tau xx}^{-1}\hat{S}_{\tau xz})$ and then apply the reduction function to update β:
$\beta^{(t+1)} = (\hat{\alpha}/\hat{\sigma})\hat{\beta}^*$

With unknown number of degrees of freedom v, the ECME algorithm is then extended to the following PX-ECME algorithm. At iteration $t + 1$ with input $\theta^{(t)} = (\beta^{(t)}, v^{(t)})$,

E-step of PX-ECME. The same as the E-step of PX-EM, just conditioning on the parameter estimates, $\theta^{(t)} = (\beta^{(t)}, v^{(t)})$.

CM-step 1 of PX-ECME. The same as the M-step of PX-EM.

CM-step 2 of PX-ECME. The same as the CM-step 2 of ECME.

21.6   A numerical example

We analyze a data set from Finney (1947) consisting of 39 binary responses denoting the presence $(y = 1)$ or absence $(y = 0)$ of vasoconstriction of the skin of the subjects after inspiration of a volume V of air at inspiration rate R. The data were obtained from repeated measurements on three individual subjects, the numbers of observations per subject being 9, 8, and 22. Finney (1947) found no evidence of intersubject variability, treated the data as 39 independent observations, and analyzed the data using the probit regression model with V and R in the logarithm scale as covariates. These data were also analyzed by Pregibon (1982), using robust procedures (called *resistant fitting methods*) as alternatives to logistic regression.

The data are displayed in Figure 21.2. The fitted probability contours obtained from the MLE indicate that there is little difference between the fitted probit and logistic regression models. From these contours, the robit(7) and logistic models are almost identical, suggesting again the robit(7) model as an alternative to the logistic model in the sense that the robit(7) regression model provides results that

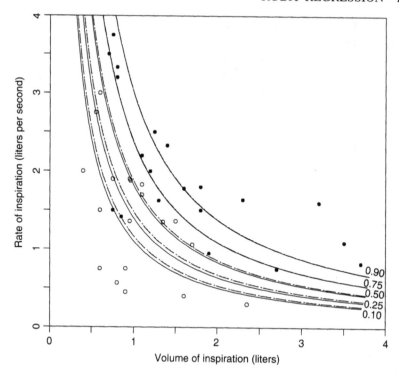

Figure 21.2   Scatterplot of the skin vasoconstriction data (with the symbols • and ○ indicating positive and negative responses respectively). The probability contours represent the probit (solid line), logistic (dotted line), and robit(7) (dashed line) models fitted by the methods of maximum likelihood.

can be understood as those from the logistic model and that the MLE of robit(7) regression model is robust.

The EM algorithm was used to choose the number of degrees of freedom. The algorithm was stopped when the likelihood increment becomes numerically instable because of the accuracy in evaluation of the probability functions of the t-distributions. The estimate of $\hat{\nu}$ is about 0.11 with the likelihood value -10.62. The fitted robit models with various numbers of degrees of freedom are represented by the probability contours in Figure 21.3. The use of a small number of the degrees of freedom is intuitively suggested by the data in which the observations with positive responses and those with negative responses can be almost separated by a line on the plane of $\log(V)$ and $\log(R)$ except for the three observations with $i = 4$, 18, and 24. These three observations are identified from the fitted individual weights. Pregibon (1982) also found that these three observations are influential to the ML estimation of the logistic model. The fitted 0.1, 0.5, and 0.9 contours by Pregibon are similar to those obtained from the robit model with about $\nu = 2$ degrees of freedom.

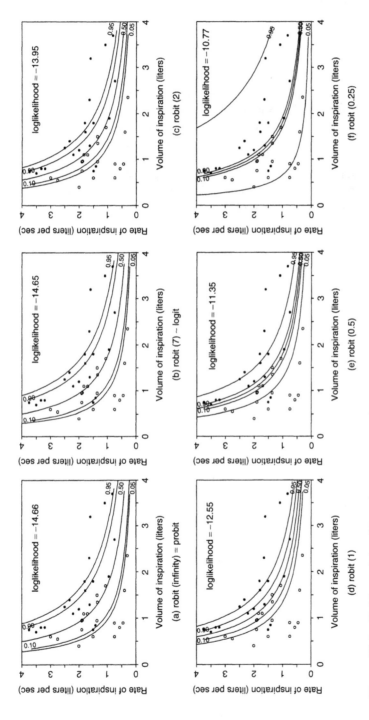

Figure 21.3 Robit models with various numbers of degrees of freedom fitted to the skin vasoconstriction data using the methods of maximum likelihood.

21.7   Conclusion

We have shown that the robit model is a usable robust alternative to the probit and logistic models for analyzing binary response data. The advantages of using the robit model include (1) inference based on the robit model is robust to the presence of outlying observations, and (2) computation for a Bayesian robit regression model using Markov chain Monte Carlo (MCMC) methods is simpler than that for the logistic model (see, for example, Zeger and Karim (1991)). Since robit (v) with small v gives more weight to the observations that are close to the dividing line $\Pr(y = 1|x) = (\Pr(y = 0|x) = 1/2$ when they agree with the fitted model, the robit model with a small number of degrees of freedom should also be useful in classification. In addition, as with the probit model (e.g., Albert and Chib, 1993; Chib and Greenberg, 1998), the extension of the robit model to correlated multivariate binary responses is straightforward.

22

Using EM and data augmentation for the competing risks model

Radu V. Craiu and Thierry Duchesne[1]

We consider a survival analysis problem in which items are subject to failure from competing risks. For some of the items, the failure cause is known only to belong to a subset of the set of all possible causes, while for the remaining items the cause of death is known precisely. In this chapter, we investigate two complementary analyses based on models in which the hazard rates are assumed piecewise constant. The approaches proposed rely on the EM algorithm and its Bayesian counterpart, the data augmentation (DA) algorithm. An example is used to illustrate the advantages of each analysis.

22.1 Introduction

In situations in which the survival data involve several different failure types, the analysis is performed using the theory of competing risks. In most medical and industrial applications, the data includes the time of censoring or failure and an indicator of the failure cause for each item/patient. However, it is often the case, especially with modular systems, that for a certain subset of the items the true cause of failure is not known exactly. Such items are said to have a *masked cause*

[1]Department of Statistics, University of Toronto, Ontario, and Universite Laval, Departement de Mathematiques et de Statistique, Quebec.

Applied Bayesian Modeling and Causal Inference from Incomplete-Data Perspectives.
Edited by A. Gelman and X-L. Meng © 2004 John Wiley & Sons, Ltd ISBN: 0-470-09043-X

of failure. While in some cases the failure can be isolated down to a subset of causes, without any such additional information the masking group is considered to be the entire set of causes. In certain experiments, a second-stage analysis can be conducted so that part of the items with a masked cause of failure are investigated and an exact diagnostic of the failure cause is obtained.

The literature on competing risks with masked causes of failure has grown greatly in the recent years. In the context of carcinogenicity studies, Racine-Poon and Hoel (1984) establish a nonparametric estimate of the survival function, while Dinse (1986) proposes nonparametric maximum likelihood estimators of prevalence and mortality. Several other authors also discuss the problem of missing cause-of-death in carcinogenicity studies (Kodell and Chen, 1987; Lagakos, 1982; Lagakos and Louis, 1988). Goetghebeur and Ryan (1990), and Dewanji (1992) construct a log-rank test to assess the difference between survival functions for subgroups of the population under study in the presence of covariates. Goetghebeur and Ryan (1995) subsequently generalize the approach to proportional cause-specific hazards regression models. Flehinger, Reiser, and Yashchin (1998, 2002) consider the analysis of data sets in which there are second-stage data. They propose maximum likelihood estimation using a model with nonparametric proportional cause-specific hazards (Flehinger, Reiser, and Yashchin, 1998) and a model with completely parametric cause-specific hazards (Flehinger, Reiser, and Yashchin, 2002). The literature regarding the Bayesian analyses of this problem is reviewed at the beginning of Section 22.4.

Proportionality between the cause-specific hazards or their complete parametric specification are assumptions that do not always mirror reality. In this chapter, we propose two approaches, both based on piecewise constant hazards. We assume no proportionality between the hazards and only weak parametric assumptions are made, namely no particular shape is imposed on the hazards. The model is defined in Section 22.2. In Section 22.3, we briefly describe an EM-based approach, which is analyzed in detail by Craiu and Duchesne (2004). The model described in Section 22.2 is the backbone of the Bayesian analysis presented in Section 22.4, which represents the main contribution of this chapter. While pros and cons are discussed for each analysis, we hope that the illustration from Section 22.5 will emphasize the advantages of each approach as well as the potential for combining their strengths. Conclusions and further work are in Section 22.6.

22.2   The model

We consider a situation in which n independent items are observed in the time interval $[0, T_{max}]$ and each of them can fail because of exactly one of J possible causes. The data are collected in two stages. In the first stage, we observe for each item its failure time, which may be censored if at time T_{max} the item was still functioning. For those items that have failed while in the study, we can observe one of the following two situations: (1) item i fails due to cause j at time t, (2) item i fails due to an unknown cause of failure, which is known to belong to a group of failure causes $g(i) \subset \{1, \dots, J\}$. The items that belong to the second situation have a masked failure cause. In the second stage, a subset of the masked items is sent for further analysis and the precise cause of failure is then determined. It is intuitive that the masking parameters shall be estimated using those items that are sent to the second stage of the experiment. In fact, if all the items were sent to the second stage, then all the information needed for estimation would be available and no missing data procedure would be necessary. Hence, we get a natural definition of the complete data as the data set that we would obtain if every masked item with an uncensored failure time were sent to a second-stage analysis. Suppose there are M masking groups in the data set (including the groups consisting of the individual failure causes). The observation for item i in the complete data set would be $(t_i, \gamma_{ig_1}, \dots, \gamma_{ig_M}, \delta_{i1}, \dots, \delta_{iJ})$, where γ_{ig} is the indicator that item i's failure cause was masked to group g at the first stage (if the failure cause is known to be j at the first stage, then we say that it is masked to $g = \{j\}$), δ_{ij} is the indicator that item i's actual failure cause is j (if an item is right-censored, then all the indicators δ_{ij}, $j = 1, \dots, J$, take on the value 0). The groups containing more than one cause are called *proper*.

Here is a short example to set the notation straight. Suppose that we have two potential causes of failure, say causes 1 and 2. Let us assume that at the first stage we either identify the cause of failure directly (in which case we say that it is masked in group $\{1\}$ or $\{2\}$ accordingly) or we only know that failure is due to one of causes 1 or 2 (in which case we say that failure is masked in group $\{1, 2\}$). For item 1, we have failure at time 2.4 masked in group $\{1, 2\}$ at stage 1 with no second stage. Item 2 fails at time 6.3 of a cause masked in group $\{1, 2\}$ and it is found in a second-stage analysis that failure was actually due to cause 2. Item 3 is right-censored at time 4.1, and item 4 fails at time 7.2 and its failure is diagnosed in stage 1 as being due to the first cause. These four observations would be coded as

$$(t_1, \gamma_{1\{1\}}, \gamma_{1\{2\}}, \gamma_{1\{12\}}, \delta_{11}, \delta_{12}) = (2.4, 0, 0, 1, \cdot, \cdot)$$

$$(t_2, \gamma_{2\{1\}}, \gamma_{2\{2\}}, \gamma_{2\{1,2\}}, \delta_{21}, \delta_{22}) = (6.3, 0, 0, 1, 0, 1)$$

$$(t_3, \gamma_{3\{1\}}, \gamma_{3\{2\}}, \gamma_{3\{1,2\}}, \delta_{31}, \delta_{32}) = (4.1, \cdot, \cdot, \cdot, 0, 0)$$

$$(t_4, \gamma_{4\{1\}}, \gamma_{4\{2\}}, \gamma_{4\{1,2\}}, \delta_{41}, \delta_{42}) = (7.2, 1, 0, 0, 1, 0)$$

where "$\cdot$" represents missing data. We denote by M_2 all masked items that have not been sent to a second-stage analysis and by G_j the set of all masking groups containing cause j. The number of elements in G_j is denoted L_j and we define $G_j^* = G_j \backslash \{j\}$.

The statistical model has a part involving the competing-risk aspect (failure times, hazard rates) and a part due to masking (masking probabilities). If T^* and

J^* are random variables that represent the failure time and the cause of failure respectively, then the cause-specific hazards are

$$\lambda_j(t) = \lim_{h \downarrow 0} \frac{\Pr[t < T^* \le t + h, \, J^* = j | T^* \ge t]}{h}, \quad j = 1, \dots, J. \quad (22.1)$$

In this chapter, we suppose that the cause-specific hazard functions are piecewise constant, that is, there exists a partition of the time interval $[0, T_{\max}]$ given by $0 = a_0 < a_1 < \cdots < a_K = T_{\max}$ such that, if $1_k(t)$ is the indicator that $t \in (a_{k-1}, a_k]$, then

$$\lambda_j(t) = \sum_{k=1}^{K} \lambda_{jk} 1_k(t). \quad (22.2)$$

The choice of the same endpoints for the hazard intervals $(a_{k-1}, a_k]$ is justified because it allows testing for the proportionality of cause-specific hazards and symmetry, as shown in Craiu and Duchesne (2004). However, if no such tests are necessary, the analysis described here can be carried on even if the intervals have different lengths for different cause-specific hazards. In such a situation, the notation for the endpoints would have to include a second index, j, to show their dependence on the cause. Of ultimate interest are the *diagnostic probabilities*

$$\pi_{j|g(i)}(t_i) = \Pr[\text{item } i \text{ failed of } j | \text{failed at } t_i \text{ and was masked in } g(i)],$$

for all masked items i and all causes $j \in g(i)$. In order to compute $\pi_{j|g}(t)$, we need the *masking probabilities*,

$$p_{g|j} = \Pr[\text{cause masked to group } g \text{ at stage 1} | \text{actual failure cause is } j], \quad j \in g.$$

With Bayes' rule we obtain

$$\pi_{j|g}(t) = \frac{\lambda_j(t) p_{g|j}}{\sum_{l \in g} \lambda_l(t) p_{g|l}}. \quad (22.3)$$

If θ is the vector of parameters that contains λ_{jk}, $j = 1, \dots, J$, $k = 1, \dots, K$ and $p_{gm|j}$, $j = 1, \dots, J$, $m = 1, \dots, M$, then the log-likelihood function under complete data is

$$\log p_C(\theta) = \sum_{i=1}^{n} \sum_{j=1}^{J} \left\{ \left[\delta_{ij} \log \sum_{k=1}^{K} \lambda_{jk} 1_k(t_i) - \sum_{k=1}^{K} \lambda_{jk} \int_0^{t_i} 1_k(u) \, du \right] \right.$$

$$\left. + \delta_{ij} \left[\left(1 - \sum_{g \in G_j^*} \gamma_{ig} \right) \log \left(1 - \sum_{g \in G_j^*} p_{g|j} \right) + \sum_{g \in G_j^*} \gamma_{ig} \log p_{g|j} \right] \right\}. $$

$$(22.4)$$

The likelihood (22.4) contains a competing-risk part that involves the failure times and failure causes (first line), and a masking part that involves the masking probabilities (second line). Under complete data, these two parts would be maximized separately making the maximum likelihood estimates of the masking probabilities robust to the specification of hazard intervals. One can notice that for right-censored observations the term on the second line of equation (22.4) vanishes and for such items there is no need to know γ_{ig}.

22.3   EM-based analysis

The EM algorithm (Dempster, Laird, and Rubin, 1977) has become a classic among the methods designed to handle the maximization of intractable likelihood functions. The use of EM to maximize (22.4) is recommended since the log-likelihood is linear in the missing data $\{\delta_{ij} : i \in M_2, 1 \leq j \leq J\}$ and the maximization required in the M-step can be performed in closed form, as shown below.

The algorithm

For each $i \in M_2$ with uncensored failure time t_i and with a failure cause masked in $g(i)$, we have that

$$E[\delta_{ij}|Y_{\text{OBS}}, \theta] = \hat{\pi}_{j|g(i)}(t_i) = \frac{\hat{\lambda}_j(t_i)\hat{p}_{g(i)|j}}{\sum_{l \in g_i} \hat{\lambda}_l(t_i)\hat{p}_{g(i)|l}}.$$

Since the complete-data log-likelihood (22.4) is linear in the missing δ_{ij}, substitution of the missing δ_{ij} with $E[\delta_{ij}|Y_{\text{OBS}}, \theta]$ constitutes the E-step of the algorithm. In addition, if we let

$$e_k = \sum_{i=1}^{n} \int_0^{t_i} 1_k(u)\, du \qquad (22.5)$$

denote the k-th interval *exposure*, that is, the total time lived by all items in the interval $(a_{k-1}, a_k]$, then one easily obtains that (22.4) is maximized when

$$\hat{\lambda}_{jk} = \frac{\sum_{i=1}^{n} \delta_{ij} 1_k(t_i)}{e_k} \quad \text{and} \quad \hat{p}_{g|j} = \frac{\sum_{i=1}^{n} \delta_{ij} \gamma_{ig}}{\sum_{i=1}^{n} \delta_{ij}}.$$

Hence, once the starting points have been chosen, the algorithm iterates between the E-step described above and the M-step given by

$$\hat{\lambda}_{jk}^{(l)} = \frac{\sum_{i=1}^{n} E_{\hat{\theta}^{(l-1)}}[\delta_{ij}|Y_{\text{OBS}}] 1_k(t_i)}{e_k} \quad \text{and} \quad \hat{p}_{g|j}^{(l)} = \frac{\sum_{i=1}^{n} E_{\hat{\theta}^{(l-1)}}[\delta_{ij}|Y_{\text{OBS}}] \gamma_{ig}}{\sum_{i=1}^{n} E_{\hat{\theta}^{(l-1)}}[\delta_{ij}|Y_{\text{OBS}}]}.$$
$$(22.6)$$

The algorithm can be easily extended to include time-varying masking probabilities $p_{g|j}(t)$ (see Craiu and Duchesne, 2004).

In all situations encountered with relatively large sample sizes and a 30 to 50% percentage of masked items sent to the second-stage analysis, the algorithm converges in less than 10 to 20 iterations. Caution is required in situations in which there are no data collected in the second stage and the cause-specific hazard rates are proportional. In such a case, the parameters are unidentifiable conditional on the observed data (Flehinger, Reiser, and Yashchin, 1998) but are identifiable given the complete data. If the parameters are identifiable only in the complete-data model, there is a ridge of local maxima in the likelihood surface and the EM algorithm will converge to one of the points on the ridge, depending on the starting point. The erratic behavior of the EM can be detected by using multiple starting points. Previous authors (Goetghebeur and Ryan, 1990; Dewanji, 1992; Lo, 1991) propose a working hypothesis of *symmetry* to reduce the number of parameters and obtain identifiability. The symmetry assumption states that the masking probabilities $p_{g|j}$ does not depend on the cause j, that is, $p_{g|j} = p_g$ for any group g and any $j \in g$.

Craiu and Duchesne (2004) prove results regarding the convergence of the EM algorithm, develop inference methods such as likelihood ratio tests for the assumptions of symmetry and proportionality of hazards, and apply the supplementary EM (SEM) algorithm (Meng and Rubin, 1991) for the estimation of the asymptotic variance matrix of the maximum likelihood estimators.

However, even if the cause-specific hazards are not proportional, with little or nonexistent second-stage data, the information about the $p_{g|j}$'s is obtained via the hazard rate estimates, which are time dependent. As a result, if the intervals for the hazards are misspecified, then the maximum likelihood estimates can be far from the true values. Equations (22.6) require that the hazard intervals are chosen so that for each interval $1 \le k \le K$ and for each failure cause $1 \le j \le J$, there exists an i such that $j \in g_i$ and $1_k(t_i) = 1$. In most cases, this implies that the intervals for the piecewise hazards are fairly large, leading naturally to misspecification. We expect that combining the previous approach with the Bayesian analysis proposed in the next section will remedy this problem since, owing to the prior specifications, there are no restrictions on the number and size of intervals for each cause-specific hazard.

22.4 Bayesian analysis

Most of the Bayesian inferences presented in the literature of competing hazards allow parametric models for the hazard rates. Reiser et al. (1995) assume that the component lifetimes are exponentially distributed, Kuo and Yang (2000) consider also Weibull-distributed lifetimes, while Basu et al. (2003) incorporate in their analysis all commonly used parametric distributions. In recent years, the nonparametric Bayesian analysis of survival models has spurred a lot of work. Following the ideas of beta and gamma processes devised by Hjort (1990), Kalbfleisch (1978), Dykstra and Laud (1981), statisticians have increased the complexity of the prior elicitation for the hazards rates in the competing-risks models. We refer the reader to

Arjas and Gasbarra (1994), Walker and Mallick (1997), Gasbarra and Karia (2000), Salinas-Torres, Pereira, and Tiwari (2002), Nieto-Barajas and Walker (2002), and Ibrahim, Chen, and Sinha (2001).

While the EM-based inference offers simplicity and robustness to the misspecification of the hazards rate when there are enough second-stage data, it can also produce the wrong estimates when there is not enough information (sparse second-stage data, large percentage of masked items, etc.). It is therefore important to be able to incorporate in the model the knowledge accumulated from past similar experiments. In addition, the performance may be improved with a more flexible choice of the intervals for the hazards.

In the following, we construct a Bayesian analysis structured on the model (22.4) that uses the work of Nieto-Barajas and Walker (2002) to define the prior distribution on the hazard rates. More precisely, their discrete gamma process is used to model piecewise constant hazard rates as we adapt their method to the context of competing proportional cause-specific hazards.

Prior distributions

As before, assume that for each cause $j \in \{1, \ldots, J\}$, we define K intervals on which the j-th hazard is constant and equal to λ_{jk}, $1 \leq k \leq K$. If we consider these intervals to be shorter, then it is likely that the values of the hazards in two successive pieces are not independent. We follow Nieto-Barajas and Walker (2002) and assume a latent process u_{jk} so that for each cause j, there is a Markovian dependence summarized by the graph

$$\lambda_{j1} \to u_{j1} \to \lambda_{j2} \to \ldots \to u_{jK-1} \to \lambda_{jK}.$$

Adding the latent variables u_{jk} allows one to model and control the dependence between values taken by one cause-specific hazard rate on adjacent intervals. Such dependence is important in situations in which we choose the intervals without a good knowledge of the underlying process (as is usually the case in practice). Alternatively, one may interpret the u_{jk}'s as virtual failures of a process identical in nature to the one under study; this point of view is attractive as it allows an intuitive interpretation of the model.

Formally, take the following conditional distributions

$$\lambda_{j1} \sim \text{Gamma}(\alpha_{j1}, \beta_{j1}),$$

$$u_{jk} | \lambda_{jk} \sim \text{Poisson}(c_{jk} \lambda_{jk})$$

$$\lambda_{j,k+1} | u_{jk} \sim \text{Gamma}(\alpha_{jk+1} + u_{jk}, \beta_{jk+1} + c_{jk}) \tag{22.7}$$

with $\alpha_{jK+1} = \beta_{jK+1} = 0$ for all $1 \leq j \leq J$ and $1 \leq k \leq K$. The c_{jk} regulates the smoothing of the hazard λ_j so that if $c_{jk} = 0$ then λ_{jk} and λ_{jk+1} are independent. In general, $10 \leq c_{jk} \leq 20$ is enough to produce smoother hazards, while taking $c = 0$ will result in approximately the same inference as the EM-based one. The choice of the c_{jk}'s has to be done in connection with the width of the intervals, for example, a succession of larger intervals requires a smaller value of the smoothing parameter. One can also let the data decide by considering the c's as part of the parameter vector and assigning exponential priors to them as suggested in Nieto-Barajas and Walker (2002). In the absence of prior information, the α_{jk} and β_{jk} are recommended to be small. If we have prior information regarding the process λ_j, say we know $E[\lambda_{jk}] = \psi_{jk}$, then we can choose the α_{jk}, β_{jk}, and c_{jk} such that

$$\frac{\alpha_{jk+1}}{\beta_{jk+1}} = \frac{\psi_{jk+1} - \xi_{jk+1}\psi_{jk}}{1 - \xi_{jk+1}},$$

where $\xi_{jk+1} = c_{jk}/(\beta_{jk+1} + c_{jk})$. We refer to Nieto-Barajas and Walker (2002) for other properties of the gamma process prior.

A natural conjugate prior assigned to the masking probabilities is

$$(p_{g_1|j}, p_{g_2|j}, \ldots, p_{g_{L_j}|j}) \sim \text{Dirichlet}(\eta_{1j}, \ldots, \eta_{L_j j}), \tag{22.8}$$

for all $1 \leq j \leq J$ causes. Lack of information on the masking probabilities will produce $\eta_{ij} = \text{constant}$ for all $1 \leq i \leq L_j$ and all causes j, while prior information can be included as $E[p_{g_i|j}] = \eta_{ij}/\sum_{h=1}^{L_j} \eta_{hj}$.

Data augmentation algorithm

There are two sets of latent variables in the model. For each item $i \in M_2$, there are J unobserved random variables $(\delta_{i1}, \ldots, \delta_{iJ})$. In addition, the prior (22.7) introduces $K - 1$ additional latent variables, $(u_{j1}, \ldots, u_{jK-1})$, for each cause j. In the initialization step, we need to input initial guesses for all the latent variables. For the set of δ's, one can use the output from the EM algorithm described in the previous section. Although the δ_{ij}^{EM} computed in the E-step are not integers, we can choose for each $i \in M_2$ the j_0 with the largest δ_{ij}^{EM} and assign $\delta_{ij_0}^{(0)} = 1$, $\delta_{ij}^{(0)} = 0$, $j \neq j_0$. In our applications, we use $u_{jk}^{(0)} = 1$ for all j, k.

The data augmentation algorithm (Tanner and Wong, 1987) consists in the following steps at iteration t:

Masking probabilities For each $j \in \{1, \ldots, J\}$ sample

$$\left(p_{g_1|j}^{(t)}, \ldots, p_{g_{L_j}|j}^{(t)}\right) \sim \text{Dirichlet}\left(\eta_{1j} + \sum_{i=1}^{N} \gamma_{ig_1}\delta_{ij}^{(t-1)}, \ldots, \eta_{L_j j} + \sum_{i=1}^{N} \gamma_{ig_{L_j}}\delta_{ij}^{(t-1)}\right).$$

Hazard rates For each $j \in \{1, \ldots, J\}$

$$\lambda_{j1}^{(t)} \sim \text{Gamma}\left(\alpha_{j1} + u_{j1}^{(t-1)} + n_{j1}, \beta_{j1} + c_{j1} + e_1^{(t-1)}\right),$$

$$\lambda_{jk}^{(t)} \sim \text{Gamma}\left(\alpha_{jk} + u_{jk-1}^{(t-1)} + u_{jk}^{(t-1)} + n_{jk}^{(t-1)}, \beta_{jk} + c_{jk-1} + c_{jk} + e_k^{(t-1)}\right),$$

where $n_{jk}^{(t-1)}$ is the number of items that fail in the k-th interval due to cause j, $e_k^{(t)}$ is defined by equation (22.5), and $c_{jK} = u_{jK} = 0$. The superindex $t - 1$ means that these numbers are estimated using the latent variables imputed at step $t - 1$.

Latent variables For each item $i \in M_2$,

$$
\left(\delta_{i1}^{(t)}, \ldots, \delta_{iJ}^{(t)} \right) \sim \text{Multin} \left(1, \frac{p_{g(i)|1}^{(t)} \lambda_1^{(t)}(t_i)}{\sum_{j \in g(i)} p_{g(i)|j}^{(t)} \lambda_j^{(t)}(t_i)}, \ldots, \frac{p_{g(i)|J}^{(t)} \lambda_J^{(t)}(t_i)}{\sum_{j \in g(i)} p_{g(i)|j}^{(t)} \lambda_j^{(t)}(t_i)} \right).
$$

For each cause $j \in \{1, \ldots, J\}$ and for each interval $k \in \{1, \ldots, K\}$,

$$
\Pr \left(u_{jk}^{(t)} = u \right) \propto \frac{[c_{jk}(c_{jk} + \beta_{jk+1}) \lambda_{jk}^{(t)} \lambda_{jk+1}^{(t)}]^u}{\Gamma(u+1) \Gamma(\alpha_{jk} + u)}.
$$

The proportional hazards case

The assumption of proportional cause-specific hazards, here denoted A_{PH}, is recurrent in the literature of competing risks. However, tests to assess the correctness of such a hypothesis are rare. Craiu and Duchesne (2004) develop a likelihood ratio test for the hypothesis A_{PH}. In the present context, one first needs to construct a data augmentation algorithm to sample from the parameter subspace defined by the constraints

$$
A_{\text{PH}} : \lambda_{jk} = \phi_j \lambda_{1k}
$$

for all $2 \leq j \leq J$ and all $1 \leq k \leq K$. The masking part of the model as well as the prior specification of λ_1 remain the same. For each $j \geq 2$, the prior distribution of ϕ_j is Gamma(v_j, χ_j). The DA algorithm for the unrestricted model changes in that only the $\{u_{1k} : 1 \leq k \leq K - 1\}$ is imputed and the hazard rates step becomes:

Hazard rates

$$
\lambda_{11}^{(t)} \sim \text{Gamma} \left(\alpha_{11} + u_{11}^{(t-1)} + n_{11}, \beta_{11} + c_{11} + e_1^{(t-1)} \right),
$$

$$
\lambda_{1k}^{(t)} \sim \text{Gamma} \left(\alpha_{1k} + u_{1k-1}^{(t-1)} + u_{1k}^{(t-1)} + n_{1k}^{(t-1)}, \beta_{1k} + c_{1k-1} + c_{1k} + e_k^{(t-1)} \right),
$$

$$
\phi_j^{(t)} \sim \text{Gamma} \left(v_j + \sum_{i=1}^{N} \delta_{ij}, \chi_j + \sum_{k=1}^{K} \lambda_{1k}^{(t)} e_k \right).
$$

We denote A_{PC} the general model with piecewise constant hazards. In assessing the validity of A_{PH} of interest is the Bayes factor

$$
B_{\text{PH}} = \frac{p(Y_{\text{OBS}}|A_{\text{PC}})}{p(Y_{\text{OBS}}|A_{\text{PH}})}. \tag{22.9}
$$

It is well known (for example, Kass and Raftery, 1995; Meng and Wong, 1996; Chen, Shao, and Ibrahim, 2000) that

$$p(Y_{\text{OBS}}|A_{\text{PH}}) = \int p(Y_{\text{OBS}}|\theta_{\text{PH}}, A_{\text{PH}}) p(\theta_{\text{PH}}|A_{\text{PH}}) \, d\theta_{\text{PH}}$$

is just the normalizing constant of the posterior density $p(\theta_{\text{PH}}|Y_{\text{OBS}}, A_{\text{PH}})$. As a result, the estimation of (22.9) is equivalent to the estimation of a ratio of two normalizing constants. The latter problem has been intensively studied in the last years, particularly in situations in which only samples (independent or dependent) from the two distributions of interest are available to the analyst. A simple but often highly variable solution is based on importance sampling (Geweke, 1989). Meng and Wong (1996) develop bridge sampling as a generalization of importance sampling that exploits optimally the overlap between the supports of the distributions. Recently, Gelman and Meng (1998) introduced path sampling as the limit of an infinite sequence of bridge samplers. While the theory of bridge sampling has been developed for situations in which independent realizations from each distribution are available, subsequent applications and studies (Servidea, 2002) have shown that the method also works well with dependent samples. In the context of the present analysis, we have reasonable confidence that the two models have a significant overlap since the two parameter spaces share the subset of masking probabilities. It is worth adding that the unnormalized posterior density can be computed at any point because $p(Y_{\text{OBS}}|\theta_{\text{PH}}, A_{\text{PH}})$ and $p(Y_{\text{OBS}}|\theta, A_{\text{PC}})$ can be expressed in closed form (Craiu and Duchesne, 2004).

22.5   Example

To assess the importance of transferring information between adjacent intervals, we consider a simulation example in which the hazards rates are Weibull distributed and there is no proportionality among them. There are 300 observations with times of failure between 0 and 15. Only 20% of the masked items are sent to a second-stage analysis. There are three possible causes of failure and there are three masking groups: $g_1 = \{1, 2\}$, $g_2 = \{1, 3\}$, and $g_3 = \{1, 2, 3\}$. The probability of having the cause masked is: 60% for cause 1, 80% for cause 2, and 70% for cause 3. We implement the Bayesian analysis with 10 or 20 intervals. In the absence of prior information, we take $c_{jk} = C, \alpha_{jk} = \beta_{jk} = 0.001$ for all causes j and all intervals k.

The DA algorithm has been used to generate 4,000 iterations out of which we used the last 2,000 for estimation. The convergence assessment has been done following the ideas of Gelman and Rubin (1992) using Andrew Gelman's *itsim* function in S-plus applied to four parallel chains. The simulation lasted approximately one hour. Figures 22.1 and 22.2 illustrate the effect of increasing the "smoothing parameter" C from 0 to 10 when the number of intervals is relatively

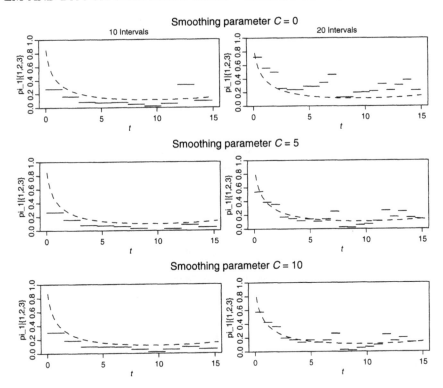

Figure 22.1   Plot of the posterior mean of the diagnostic probability $\pi_{1|\{1,2,3\}}$ against time as the number of intervals and the value of the smoothing parameter C vary. The true curve is represented by the dashed line, and the estimates within each interval are rendered with the solid lines.

moderate (10–20). Under consideration are the posterior means of the estimators for the diagnostic probability $\hat{\pi}_{1|\{1,2,3\}}(t)$ and $\hat{\pi}_{2|\{1,2,3\}}(t)$. Each plot shows in solid line the true value, and in dotted line the piecewise constant estimator. The $C = 0$ value corresponds roughly to the EM-based inference. It is seen here that if we increase the number of intervals, the EM estimator is too rough due to the lack of sufficient data in some of the intervals. Raising the value of the smoothing parameter noticeably increases the precision of the estimate. It can also be seen from the plot that the difference between the estimators obtained for $C = 5$ and $C = 10$ is quite small.

   The Bayes factor (22.9) can be calculated following the iterative construction of Meng and Wong (1996). This calculation is possible since one can compute the observed likelihood in any point as shown in Craiu and Duchesne (2004). With any of the above values for C, (22.9) ranges between 25 and 40 and shows no support for A_{PH}.

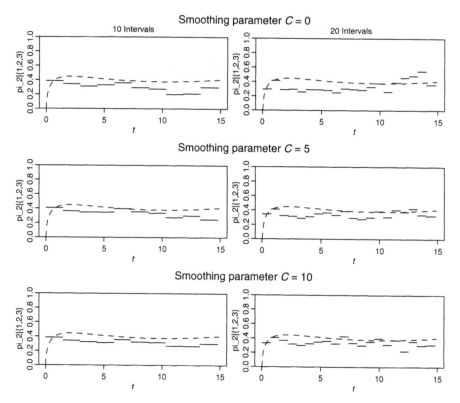

Figure 22.2   Plot of the posterior mean of the diagnostic probability $\pi_{2|\{1,2,3\}}$ against time as the number of intervals and the value of the smoothing parameter C vary. The true curve is represented by the dashed line, and the estimates within each interval are rendered with the solid lines.

22.6   Discussion and further work

The two methods presented are complementary and should be used together to increase the strength of the analysis. While the EM analysis produces robust inference of the masking probabilities and can be used to test for the symmetry and proportional hazards assumptions, it can also be used to determine the posterior modes for some or all of the model parameters as suggested in Gelman, Carlin, Stern, and Rubin (2003, Chapter 12). The Bayesian analysis is particularly useful in producing more sensible estimates of the hazard rates when the data are sparse. In addition, the calculation of the posterior variance of the diagnostic probabilities is more straightforward once it is possible to sample from the posterior distribution of the parameters.

Within the Bayesian framework, it may be of interest to produce an automatic sequential design procedure to help the experimenter decide which masked items should be sent to the second-stage analysis so that a certain given utility function is maximized.

We would like to enrich the class of possible models by relaxing the condition of piecewise linearity of the hazards. However, the computation complexity increases rapidly once we give up linearity and needs further investigation.

23

Mixed effects models and the EM algorithm

Florin Vaida, Xiao-Li Meng, and Ronghui Xu[1]

23.1 Introduction

Random effects models are a powerful and popular tool for clustered data. In this chapter, we show how to use the Monte Carlo EM algorithm (MCEM, Wei and Tanner, 1990) for maximum likelihood inference in the generalized linear mixed effects model (GLMM), and the proportional hazards mixed effects model (PHMM). The computation of the maximum likelihood estimator (MLE) for these models is complex, due to the analytically intractable marginal likelihood. As for linear mixed-effects models (Laird and Ware, 1982; Meng and van Dyk, 1998), we treat the random effects as "missing data." The expectation of the conditional log-likelihood at the E-step is analytically intractable for GLMM and PHMM, and is computed via the Monte Carlo simulation.

The GLMM has received increasing attention (e.g., Breslow and Clayton, 1993; Diggle and Kenward, 1994; Lee and Nelder, 1996; Chan and Kuk, 1997). When the random effects have a crossed design, the data cannot be reduced to small independent clusters, and therefore numerical methods are impractical; this is the case of interest here. For the computation of the E-step, several methods have been proposed: a Metropolis–Hastings algorithm (McCulloch, 1997); an independent

---

[1] Department of Biostatistics and Department of Statistics, Harvard University, Boston, Mass.

Applied Bayesian Modeling and Causal Inference from Incomplete-Data Perspectives.
Edited by A. Gelman and X-L. Meng © 2004 John Wiley & Sons, Ltd ISBN: 0-470-09043-X

sampler based on either multivariate importance sampling or rejection sampling (Booth and Hobert, 1999); and a Gibbs sampler (McCulloch and Searle, 2001, for probit link). Bayesian analyses for GLMM include Karim and Zeger (1992), Clayton (1996), Damien, Wakefield, and Walker (1999). In this chapter, by invoking a data-augmentation scheme larger than the one used by the M-step of EM, we obtain a straightforward slice sampler (Neal, 2003) for the E-step, which naturally accommodates any link function and distribution of random effects. Secondly, we propose a new EM scheme for fitting GLMM, in which the missing data are the variance-standardized random effects. Using the well-known salamander mating data of McCullagh and Nelder (1989, p. 439), we compare the two slice-EM algorithms with other methods in the literature.

The second half of the chapter is dedicated to the PHMM. In survival analysis, the random effects have traditionally been incorporated in the proportional hazards model (Cox, 1972) through the "frailty" term, a univariate gamma-distributed random effect that factors in the hazard function (Vaupel, Manton, and Stallard, 1979; Nielsen, Gill, Andersen, and Sorensen, 1992; Klein, 1992). Recent work includes Ripatti and Palmgren (2000) and O'Quigley and Stare (2002). Bayesian analysis of PHMM-related models include Gray (1994), Sargent (1998), and Carlin and Hodges (1999). Following Vaida and Xu (2000), we propose here a general PHMM with a linear mixed-effects predictor acting on the log-hazard scale, for the analysis of censored clustered survival data. The model is applied in the analysis of a lung cancer data set.

23.2   Binary regression with random effects

In the salamander mating data, 60 females and 60 males of two species of salamander, the Rough Butt (R), and White Side (W) were paired following a crossed, blocked, and incomplete design, in an experiment studying whether the two species have developed genetic mechanisms that would prevent interbreeding. The response is binary—successful ($y_{ij} = 1$) or unsuccessful ($y_{ij} = 0$) mating between female i and male j. We adopt the model

$$\text{logit Pr}(y_{ij} = 1 | \mathbf{u}, \beta) = \beta_{IJ} + u_i^F + u_j^M, \qquad (23.1)$$

where β_{IJ} is the fixed effect corresponding to the species combination of the $\{i, j\}$-pair of salamanders with $\beta = (\beta_{RR}, \beta_{RW}, \beta_{WR}, \beta_{WW})$; $\mathbf{u} = (\mathbf{u}^F, \mathbf{u}^M)$ is the vector of female and male random effects respectively, for which it is assumed that, independently, $u_i^F \overset{iid}{\sim} N(0, \sigma_F^2)$, $u_j^M \overset{iid}{\sim} N(0, \sigma_M^2)$, $i, j = 1 \ldots 60$. Each animal participates in six matings. The experiments yielded the female–male mating proportions: R–R = 60/90, R–W = 50/90, W–R = 19/90, W–W = 60/90. We only focus here on computing MLEs of the parameters β and $\delta = (\sigma_F, \sigma_M)$ under the simple model (23.1). The proposed methods are applicable to more sophisticated models, such as those that allow correlated, species-specific, and/or experiment-specific random effects (e.g., McCullagh and Nelder, 1989; Karim and Zeger, 1992; Chan and Kuk, 1997).

The main interest is on $\theta = (\beta, \delta)$, which determines the probability of a successful mating for each crossing, $\pi_{IJ} = E[G(\beta_{IJ} + u^F + u^M)]$, where G is the inverse logit function, $G(\eta) = (1 + \exp(-\eta))^{-1}$, and the expectation is over $\mathbf{u}$. The random effects may be used to estimate the mating propensity of individual animals. Conditional on $\mathbf{u}$, the likelihood for β is

$$p(\mathbf{y}|\beta, \mathbf{u}) = \frac{\exp\left(\sum_{I,J} y_{IJ} \beta_{IJ} + \sum_i y_{i.} u_i^F + \sum_j y_{.j} u_j^M\right)}{\prod_{i,j} \left\{1 + \exp(\beta_{IJ} + u_i^F + u_j^M)\right\}},$$

where $y_{IJ} = \sum y_{ij}$ and the sum extends over all observations from the crossing (I, J); $y_{i.}$ and $y_{.j}$ are the total number of successful matings for the ith female and jth male respectively (between 0 and 6); and i, j extend over all females and males, respectively, in the experiment. The marginal likelihood to maximize is $p(\mathbf{y}|\theta) = E[p(\mathbf{y}|\beta, \mathbf{u})|\delta]$. This 120-dimensional integral can be decomposed into a product of six 20-dimensional integrals, which cannot be reduced any further.

Slice-EM algorithms for GLMM

The EM algorithm for GLMM solves iteratively Fisher's equation $s(\theta; \mathbf{y}) = E[s(\theta; \mathbf{y}, \mathbf{u})|\mathbf{y}, \theta] = 0$, where $s(\theta; \mathbf{y})$ and $s(\theta; \mathbf{y}, \mathbf{u})$ are the observed-data and augmented-data score functions respectively. Operationally, the E-step computes $s(\theta|\theta^{(t)}) \equiv E[s(\theta; \mathbf{y}, \mathbf{u})|\mathbf{y}, \theta^{(t)}]$, and then the M-step solves $s(\theta|\theta^{(t)}) = 0$ for θ to determine $\theta^{(t+1)}$. The algorithm is iterated to convergence.

For a general GLMM with binary response y_i, linear predictor $\eta_i = \mathbf{x}_i^\top \beta + \mathbf{z}_i^\top \mathbf{u}$, and mean response $E(y_i) = G(\eta_i) = G_i$, $s(\theta|\tilde{\theta})$ conveniently separates the fixed effect β from the variance parameter δ: $s(\theta|\tilde{\theta}) = s(\beta|\tilde{\theta}) + s(\delta|\tilde{\theta})$,

$$s(\beta|\tilde{\theta}) = \sum_{i=1}^{n} \mathbf{x}_i E\left\{\left.\frac{G_i'(y_i - G_i)}{G_i(1 - G_i)}\right| \mathbf{y}, \tilde{\theta}\right\}, \tag{23.2}$$

$$s(\delta|\tilde{\theta}) = E\left\{\left.\frac{\partial}{\partial \delta} \log p(\mathbf{u}|\delta)\right| \mathbf{y}, \tilde{\theta}\right\}, \tag{23.3}$$

where $G_i' = dG/d\eta_i$. Using the Monte Carlo simulation to compute (23.2) and (23.3), at the M-step we then solve

$$\hat{s}_m(\beta) = \sum_{i=1}^{n} \mathbf{x}_i \frac{1}{m} \sum_{k=1}^{m} \frac{G_{ik}'(y_i - G_{ik})}{G_{ik}(1 - G_{ik})} = 0 \tag{23.4}$$

and similarly for $s(\delta|\tilde{\theta})$, where $\mathbf{u}_1 \ldots \mathbf{u}_m$ are draws from $p(\mathbf{u}|\tilde{\theta}, \mathbf{y})$, G_{ik} and G'_{ik} are obtained by substituting $\eta_{ik} = \mathbf{x}_i^\top \beta + \mathbf{z}_k^\top \mathbf{u}_k$ for η_i in G_i and G'_i, respectively. In (23.4), $\hat{s}_m(\beta)$ is the score function of a GLM with mn observations y_i (repeated m times) and linear predictor η_{ik}, and the M-step for β is the estimation of a GLM with offsets $\mathbf{u}_k$. The M-step for δ is the same as finding MLE for δ under $p(\mathbf{u}|\delta)$ based on perceived i.i.d. observations $\mathbf{u}_1 \ldots \mathbf{u}_m$. For the salamanders mating data, the solutions were $\{\sigma_F^2\}^{(t+1)} = (Im)^{-1} \sum_{i=1}^{I} \sum_{k=1}^{m} (u_{ik}^F)^2$ and $\{\sigma_M^2\}^{(t+1)} = (Jm)^{-1} \sum_{j=1}^{J} \sum_{k=1}^{m} (u_{jk}^M)^2$, where $I = J = 60$ and u_{ik}^F's and u_{jk}^F's are the output from the E-step sampler.

We sample from $p(\mathbf{u}|\mathbf{y}, \tilde{\theta})$ via a slice sampler (Neal, 2003), which produces a data-augmentation scheme for the E-step larger than the one for the M-step (see also Meng and van Dyk, 1997). Specifically, we further augment $\{\mathbf{u}, \mathbf{y}\}$ to $\{\mathbf{u}, \mathbf{y}, \mathbf{v}\}$, where $\mathbf{v} = (v_1 \ldots v_n)$ are an i.i.d. sample from the uniform distribution on $[0,1]$; $\mathbf{v}$ is independent of $\mathbf{u}$ and is connected to $\mathbf{y}$ via the threshold representation,

$$ y_i = I[v_i \leq G(\eta_i)], \qquad i = 1 \ldots n, \tag{23.5} $$

where $I[\cdot]$ is the indicator function. We sample $\mathbf{v}$ from $p(\mathbf{v}|\mathbf{u}, \mathbf{y})$ and $\mathbf{u}$ from $p(\mathbf{u}|\mathbf{v}, \mathbf{y})$. The two distributions are proportional to the joint distribution restricted by a set of linear inequalities (both also condition on θ): $p(\mathbf{v}|\mathbf{u}, \mathbf{y}) \propto I_{\mathcal{R}(\mathbf{y})} p(\mathbf{v})$, $p(\mathbf{u}|\mathbf{v}, \mathbf{y}) \propto I_{\mathcal{R}(\mathbf{y})} p(\mathbf{u})$, where $\mathcal{R}(\mathbf{y})$ is the set of all vectors $(\mathbf{u}, \mathbf{v})$ for which (23.5) holds. The distribution $p(\mathbf{v}|\mathbf{u}, \mathbf{y})$ is truncated uniform on the unit hypercube, and $p(\mathbf{u}|\mathbf{v}, \mathbf{y})$ is a truncated $p(\mathbf{u})$. Slice sampling is a general method for constructing useful Gibbs samplers (Damien, Wakefield, and Walker, 1999), and it has good convergence properties (Mira and Tierney, 2002).

An alternative slice-EM algorithm is obtained by writing equation (23.1) as logit $\Pr(Y_{ij} = 1|w_i^F, w_j^M) = \beta_{IJ} + \sigma_F w_i^F + \sigma_M w_j^M$, with $w_i^F, w_j^M \overset{iid}{\sim} N(0, 1)$ for all i, j, that is, σ_F, σ_M become the regression coefficients of the standardized random effects w_i^F, w_j^M. The E-step remains essentially unchanged, but the key difference is that the variance parameters are now part of the mean parameter to be estimated at the M-step, $\theta = (\beta, \sigma_F, \sigma_M)$. This leads to faster convergence for σ_F, σ_M, for reasons similar to those given in Meng and van Dyk (1997, 1998).

The standard errors (SEs) of the estimates are computed from the Fisher information matrix, as a by-product of the MCEM using the formula (Orchard and Woodbury, 1972; Louis, 1982):

$$ I_\mathbf{y}(\hat{\theta}) = \mathrm{E}\left[-s'(\hat{\theta}; \mathbf{y}, \mathbf{u}) \mid \mathbf{y}, \hat{\theta} \right] - \mathrm{E}\left[s(\hat{\theta}; \mathbf{y}, \mathbf{u}) s(\hat{\theta}; \mathbf{y}, \mathbf{u})^\top \mid \mathbf{y}, \hat{\theta} \right], \tag{23.6} $$

where $\hat{\theta}$ is the MLE of θ, and $s' = ds/d\theta$. The right side is estimated via Monte Carlo averages of $-s'(\hat{\theta}; \mathbf{y}, \mathbf{u})$ and $s(\hat{\theta}; \mathbf{y}, \mathbf{u})s(\hat{\theta}; \mathbf{y}, \mathbf{u})^\top$. The second slice-EM has the added appeal that the augmented-data score and Fisher information have standard GLM forms, regardless of the distribution of the random effects.

Implementation and results of the slice-EM

In MCEM, the E-step computation is not exact, the EM sequence of likelihood values is no longer guaranteed to be monotone, and the convergence to the MLE is stochastic. Chan and Ledolter (1995), Biscarat (1994), Vaida (1998) establish the convergence of MCEM under suitable conditions. Practical methods of monitoring convergence of MCEM include graphical monitoring of some sequences of parameters (Chan and Ledolter, 1995); monitoring the likelihood ratio via bridge sampling (Meng and Schilling, 1996); stopping when the Monte Carlo error is small relative to the statistical error (Booth and Hobert, 1999). Our strategy is to implement an MCEM in three stages: (1) The burn-in, where the starting point is "forgotten." We use a small sample size m; our goal is just to approach the region of convergence. (2) The transition stage, where m is increased gradually (e.g., linearly). (3) The plateau, where the algorithm is run with large m to achieve a small Monte Carlo error.

The starting point was $\beta = 0$, $\sigma_F^2 = \sigma_M^2 = 1$. The burn-in had 50 steps at $m = 100$. Figure 23.1 shows that this was enough to approach stationarity. The transition had 20 steps, with m increased linearly to 10,000, ensuring a smooth transition to the plateau. The plateau stage, with $m = 10,000$ for 50 steps, showed the stationarity of the process, and gave more precise MLE. The total running time was less than 30 min (Slice-EM1 and Slice-EM2 took about same time per iteration), with 20-s burn-in and 25-min plateau. All programs were implemented in C and were run on a Sun Ultra 30 workstation. The MLE was the average of the MCEM iterates from the plateau stage. Table 23.1 compares the point and interval estimates from the two algorithms, and the error due to the simulation. The latter was estimated on the basis of an AR(1) approximation to $\theta^{(t)}$ during the plateau stage (see Chan and Ledolter, 1995; Vaida, 1998). Slice-EM2 has uniformly lower MCEM error than Slice-EM1, with small improvements for β, but 50% reduction for the variance estimates. In general, the MCEM error is negligible compared to the standard error of the MLE.

The population-level probabilities of mating, π_{IJ} are given by $\pi_{IJ} = G(\beta_{IJ}/\sqrt{1+c^2\sigma^2})$, where $c^2 = (16\sqrt{3}/(15\pi))^2 \approx 0.346$ and G is the inverse logit function (Zeger, Liang, and Albert, 1988). They are reported in Table 23.2. Interval estimates for these were obtained by simulating the 90% highest posterior density interval for π_{IJ} as a function of $(\beta, \log\sigma_F^2, \log\sigma_M^2)$. The mating probabilities are large and very similar for same-species matings, $\pi_{WW} = .676$, $\pi_{RR} = .673$, but very low for WR, $\pi_{WR} = .197$. To test whether the mating between species is less probable than within the same species, we calculated the marginal odds ratios (OR) relative to RR for the three other crossings. There is strong evidence of a smaller probability of mating for WR: $OR = .12$, 95% $CI = (.05, .26)$. For the other two comparisons, the 95% CI include $OR = 1$.

Table 23.2 compares the mating probabilities π_{IJ} and 90% intervals estimates from four different models/methods: GLMM (computed via Slice-EM2), a

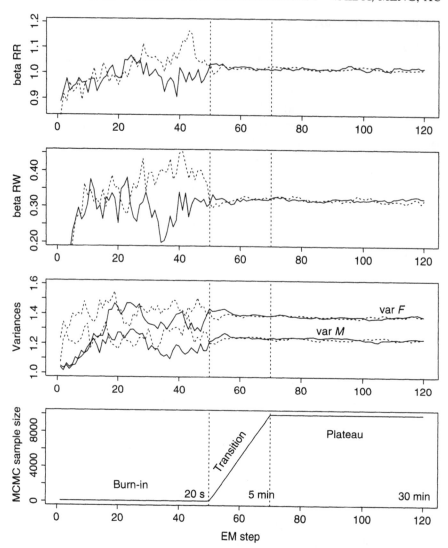

Figure 23.1   Convergence and comparison of algorithms for the salamander data: Slice-EM1—; Slice-EM2 ⋯. (a) Two mean parameters, β_{RR}, (b) β_{RW}, and (c) the variance parameters σ_F^2, σ_M^2 are shown. (d) includes the MCMC sample size m as a function of the phase and step of the algorithm, and computation time.

fixed-effects GLM, the Bayesian model of Karim and Zeger (1992), and penalized quasi-likelihood (PQL, Breslow and Clayton, 1993). GLMM and the Bayesian method produce numerically identical results. The GLM probabilities are also very similar to GLMM; however, the confidence intervals from GLM are slightly narrower due to the absence of the random effects.

Method	Estimate	MCEM St Error	MLE St Error	95% Conf Int	
β_{RR} Slice-EM1	1.019	.0016	.415	(0.19,	1.85)
β_{RR} Slice-EM2	1.018	.0013	.407	(0.20,	1.83)
β_{RW} Slice-EM1	.321	.0015	.393	(−0.47	1.11)
β_{RW} Slice-EM2	.320	.0013	.389	(−0.46,	1.10)
β_{WR} Slice-EM1	−1.940	.0019	.475	(−2.89	−0.99)
β_{WR} Slice-EM2	−1.941	.0013	.473	(−2.89,	−0.99)
β_{WW} Slice-EM1	.997	.0016	.415	(0.17	1.83)
β_{WW} Slice-EM2	.994	.0014	.417	(0.16,	1.83)
σ_F^2 Slice-EM1	1.384	.0044	.658	(0.54	3.58)
σ_F^2 Slice-EM2	1.385	.0016	.626	(0.56	3.42)
σ_M^2 Slice-EM1	1.238	.0039	.583	(0.48	3.18)
σ_M^2 Slice-EM2	1.234	.0016	.580	(0.48	3.16)

Table 23.1 Maximum likelihood estimates for the salamander data from the two Slice-EM algorithms, MCEM error, standard error of the MLE, and approximate 95% confidence intervals.

	GLMM			GLM			Bayes			PQL
π_{RR}	.68	(.56	.77)	.67	(.59	.75)	.67	(.56	.77)	.66
π_{RW}	.56	(.44	.66)	.56	(.47	.64)	.56	(.44	.66)	.55
π_{WR}	.20	(.13	.30)	.21	(.14	.28)	.20	(.13	.30)	.22
π_{WW}	.67	(.56	.77)	.67	(.59	.75)	.68	(.56	.77)	.66

Table 23.2 Marginal probabilities and 90% intervals from GLMM, GLM, and Karim and Zeger's Bayesian model, and marginal probabilities from PQL.

23.3 Proportional hazards mixed-effects models

A natural extension of the proportional hazards model (Cox, 1972) to clustered survival data is to incorporate the random effects in the log relative risk:

$$\lambda_{ij}(t) = \lambda_0(t) \exp({'\mathbf{z}_{ij}\beta' + \mathbf{w}_{ij}'\mathbf{b}_i}), \qquad (23.7)$$

where $\lambda_{ij}(t)$ is the hazard function of the jth observation for the ith cluster ($i = 1 \ldots n$, $j = 1 \ldots n_i$), $\mathbf{b}_i$ is the vector of random effects from the ith cluster, and $\mathbf{z}_{ij}, \mathbf{w}_{ij}$ are the covariate vectors for the fixed and random effects. Often $\mathbf{w}_{ij}$ is a submatrix of $\mathbf{z}_{ij}$, apart from possibly a '1,' which multiplies the cluster effect on the baseline hazard. In the following, we assume that $\mathbf{b}_i \overset{iid}{\sim} N(0, V)$. The typical frailty model (Vaupel, Manton, and Stallard, 1979) corresponds to a univariate random effect in (23.7) with $w_{ij} = 1$ and $\exp(\mathbf{b}_i) \sim \text{gamma}(\sigma^{-2}, \sigma^2)$.

Model (23.7) reflects the fact that some of the regression parameters in the proportional hazards model are cluster-dependent and that they may be treated as random. As an example, in a lung cancer trial conducted by the Eastern Cooperative Oncology Group (EST 1582) to compare two different chemotherapy regimens, there was evidence that the treatment effect varied substantially across the 31 participating institutions (Gray, 1994, 1995). Vaida and Xu (2000) apply PHMM to recurrent events data, and I. Liu (2003) to the genetic epidemiology of alcoholism.

Let γ be the vector of parameters in V. In certain cases, with proper parameterization of the random effects, the covariance matrix V can be chosen to be diagonal ($\gamma = \text{diag}(V)$). Let $\theta = (\beta, \gamma, \lambda_0)$ denote the "vector" of parameters. While the random effects $\mathbf{b}_i$ are not parameters in the strict sense, they are estimable quantities. The interest and focus on the unknown quantities β, $\mathbf{b}_i$, or γ depend on the application at hand: in a clinical trial, the fixed treatment effect is usually of importance; in a biological application where the cluster of observations corresponds to a certain animal, or breed, the interest is in ranking the random effects and selecting the best animal or breed; for genetic data, the focus is on γ, which contains the components of genetic variability. The baseline hazard $\lambda_0(t)$ is needed for the estimation and prediction of the survival probabilities. As will be seen, our method generates the nonparametric maximum likelihood estimate (NPMLE) of $\lambda_0(\cdot)$ and its estimated variance as a by-product.

Each observation of the data can be written as $y_{ij} = (t_{ij}, \delta_{ij}, \mathbf{z}_{ij}, \mathbf{w}_{ij})$, where t_{ij} is the failure time, possibly censored, and δ_{ij} is the event indicator (1 for an observed failure and 0 otherwise). Let $\mathbf{y}_i = (y_{i1} \dots y_{in_i})$, that is, the data for cluster i. Our inference is based on the full likelihood from the observed data. For cluster i, conditional on the random effect, the (full) log-likelihood is

$$\log p(\mathbf{y}_i | \mathbf{b}_i) = \sum_{j=1}^{n_i} \{\delta_{ij} \log \lambda_0(t_{ij}) + \delta_{ij}(\mathbf{z}_{ij}\beta' + \mathbf{w}'_{ij}\mathbf{b}_i) - \Lambda_0(t_{ij})e^{\mathbf{z}_{ij}\beta' + \mathbf{w}'_{ij}\mathbf{b}_i}\},$$

(23.8)

where $\Lambda_0(t) = \int_0^t \lambda_0(s)\,ds$ is the cumulative baseline hazard. The likelihood based on the observed data is then

$$p(\mathbf{y}|\theta) = \int p(\mathbf{y}|\theta, \mathbf{b})p(\mathbf{b}|\gamma)\,d\mathbf{b} = \prod_{i=1}^{n} \int p(\mathbf{y}_i|\beta, \lambda_0, \mathbf{b}_i)p(\mathbf{b}_i|\gamma)\,d\mathbf{b_i}, \quad (23.9)$$

where $p(\mathbf{y}|\theta, \mathbf{b}) = \prod_i p(\mathbf{y}_i|\beta, \lambda_0, \mathbf{b}_i)$ is the likelihood conditional on the random effects $\mathbf{b} = (\mathbf{b}_1 \dots \mathbf{b}_n)$. Usually, no closed-form expression is available for $p(\mathbf{y}|\theta)$ and its calculation involves d-dimensional integration.

In the fixed-effect proportional hazards model, the partial likelihood is the profile likelihood obtained by maximizing out the baseline hazard function within the family of nonparametric discrete hazards (Johansen, 1993), which amounts to replacing λ_0 and Λ_0 by their Nelson–Aalen estimators, respectively. In the random effects case, the NPMLE $\hat{\theta} = (\hat{\beta}, \hat{\gamma}, \hat{\lambda}_0)$ maximizes the likelihood (23.9), and can be obtained using the EM algorithm (Gill, 1985). For the univariate gamma-frailty model, the algorithm was described in Klein (1992) and a corrected variance estimator was later given in Andersen et al. (1997) following Parner (1998). Asymptotic theory for gamma-frailty models was developed in Murphy (1994, 1995) and Parner (1998). They showed the consistency and asymptotic normality of the NPMLE and that the asymptotic variance can be consistently estimated by the inverse of a discrete observed information matrix Parner (1998).

The EM algorithm for the PHMM

For the EM algorithm, the augmented data is $(\mathbf{y}_i, \mathbf{b}_i)$. Conditional on the current parameter $\tilde{\theta}$ and the observed data, the expected log-likelihood is

$$Q(\theta) = \mathrm{E}[\log p(\mathbf{y}, \mathbf{b}|\theta) \mid \mathbf{y}, \tilde{\theta}] = \mathrm{E}[\log p(\mathbf{y}|\beta, \lambda, \mathbf{b}) \mid \mathbf{y}, \tilde{\theta}] + \mathrm{E}[\log p(\mathbf{b}|\gamma) \mid \mathbf{y}, \tilde{\theta}].$$
(23.10)

Denote the two terms on the right side of the above equation by $Q_1(\beta, \lambda)$ and $Q_2(\gamma)$. Then

$$Q_1(\beta, \lambda) = \sum_{i=1}^{n} \sum_{j=1}^{n_i} \left\{ \delta_{ij} (\log \lambda_0(t_{ij}) + \mathbf{z}_{ij}\beta' + \mathbf{w}'_{ij}\mathrm{E}[\mathbf{b}_i]) - \Lambda_0(t_{ij})e^{\mathbf{z}_{ij}\beta' + u_{ij}} \right\},$$
(23.11)

where $u_{ij} = \log \mathrm{E}[e^{\mathbf{w}'_{ij}\mathbf{b}_i}]$, and $Q_2(\gamma) = \sum_{i=1}^{n} \mathrm{E}[l(\gamma; \mathbf{b}_i)]$. For normal random effects with diagonal V, we have

$$Q_2(\gamma) = -\sum_{g=1}^{d}(n \log \sigma_g^2 + \sigma_g^{-2} \sum_{i=1}^{n} \mathrm{E}[b_{ig}^2])/2;$$
(23.12)

a similar formula involving the expectation of cross-products $b_{ig}b_{i'g}$ is obtained for a general unconstrained V.

The conditional expectations in (23.11) and (23.12) can be obtained by numerical methods (Xue and Brookmeyer, 1996, for $d = 2$) or Monte Carlo simulation. For higher dimensions, we propose computing the E-step expectations based on a simulated sample from $p(\mathbf{b}_i|\mathbf{y}_i)$. For log-concave "prior" distributions of the random effects $p(\mathbf{b}_i)$, $p(\mathbf{b}_i|\mathbf{y}_i)$ is also log-concave, and a Gibbs sampler may be implemented on the basis of the adaptive rejection sampling algorithm of Gilks and Wild (1992). Most commonly used distributions for the random effects are log-concave; the multivariate t is a notable exception, but a special algorithm is available in this case.

The M-step conveniently separates the estimation of the parameters β and λ from the variance components γ. The formula (23.11) has the same form as the log-likelihood in a Cox regression model with known offsets u_{ij}, for which standard software is available. Q_2 is equivalent to the log-likelihood corresponding to n

independent observations from the "prior" random effects distribution $p(\mathbf{b}_i)$, where the standard sufficient statistics are replaced with their conditional expectations. For diagonal V with (23.12), the estimates are $\hat{\sigma}_g^2 = n^{-1} \sum_{i=1}^n \mathrm{E}[b_{ig}^2]$ for $g = 1, \ldots, d$. If V is unconstrained, it is maximized by $\hat{V} = n^{-1} \sum_{i=1}^n \mathrm{E}[\mathbf{b}_i \mathbf{b}_i']$.

A good starting point for β and λ in the EM algorithm is given by a usual Cox regression with no random effects. The initial value for V can be taken as the identity matrix. As in the GLMM case, the convergence of the MCEM is only approximately monotone, and special stopping rules are needed.

In the Cox random effects model, the EM algorithm has the particular feature that the baseline hazard function is maximized nonparametrically; for this reason the algorithm has sometimes been called a "modified EM" (Klein, 1992). However, the theory of the parametric EM algorithm applies to this case. It may be showed with similar methods as Johansen (1993) that the NPMLE for $\lambda_0(\cdot)$ is concentrated at the observed failure times. The problem is now equivalent to maximizing a parametric likelihood with a parameter $\lambda = (\lambda_1, \ldots, \lambda_s)'$ for the baseline hazard, where $\lambda_i = \lambda_0(t_i)$ and $t_1, \ldots, t_s$ are the distinct uncensored failure times. Therefore, the standard convergence properties of the EM algorithm apply.

The variance matrix of the NPMLE $\hat{\theta} = (\hat{\beta}, \hat{\lambda}, \hat{\gamma})$ is estimated using Louis' formula, as described for GLMM. Parner (1998) showed consistency of such a variance estimator in the gamma-frailty case. See Vaida and Xu (2000) for details.

An important advantage of random effects models over the marginal, or GEE-type models, is that they allow inference of the cluster-specific random effects, and therefore a better understanding and interpretation of the variability in the data. Estimation, or prediction, of the random effect $\mathbf{b}_i$ is based on its posterior, or "empirical Bayes," distribution $p(\mathbf{b}_i|\mathbf{y}_i, \hat{\theta})$ (Morris, 1983; Carlin and Louis, 1996). Conditional on the estimated parameters $\hat{\theta}$, the point estimate for $\mathbf{b}_i$ is $\hat{\mathbf{b}}_i = \mathrm{E}(\mathbf{b}_i|\mathbf{y}_i, \hat{\theta})$, with variance $\hat{\mathbf{v}}_i = \mathrm{var}(\mathbf{b}_i|\mathbf{y}_i, \hat{\theta})$. These quantities are easily computed as by-products of the EM algorithm at convergence. We illustrate inference for the random effects in the following example.

A lung cancer clinical trial

The lung cancer trial EST 1582 compared two different chemotherapy regimens: a standard (CAV) and an alternating regimen (CAV-HEM), where cycles of CAV were alternated with HEM. The endpoint is overall survival. There were 31 institutions with a total of 579 patients. Gray (1995) found significant variation in treatment effects among institutions, but no significant variation in the baseline hazards (corresponding to the random intercept).

We fit model (23.7) to this data set with all 579 patients. The covariates are treatment, presence or absence of bone metastases, presence or absence of liver metastases, performance status at entry, and whether there was weight loss prior to entry. We first modeled all five covariates with two independent normal random effects for the intercept and treatment. Unsurprisingly, the standard deviations of the random effect on the baseline hazard converged to zero. We therefore only

		$d = 0$	$d = 1$	$d = 2$
Treatment	β	−0.25 (0.09)	−0.25 (0.10)	−0.25 (0.12)
Bone		0.22 (0.09)	0.21 (0.10)	0.23 (0.14)
Liver		0.43 (0.09)	0.42 (0.09)	0.39 (0.09)
ps		−0.60 (0.10)	−0.64 (0.11)	−0.65 (0.13)
Weight loss		0.20 (0.09)	0.22 (0.09)	0.21 (0.09)
Treatment	σ	−	0.27 (0.13)	0.21 (0.43)
Bone		−	−	0.36 (0.12)

Table 23.3 Estimates (and SE's) for the mean parameters β and random effects standard deviations σ from the lung cancer trial. Proportional hazards model ($d = 0$); PHMM with treatment random effect ($d = 1$); PHMM with treatment and bone random effects ($d = 2$).

kept the treatment random effect; see Table 23.3 ($d = 1$). (The SE's of the σs should not be used for testing $\sigma = 0$ since the null hypothesis lies on the boundary of the parameter space and the normal approximation of the null test statistic is no longer appropriate.) The EM sequence achieved satisfactory convergence after 25 iterations. The results with $d = 1$ were compared to the fit from the usual Cox regression without any random effects (Table 23.3, $d = 0$).

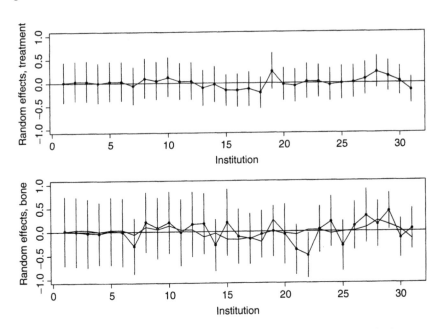

Figure 23.2 Multivariate random effects and 95% confidence intervals, lung cancer data. The institutions are ordered by increasing sample size (from 1 to 56).

When we fit the data set with independent random effects for all five covariates, three of the random-effects variances, corresponding to liver metastases, performance status, and weight loss, converged to zero. Finally, we fit a model with all five covariates and two independent random effects for treatment and bone metastases, respectively (Table 23.3, $d = 2$). The results are statistically indistinguishable from the model with $d = 1$. Figure 23.2 shows the predicted random effects of treatment and bone metastases together with the 95% credibility intervals, with the institutions ordered by their sizes. The lengths of the credibility intervals decrease as the clusters become larger.

In summary, in this analysis the random effects model leads to similar conclusions regarding the treatment and the other fixed effects, but the prediction for each hospital is different, due to the presence of the random effects.

24

The sampling/importance resampling algorithm

Kim-Hung Li[1]

24.1 Introduction

The increasing use of the Bayesian methods in statistics demands powerful sampling algorithms that are workable even for very awkward distributions. The demand is especially strong when full Bayesian analysis with nonstandard combinations of distributions is of interest. Markov chain Monte Carlo (MCMC) methods are commonly used for this sampling purpose. MCMC methods are iterative. A process having the target probability density function (pdf), $f(x)$, as its unique stationary distribution is constructed and simulated. After removing a leading portion of a realization of the simulated process, the remaining portion is taken as (correlated) samples from the target distribution. The Gibbs sampler (Geman and Geman, 1984) and the Metropolis algorithms (Hastings, 1970) are well-known examples of this type.

Rubin (1983) described a noniterative method for approximately sampling from $f(x)$. He called it the sampling/importance resampling (SIR) algorithm (Rubin 1987a, 1988). The SIR algorithm is a sample filtering method. It takes a random sample of size M from an approximate distribution as input and produces a refined sample of size m as output. In the SIR algorithm, a specially designed resampling procedure that uses the importance ratios as resampling weights is used to select the sample. As expected, the output of the SIR algorithm is "good" if the input is "good" or if M/m is large.

---

[1]Department of Statistics, Chinese University of Hong Kong, Shatin, N.T., Hong Kong.

Applied Bayesian Modeling and Causal Inference from Incomplete-Data Perspectives.
Edited by A. Gelman and X-L. Meng © 2004 John Wiley & Sons, Ltd ISBN: 0-470-09043-X

The SIR algorithm has been used in different occasions. It was used as a method to generate a starting value for the MCMC procedure (Gelman and Rubin, 1992), as an adjusting method for the weighted likelihood bootstrap (Newton and Raftery, 1994), as an updating method in the bootstrap filter (Gordon, Salmond, and Smith, 1993), and as a sampling method in the inverse Bayes formulae sampling (Tan, Tian, and Ng, 2003).

This chapter is organized as follows. The SIR algorithm is presented in Section 24.2. Then we discuss several fundamental problems of the SIR algorithm. In Section 24.3, the selection of M is studied. It is shown that when the importance ratios follow a Gamma distribution, a slightly nonlinear relation between M and m is preferred. We also show how M affects the fitness of the distribution of each individual output to $f(x)$. In Section 24.4, we suggest a selection criterion of the importance sampling pdf. In Section 24.5, we consider different resampling algorithms. The idea of the SIR algorithm is extended to yield an iterative sampling algorithm. Finally, a discussion is given in Section 24.6.

24.2 SIR algorithm

The SIR algorithm consists of two steps: a sampling step and an importance resampling step as given below:

Step 1. (Sampling step) Generate $X_1, \ldots, X_M$ independently and identically distributed (iid) from $h(x)$, the support of which includes that of $f(x)$.

Step 2. (Importance resampling step) Draw a weighted sample of size $m (m \leq M)$, say $\{Y_1, \ldots, Y_m\}$, from $\{X_1, \ldots, X_M\}$ with weight assigned to X_i being $\omega(X_i) \propto f(X_i)/h(X_i)$ for all i.

We call $\{X_1, \ldots, X_M\}$ a pool of candidate values, $\{Y_1, \ldots, Y_m\}$ a resample, $h(x)$ the importance sampling pdf, and $\omega(X)$ the importance ratio (also known as importance weight) of X. Smith and Gelfand (1992) used simple weighted random sampling with replacement in the importance resampling step and called their method the *weighted bootstrap*. Throughout this chapter, we assume that $\omega(X)$ has finite second moment. Write $\mu_k(\omega) = \mathrm{E}(\omega^k(X))$ for $k = 1$ and 2.

The SIR algorithm is most useful when $h(x)$ is a good approximation of $f(x)$. This characteristic makes the SIR algorithm attractive in sensitivity studies (Smith and Gelfand, 1992) because once a sample from a distribution is generated, we can easily resample it to give a sample from similar distributions using the SIR algorithm. Lancaster (1997), and Koop and Poirier (2001) studied two hypotheses. They generated a sample under one hypothesis and resampled it using the SIR algorithm to give a sample under the other hypothesis.

The SIR algorithm is closely related to the well-known importance sampling (IS) method (Srinivasan, 2002). Suppose $\mathrm{E}(g(X)|X \sim f(x))$ is of interest. Given

$X_1, \ldots, X_M$ an iid sample from $h(x)$, the IS method uses

$$\sum_{i=1}^{M} g(X_i)\omega(X_i) \Big/ \sum_{i=1}^{M} \omega(X_i)$$

to estimate the expectation. Instead of using $\omega(X)$ as the weight in the weighted average, the SIR algorithm uses it as the weight for resampling, and estimates the expectation by $\sum_{i=1}^{m} g(Y_i)/m$. Clearly the IS method produces a better estimator than the SIR algorithm because resampling introduces noises. Estimate produced by the SIR algorithm can be viewed as a Monte Carlo estimate of the IS estimate, just like the Monte Carlo approximation of the bootstrap distribution.

The introduction of the randomness in the importance resampling step is in fact a trade-off for greater flexibility. A well-tuned sample is preferred to a single estimate or several estimates in many occasions. A sample provides a more complete picture of the distribution, and is an efficient way to represent a distribution. For example, in bootstrap filter (Gordon, Salmond, and Smith, 1993), the SIR algorithm is used to refine the sample because keeping track of how a sample evolutes is much easier than keeping track of how a density evolutes. It explains the attractiveness of the sample-based approach (Gelfand and Smith, 1990; Smith and Gelfand, 1992).

24.3   Selection of the pool size

It was pointed out in Section 24.1 that the SIR algorithm is good when M/m is large. It means that controlling the M value alone is good enough to achieve the following two purposes: (i) avoid positive dependence among Y_i's (negative dependence may be desirable) and (ii) the distribution of any Y_i in the resample is close to $f(x)$. Choosing an appropriate value of M to achieve the two purposes is fundamental in the application of the SIR algorithm. If $\omega(x)$ is not a constant function, elements in the resample are dependent. The relative magnitude of M and m monitors the dependence among $\{Y_i\}$. On the other hand, in order to control the closeness of the distribution of each Y_i to $f(x)$, our choice of M does not depend on m as the latter plays no role in this aspect. We will discuss each of these in the following subsections.

Relative magnitude of the pool and resample sizes

The SIR algorithm generates iid samples from $f(x)$ when M/m tends to infinity. We want M/m not to be small so as to avoid a lot of duplicates in the resample. Rubin (1987) suggested a linear rule, $M/m = 20$. Smith and Gelfand (1992) recommended $M/m \geq 10$. Theoretically, the relation between M and m should depend on the distribution of the importance ratio.

Let q_i be the count of X_i in the resample $\{Y_1, \ldots, Y_m\}$. Given the pool $\{X_1, \ldots, X_M\}$, we call a resampling algorithm a *sampling without-replacement algorithm* if q_i can only be either 0 or 1 with probability one; otherwise the algorithm is called *sampling with replacement*. We want the resampling algorithms to satisfy the following weighting condition:

$$E(q_i | X_1, \ldots, X_M) \propto \omega(X_i) \quad \text{for } i = 1, \ldots M.$$

If the resample size $m (= \sum_{i=1}^{M} q_i)$ is fixed, it becomes

$$E(q_i | X_1, \ldots, X_M) = m\omega(X_i) \Big/ \sum_{j=1}^{M} \omega(X_j) \quad \text{for } i = 1, \ldots M. \qquad (24.1)$$

Given M, m, and $\{X_1, \ldots, X_M\}$, a weighted sample without replacement exists if and only if

$$m\omega(X_i) \Big/ \sum_{j=1}^{M} \omega(X_j) \leq 1 \quad \text{for } i = 1, \ldots, M. \qquad (24.2)$$

As expected, condition (24.2) implies that m is less than or equal to M.

Define $[\lceil c \rceil]$ to be the ceiling of c, $\lfloor c \rfloor$ to be the floor of c, and $[c] = c - \lfloor c \rfloor$. From (24.1), the tightest possible bounds for q_i are

$$\left\lfloor m\omega(X_i) \Big/ \sum_{j=1}^{M} \omega(X_j) \right\rfloor \leq q_i \leq \left\lceil m\omega(X_i) \Big/ \sum_{j=1}^{M} \omega(X_j) \right\rceil. \qquad (24.3)$$

An algorithm is called *tight* if (24.3) holds for all i with probability one. We call a sample that satisfies (24.3) tight. A tight sampling algorithm will be sampling without replacement whenever condition (24.2) holds.

Large variation of q_i is clearly undesirable because it usually means large variation in estimation. For this reason, given m, M, and a pool $\{X_1, \ldots, X_M\}$, a resampling algorithm should be chosen to minimize the variability of q_i's. Clearly any tight sampling algorithm is optimal with respect to this criterion.

The count q_i relates directly to the positive dependence among elements in the resample. We want M large enough for the following condition to hold:

$$\Pr(q_i \leq b \text{ for all } i) \geq 1 - \gamma, \qquad (24.4)$$

for a given positive integer $b (b < m)$, and a small positive value γ. For a tight sampling algorithm, (24.4) means

$$\Pr\left(m \max_{i=1,\ldots,M} \omega(X_i) \Big/ \sum_{j=1}^{M} \omega(X_j) \leq b \right) \geq 1 - \gamma. \qquad (24.5)$$

Insights into the requirement in (24.5) on M can be obtained if the importance weights are iid from a Gamma(α, β) distribution, where α is the shape parameter and β is the scale parameter. Under this distribution assumption, $(\omega(X_1), \ldots, \omega(X_M))/\sum_{j=1}^{M} \omega(X_j)$ follows the Dirichlet$(\alpha, \ldots, \alpha)$ distribution.

$$\Pr(q_i \leq b \text{ for all } i)$$

$$= \Pr\left(m \max_{j=1,\ldots,M} Z_j \leq b \,\Big|\, (Z_1, \ldots, Z_M) \sim \text{Dirichlet}(\alpha, \ldots, \alpha)\right)$$

$$\geq 1 - M \Pr(Z > b/m | Z \sim \text{Beta}(\alpha, (M-1)\alpha)).$$

Condition (24.4) is fulfilled if

$$M \Pr(Z > b/m | Z \sim \text{Beta}(\alpha, (M-1)\alpha)) \leq \gamma. \tag{24.6}$$

When m, b, α, and γ are given, we should choose the smallest M value satisfying (24.6). Suppose that $m = 1,000$, $b = 1$, $\alpha = 2$, and $\gamma = 0.05$. Equation (24.6) suggests $M = 7,320$. Suppose that $h(x)$ is close to $f(x)$, so that α is large, say $\alpha = 1,000$. If all other values remain unchanged, from (24.6), the required $M = 1,129$, which is close to m.

To investigate how M and m are related, we show that for large M,

$$M \Pr(Z > b/m | Z \sim \text{Beta}(\alpha, (M-1)\alpha))$$

$$\leq \frac{M(\alpha b M/m)^{\alpha-1} \exp(-\{(M-1)\alpha - 1\}b/m)(1 + o(1))}{\Gamma(\alpha)}, \tag{24.7}$$

where $\Gamma(.)$ is the gamma function (the proof of (24.7) is available from the author). It suggests choosing M to be an integer close to the larger root of the equation

$$\gamma = M(\alpha b M/m)^{\alpha-1} \exp(-\{(M-1)\alpha - 1\}b/m)/\Gamma(\alpha). \tag{24.8}$$

This formula gives quite a good approximation to (24.6). When $m = 1,000$, $b = 1$, $\alpha = 2$, and $\gamma = 0.05$, equation (24.8) gives $M = 7,286$. If α changes to $1,000$, $M = 1,119$. They show a little downward bias when compared with the use of (24.6). If both M and m are much larger than α, equation (24.8) can be approximated by $\gamma = M(\alpha b M/m)^{\alpha-1} \exp(-\alpha b M/m)/\Gamma(\alpha)$ implying that $\alpha b M/m \approx \log(M) + (\alpha - 1)\log\log(M) - \log(\gamma\Gamma(\alpha))$. This shows that M/m increases in an order $\log(M)$.

Magnitude of the pool size

Let C be an arbitrary set in the sample space of Y. Define the indicator function of the set C by $I_C(X)$. It takes value 1 when $X \in C$ and 0 otherwise. As pointed out in Section 24.2, the estimator of $\Pr(X \in C | X \sim f(x))$ based on the resample is a Monte Carlo approximation of the IS estimator, $\sum_{i=1}^{M} I_C(X_i)\omega(X_i)/\sum_{i=1}^{M} \omega(X_i)$. Lee (1997) used M to make the mean squared error (MSE) of the IS estimator bounded above by a small value, ϵ. He showed that the MSE is asymptotically bounded above by $\mu_2(\omega)/(M\mu_1^2(\omega))$ and suggested that M should be chosen to be $\mu_2(\omega)/(\epsilon\mu_1^2(\omega))$. This choice is not very satisfactory because if ϵ is small, it asks for large M even when $h(x) = f(x)$.

Let Y be a random element from the resample $\{Y_1, \ldots, Y_m\}$. We want $\Pr(Y \in C)$ to be equal or very close to $\Pr(X \in C|X \sim f(x))$. A better measure of the performance of the algorithm is $|\Pr(Y \in C) - \Pr(X \in C|X \sim f(x))|$.

$$|\Pr(Y \in C) - \Pr(X \in C|X \sim f(x))|$$

$$= \left| E\left(\sum_{i=1}^{M} I_C(X_i)q_i \bigg/ \sum_{i=1}^{M} q_i\right) - \Pr(X \in C|X \sim f(x)) \right|$$

$$= \left| E\left(\sum_{i=1}^{M} I_C(X_i)\omega(X_i) \bigg/ \sum_{i=1}^{M} \omega(X_i)\right) - \Pr(X \in C|X \sim f(x)) \right|. \quad (24.9)$$

Equation (24.9) reveals that (i) the performance of the algorithm depends on the bias rather than the MSE of the IS estimator, and (ii) as far as (24.1) is satisfied, the quality of each individual Y_i does not depend on the resampling method used.

From (24.9), it can be shown that

$$|\Pr(Y \in C) - \Pr(X \in C \mid X \sim f(x))|$$

$$= \frac{|\mu_2(\omega)E(\omega(X)I_C(X)) - \mu_1(\omega)E(\omega^2(X)I_C(X))|}{M\mu_1^3(\omega)} + o\left(\frac{1}{M}\right)$$

$$\leq \frac{E(|\mu_2(\omega)\omega(X) - \mu_1(\omega)\omega^2(X)|)}{2M\mu_1^3(\omega)} + o\left(\frac{1}{M}\right). \quad (24.10)$$

Given a maximum tolerable error, δ, the pool size, M, should be chosen so that

$$\frac{E(|\mu_2(\omega)\omega(X) - \mu_1(\omega)\omega^2(X)|)}{2\delta\mu_1^3(\omega)} \leq M. \quad (24.11)$$

If $\omega(X)$ follows a Gamma(α, β) distribution, (24.11) becomes

$$\frac{(\alpha + 1)^{\alpha+1}\exp(-\alpha - 1)}{\delta\alpha\Gamma(\alpha + 1)} \leq M.$$

When α is large (that is, the importance sampling pdf is close to $f(x)$), we can apply the Stirling's formula for the gamma function (Abramowitz and Stegun, 1964, P. 257), and obtain the following choice of M

$$M \approx 1/(\delta\sqrt{2\pi\alpha}).$$

If α is very small (that is, the importance sampling pdf is a very poor approximation of $f(x)$), choose

$$M \approx 1/(\delta\alpha e).$$

24.4 Selection criterion of the importance sampling distribution

Like the IS method, the SIR algorithm will be inefficient if the importance sampling pdf is not close to $f(x)$. When sampling parameter from its posterior distribution, normal approximation is a good importance sampling pdf. Of course, t-distribution is an attractive alternative because of its closeness to the normal distribution and its thicker tails. We can further extend the family of distributions to a mixture of the above distributions. See for example, Gelman and Rubin (1992), Gelman et al. (2003), and Tan, Tian, and Ng (2003).

The selection of an importance sampling pdf for the IS method has attracted many discussions (Srinivasan, 2002). There is, however, a basic difference in the focus between the IS method and the SIR algorithm: The main concern in the IS method is on the variance (or MSE), but that in the SIR algorithm is on the bias as demonstrated in (24.9). Usual adaptive approach (Oh and Berger, 1992; Schmidt, Gamerman, and Moreira, 1999) chooses a parametric family of pdf $h_\lambda(x)$. An adaptive rule is then activated to improve the choice of λ as more samples are drawn. A good family of pdfs should possess the following properties: (i) it is easy to draw sample from any of its members, and (ii) the family is rich in the sense that $f(x)$ can be well approximated by at least one member in the family.

Equation (24.10) suggests that the performance of an importance sampling pdf can be measured by

$$E(|\mu_2(\omega)\omega(X) - \mu_1(\omega)\omega^2(X)|)/\mu_1^3(\omega).$$

The smaller the value, the better the $h_\lambda(x)$. Let $\omega_\lambda(X)$ be the importance ratio of X for the importance sampling pdf $h_\lambda(x)$ and $\mu_k(\omega_\lambda) = E(\omega_\lambda^k(X)|X \sim h_\lambda(x))$ be the kth moment of $\omega_\lambda(X)$. Suppose $\omega_\lambda(X) = \tau f(X)/h_\lambda(X)$, then $\mu_1(\omega_\lambda) = \tau$. Thus, $h_\lambda(x)$ is a good importance sampling pdf, if for $X \sim h_\lambda(x)$,

$$E(|\mu_2(\omega_\lambda)\omega_\lambda(X) - \tau\omega_\lambda^2(X)|) \qquad (24.12)$$

is small.

Suppose $f(x)$ is the beta-binomial distribution with $n = 5$, $\alpha = 3$ and $\beta = 2$. Take Bin(n, λ) as the importance sampling pdf. The optimal λ value that minimizes (24.12) is $\lambda = 0.581$. If $n = 5$, $\alpha = \beta = 3$, $f(x)$ is symmetric with respect to $n/2$, and (24.12) suggests $\lambda = 0.5$ as expected.

If we have $X_1, \ldots, X_T$ iid from $h_{\lambda_0}(x)$, which has equal or larger support than $h_\lambda(x)$, the quantity in (24.12) can be estimated by

$$\frac{1}{T}\sum_{j=1}^{T}\omega_{\lambda_0}(X_j)\left|\frac{1}{T}\sum_{i=1}^{T}\omega_\lambda(X_i)\omega_{\lambda_0}(X_i) - \left(\frac{1}{T}\sum_{i=1}^{T}\omega_{\lambda_0}(X_i)\right)\omega_\lambda(X_j)\right|.$$

We should look for λ, which minimizes the above estimate.

24.5   The resampling algorithms

How a resample is selected from a pool is an important problem in the SIR algorithm. When M is fixed, it is not meaningful to restrict our attention to sampling without-replacement algorithms because it is difficult to rule out the possibility that condition (24.2) fails. Two common situations that cause trouble are (i) when the importance ratio is unbounded from above and (ii) when the importance ratio is not bounded away from zero.

In Section 24.3, tight sampling algorithms are defined and recommended. They minimize the variability of each q_i, and thus usually produce an estimator with a smaller standard error. Generating a tight sample is easy once a weighted sampling without-replacement method is available. For any tight sampling algorithm, we have $q_i \geq \lfloor m\omega(X_i)/\sum_{j=1}^{M} \omega(X_j) \rfloor$. This means that at least $\lfloor m\omega(X_i)/\sum_{j=1}^{M} \omega(X_j) \rfloor$ copies of X_i must appear in the resample. We call these $\lfloor m\omega(X_i)/\sum_{j=1}^{M} \omega(X_j) \rfloor$ copies of X_i a self-selective sample. Let Ω be the set of all self-selective samples, that is, Ω contains $\lfloor m\omega(X_1)/\sum_{j=1}^{M} \omega(X_j) \rfloor$ copies of X_1, $\lfloor m\omega(X_2)/\sum_{j=1}^{M} \omega(X_j) \rfloor$ copies of X_2, and so on. Denote the size of Ω by $\nu(\Omega)$. Let Ξ contain a weighted sample from $\{X_1, \ldots, X_M\}$ of size $m - \nu(\Omega)$. The weight assigned to $X_i (i = 1, \ldots, M)$ is $[[m\omega(X_i)/\sum_{j=1}^{M} \omega(X_j)]]$. It can be shown that these revised weights satisfy condition (24.2). Therefore, Ξ can be a without-replacement sample, and $\Omega \cup \Xi$ is a tight sample of size m.

A sampling algorithm from a pool is called a *one-pass algorithm*, if each element in the pool is to be read once in order to draw the final sample. One-pass sampling algorithm has some attractive advantages. First, the pool need not to be stored, and the computer storage requirement is independent of the pool size M. Second, many one-pass algorithms do not need prior knowledge of M and thus we can choose M adaptively to control the accuracy of the method. Third, if we want to generate a sequence of $\{Y_i\}$, the sampling algorithm must be a one-pass algorithm.

The SIR algorithm generates a sample of size m. We can extend the idea of the SIR algorithm to cases when we want $n(n > 1)$ samples each of size m, and when we want a sequence of values. We will discuss each of the three cases in the following sections.

Generating one sample

In survey sampling, many sampling algorithms have been proposed to draw a weighted sample with or without replacement from a finite population. Brewer and Hanif (1983) discussed 50 of them, not to mention the significant additions after 1983. With the weights defined to be the importance ratios, many weighted sampling algorithms can be used in the importance resampling step of the SIR algorithm. In this subsection, we consider only four methods. They are the simple weighted sampling with replacement, the Yates–Grundy draw by draw procedure, the ordered systematic procedure, and Chao's algorithm. The first two are the algorithms that are currently in use with the SIR algorithm. The ordered systematic procedure is chosen because of its simplicity and its relation with another algorithm in Section 24.5. Chao's algorithm is a one-pass algorithm and can be modified to generate a tight sample.

Simple weighted sampling with replacement

It is the standard weighted sampling with replacement method. Elements are drawn independently one by one with replacement. In each draw, an X_i in the pool is selected with probability equal to $\omega(X_i)/\sum_{j=1}^{M} \omega(X_j)$. Brewer and Hanif (1983) called it the *multinomial sampling* because of its close relation to the multinomial distribution. It is a popular sampling method used in the importance resampling step of the SIR algorithm (Smith and Gelfand, 1992; Gordon, Salmond, and Smith, 1993, McAllister, Pikitch, Punt, and Hilborn, 1994; Avitzour, 1995; Gordon, Salmond, and Ewing, 1995; Lancaster, 1997; Lopes, Moreira, and Schmidt, 1999; Koop and Poirier, 2001; Tanizaki, 2001). This sampling method is simple. However, from the statistical point of view, it is usually (but not always) less efficient than the weighted sampling without-replacement algorithms when the latter is applicable.

Yates–Grundy draw by draw procedure

The Yates–Grundy draw by draw procedure (Yates and Grundy, 1953) is a weighted sampling without-replacement algorithm in sample survey. Units are drawn one by one. The first unit in the sample is selected with probability proportional to $\omega(X_i)$; the second unit, without replacement, again with probability proportional to $\omega(X_i)$; and so on until m values are drawn. It is the method used in Raghunathan and Rubin (1990), Gelman (1992), Gelman et al. (1995), and Tan, Tian, and Ng (2003). Efficient one-pass algorithm is available in Li (1994). A major drawback of this method is that condition (24.1) holds only approximately when M/m is large. It explains why it can always give a sample without replacement even if condition (24.2) is violated.

Ordered systematic procedure

The ordered systematic procedure (Hartley, 1966) is a tight sampling algorithm. To use it in the SIR algorithm, we arrange X_i's in a convenient order. Compute $S_i = \sum_{j=1}^{i} \omega(X_j)$ for $i = 1, \ldots, M$. Generate $u \sim U(0, S_M/m)$. For $i = 1, \ldots, m$, let $Y_i = X_j$, where j is the smallest integer such that $u + (i-1)S_M/m \leq S_j$. Then $\{Y_1, \ldots, Y_m\}$ is the resample. Kitagawa (1996) considered the case when u is a fixed value, and found in an example that it outperforms the simple weighted sampling with replacement method.

As we are free to determine the ordering of X_i's, we can take advantage of this flexibility. For example, if we want to estimate $E(g(X)|X \sim f(x))$, a regular sample can be obtained if X_i's are ordered according to the magnitude of $g(X_i)$'s. A disadvantage of this method is that it is not a one-pass algorithm, and the pool has to be stored in the computer.

Chao's algorithm

Chao's (1982) algorithm is a one-pass sampling method. When the algorithm is applied to the SIR algorithm, it gives a sample, which is the union of two disjoint sets, say A and B. Let $S_M = \sum_{j=1}^{M} \omega(X_j)$, which is available when Chao's algorithm ends. The set $A = \{X_i : m\omega(X_i)/S_M > 1\}$ (in Chao's notation $A = A_{M-1}$), and B is a weighted random sample without replacement from $(\{X_1, \ldots, X_M\} - A)$ of size $(m - \nu(A))$ with $\omega(X_i)$ being the weight of X_i. When condition (24.2) holds, A is empty, and B is a without-replacement sample. We can modify Chao's algorithm to yield a tight sample when A is not empty. As defined in Section 24.5, we let Ω be the set of all self-selective samples. Draw Ξ a weighted sample without replacement of size $(m - \nu(\Omega))$ from $A \cup B$ with weight $[[m\omega(X_i)/S_M]]$ for every $X_i \in A$, and with weight $m(S_M - \sum_{j \in A} \omega(X_j))/(\nu(B)S_M)$ for every $X_i \in B$. Then $\Omega \bigcup \Xi$ is a tight sample.

Generating multiple samples

Like the bootstrap, we may want n samples of size m. A simple way to generate the samples is to repeat the SIR algorithm independently n times. Each time a sample of size m is simulated. We call this method the independent multiple sampling (IMS) method. Another way is to use the SIR algorithm to generate a sample of size mn, then randomly partition the sample into n sets of m values. If a tight sampling is used in the importance resampling step, we call this method a balanced multiple sampling (BMS) method because it reduces to the balanced bootstrap (Davison, Hinkley, and Schechtman, 1986) when $M = m$ and $h(x) = f(x)$.

It is of interest to compare the BMS method with the IMS method when a tight sampling method is used. Let the pool size used in the BMS method be nM, where M is the pool size used in the IMS method. Therefore, both methods require the same number of X_i's. As the pool size used in the BMS method is nM, which is larger than the pool size M in the IMS method, from (24.10), an individual sample from the BMS method is of better quality than that from the IMS method. On the other hand, let q_i^* be the count of X_i in the resample of size mn. Assume that $\omega(X)$'s are iid Gamma(α, β) distributed. From (24.7), for the IMS method,

$$\Pr(q_i^* \le b \text{ for all } i)$$
$$\ge \{1 - M(\alpha bM/m)^{\alpha-1} \exp(-\{(M-1)\alpha - 1\}b/m)(1 + o(1))/\Gamma(\alpha)\}^n$$
$$= 1 - nM(\alpha bM/m)^{\alpha-1} \exp(-\{(M-1)\alpha - 1\}b/m)(1 + o(1))/\Gamma(\alpha).$$

For the BMS method, the chance is

$$\Pr(q_i^* \le b \text{ for all } i)$$
$$\ge 1 - nM(\alpha bM/m)^{\alpha-1} \exp(-\{(nM-1)\alpha - 1\}b/(mn))(1 + o(1))/\Gamma(\alpha)$$
$$= 1 - nM(\alpha bM/m)^{\alpha-1} \exp(-\{(M-1/n)\alpha - 1/n\}b/m)(1 + o(1))/\Gamma(\alpha).$$

The BMS method has a slightly larger lower bound than the IMS method. In summary, the BMS method is preferred to the IMS method as it has better individual property and slightly larger lower bound for the chance of having less duplicates in the resample. A drawback for the BMS method is that the mn selected values are needed to be stored.

Generating a sequence of values

Like the uniform random number generation, it is desirable if we can generate a sequence of random values. To do so, we construct and simulate a stochastic process $\{Y_i\}$. The process should have $f(x)$ as its unique stationary distribution. Gelman (1992) proposed an iterative procedure that fails to satisfy the above requirement. We are going to introduce a new iterative procedure.

Consider the following continuous time process $\{X(t): t \geq 0\}$.

Step 1: Set $r = 0$.
Step 2: Generate Z from $h(x)$, and a variable value $L(L \geq 0)$ given Z.
Step 3: Set $X(t) = Z$ for $r \leq t < r + L$ when $L > 0$.
Step 4: Increase r by L and go to Step 2.

The process $\{X(t)\}$ is a regenerative process (see for example, Shedler, 1993). Variable $X(t)$ converges in distribution to $f(x)$ if $E(L|Z) \propto \omega(Z)$, the importance ratio of Z. We call a process satisfying the above condition an importance weighted regenerative (IWR) process. To get a sequence of Y_i from a realization of an IWR process, we choose a length of the "burn-in" period $c(c \geq 0)$, and a sampling interval $k(k > 0)$. Generate a random starting point u from $U[c, c + k)$, and set $Y_i = X(u + (i - 1)k)$ for $i = 1, 2, \ldots$.

A simple IWR process is to set $L = \omega(Z)$. It relates closely to the ordered systematic sampling. The differences are (i) the sampling interval k is fixed rather than data-dependent in the ordered systematic sampling, (ii) we do not have a finite pool in the IWR process, and (iii) the IWR process has a "burn-in" period.

Another reasonable choice is to set $k = 1$, and $c = 0$ and let L be a random variable taking only nonnegative integer values. In this case, L is simply the number of copies of Z in the final sample. This form of IWR process relates to the Friedman's sampling method and the method considered in Patil and Rao (1977). For a given positive scale parameter λ, the former corresponds to the case when L takes value either $\lfloor \lambda \omega(Z) \rfloor$ or $\lceil \lambda \omega(Z) \rceil$, while the latter has L equal to either 0 or $\lceil \lambda \omega(Z) \rceil$. Both methods reduce to the acceptance–rejection method when λ is chosen such that $\lambda \omega(Z) \leq 1$ for all Z in the support of $h(x)$.

The use of the IWR process does not have the problem of determining M. However, we have to choose a sampling interval k and the length of the "burn-in" period c. The former plays a crucial role in q, the number of copies of an $X(t)$ in the output values. As $\Pr(q \leq b) \geq \Pr(L < bk)$. With a given integer b in mind, pick a k value large enough so that $\Pr(L < bk)$ is close to one.

To diminish the effect of successive identical Y's when we encounter Z values with large $\omega(Z)$, we can apply the table-shuffling method that works successfully in the combination of pseudorandom numbers (MacLaren and Marsaglia, 1965; Nance and Overstreet, 1978). First, a table of size n is initialized to store $Y_1, \ldots, Y_n$. Whenever a random sample from $f(x)$ is wanted, randomly select one element from the table as our sample. Then replace the selected element in the table by the next Y_i.

24.6 Discussion

Apart from the fundamental problems considered in this chapter, other topics about the SIR algorithm are worth discussion. The requirement that elements in the pool, $\{X_1, \ldots, X_M\}$, are iid samples is too restrictive. Lancaster (1997), and Koop and Poirier (2001) applied the SIR algorithm to optimal job-search models. However, the pool is not a simple random sample but a realization of a Gibbs sampler. A simple way to gain efficiency is to allow negative dependence among X_i's. The ideas of stratification and systematic sampling are helpful in this direction. The use of nonidentically distributed sample is also desired, especially when an adaptive rule is implemented to update our choice of the importance sampling pdf as more and more values are sampled.

Real-time control of M is also an interesting problem. Instead of fixing M before simulation, we can go on sampling from $h(X)$ until a certain stopping criterion is satisfied. A simple stopping rule is to stop simulating X_i when M is larger than a fixed value and condition (24.2) is satisfied. This stopping rule ensures that there are no duplicates in the output. Chao's algorithm is a good resampling algorithm for real-time control of M because it is a one-pass algorithm and does not require the knowledge of M before sampling.

Part IV

Applied Bayesian inference

25

Whither applied Bayesian inference?

Bradley P. Carlin[1]

In this chapter, I would like to offer some thoughts on where we have been, where we are, and where we hope to go as applied Bayesians in the next few years. Mercifully, this task is far easier than reviewing Prof. Rubin's contributions to the field (a summary of the impact of the EM algorithm and its extensions alone would easily fill my available space), or even giving a cogent overview of the applied Bayesian papers included in this part of the book (which also cover a very broad range of application). Nor will I attempt a textbook-style review of applied Bayesian methods, since that is not the purpose of this book and, in any case, many such fine treatments (including of course Gelman, Carlin, Stern, and Rubin, 2003) already exist.

25.1 Where we've been

There was a time not so long ago (say, 1960) when the term "applied Bayes" would have been viewed by virtually all practicing statisticians as inherently self-contradictory, as we would now view the terms "jumbo shrimp" or "sensible p-value." This was because, prior to modern computing, believable analyses were

[1] Division of Biostatistics, School of Public Health, University of Minnesota, Minneapolis, Minn. The author is grateful to Profs. Andrew Gelman and Xiao-Li Meng for their assistance in preparing these remarks, and to Prof. Donald Rubin for a lifetime of insightful, passionate, relentless, and irreverent devotion to Bayesian statistical science.

Applied Bayesian Modeling and Causal Inference from Incomplete-Data Perspectives.
Edited by A. Gelman and X.-L. Meng © 2004 John Wiley & Sons, Ltd ISBN: 0-470-09043-X

not possible unless one could somehow argue that the low-dimensional model and conjugate prior employed were actually reflective of real life. Of course, this is not to say that there was no applied Bayes work at all prior to this time; examples can be cited as far back as Laplace, and Bayesian methods had a robust (if primarily theoretical) following under the banner of "inverse probability." Moreover, the seeds of Bayesian thinking creeping into the applied work of mainstream statisticians were already being sown in the 1960s by James and Stein (1961), who showed that Bayesianly derived estimators could outperform traditional ones even when judged by frequentist criteria. Still, the lion's share of the history of applied Bayesian inference is irretrievably tied to the history of computing. In my applied Bayesian courses, I often refer to the "prehistoric" period in Bayesian computing as starting in 1763 (with the publication of Bayes' rule) and ending sometime during the 1960s, when compiled languages like `Fortran` coupled with Newton–Cotes type integration routines at last made "real" applied Bayesian work possible—at least for models with no more than 10 or 12 unknown parameters.

Fortunately, the publication of the EM algorithm by Dempster, Laird, and Rubin (1977) and subsequent 1980s emergence of practical Monte Carlo integration methods and 1990s popularization of the Gibbs and Metropolis samplers changed this situation forever. (These changes took place roughly in parallel with the development and popularization of the similarly computationally intensive bootstrap method within the frequentist camp.) Since the 1980s, every statistician has had enough computing power in a desktop workstation or PC to allow these algorithms to deliver good Bayesian answers for a remarkably broad array of applied problems. Starting around 15 years ago, applied Bayesians systematically conquered virtually every standard model typically used by working statisticians: linear, loglinear, nonlinear, categorical, longitudinal, latent variable, mixture, survival, spatial, and on and on. Perhaps even more remarkable, the reduced reliance of the Monte Carlo–Bayes approach on closed forms and restrictive assumptions meant that it quickly surpassed the ability of traditional likelihood methods in handling complicated or nonstandard settings. Small but annoying problems like nonnormal error distributions, unbalanced data, unequal variances across populations, and the like that greatly complicated the asymptotic theory supporting traditional methods now caused no additional complexity in the Bayesian solution. Since this solution was already the conceptually simpler one, it became the case that when the applied statistical going gets tough, the tough get Bayesian.

Of course, without access to user-friendly software, Bayesian methods might still be the purview of just a few PhD-level Bayesian statisticians with the stomach for low-level programming. But during the last decade or so, the freely available `WinBUGS` language has enabled statisticians with just an MS (or even an advanced undergraduate) understanding of probability distributions and standard models to become genuine applied Bayesians. Several other competitor programs have emerged recently, but in all cases the basic approach remains the same: code up the model (probably by modifying an existing piece of code, just as one would do in any statistical package), run a few parallel sampling chains, check that the algorithm's convergence is acceptable by looking at trace plots and maybe computing some (Gelman and Rubin, 1992) diagnostics, then create posterior summaries from a lengthy "production run" of post-burn-in MCMC samples. The particular inferential task at hand (estimation, testing, etc.) causes the "script" to diverge at this point, but not in the sense that everything still arises from the manipulation of these posterior samples.

25.2   Where we are

What then can we say about the current state of applied Bayesian statistics? Certainly there is now a widespread awareness that applied Bayesian tools are valuable in advanced modeling settings, and should be included in the kit bag of every well-educated senior statistician. Most leading statistics and biostatistics graduate programs now include courses in Bayesian methods at the PhD level, and some (including, I am proud to say, my own department) at the Master's level as well. However, the methods seem to be widely viewed as another "special topic" in both practice and education, akin to other recent methodological areas of development such as spatial statistics, genomics, or causal inference. But Bayesians do not view their approach as another methodological subcategory; there is Bayesian spatial, Bayesian genomics, and Bayesian causal. Clearly there is still work to be done in getting mainstream statistical practitioners and educators to think of our approach as a general one to be used in *all* situations, not merely when standard methods fail.

This point is related to the one made by the current ASA President, Brad Efron (2004), in his first *Amstat News* editorial, where he argued that the Bayes-frequentist controversy is at this point mostly resolved; philosophically we have all moved to the middle, and statistics is now "a unified discipline". This is certainly an entirely appropriate and upbeat view for an incoming ASA President to espouse. It is also true that traditional statisticians and other applied scientists now recognize the deficiencies in the frequentist paradigm (many medical journals now prohibit p-values, and instead insist on point estimates and confidence intervals) and incorporate many quasi-Bayesian elements in their work (SAS Proc Mixed leaps to mind). And for their part, the predominant view among Bayesians is now no longer subjectivist but objectivist, strongly respecting the data and exhibiting a healthy skepticism of "expert opinion." Despite all this, I still find Prof. Efron's view to be on the optimistic side. The problems with traditional significance testing are not going to go away, and as computing continues to advance, statisticians' comfort even with well-loved and time-honored traditional estimation methods (based as they often are on unrealistic assumptions and approximate asymptotic normality) is likely to increase.

To be fair, at this point we ought to take stock of the problems that still vex applied Bayesian inference. The first is of course the difficulty in MCMC convergence assessment. Recently, Lange (2004) somewhat darkly observed that, "Practical failures of the ergodic theorem are the Achilles heel of MCMC." While we have a large class of diagnostic plots and statistics to assist us, it is certainly true that all can be fooled if the situation is chosen carefully enough, and convergence diagnosis for arbitrary models should not be left to the inexperienced (or to the machine itself). However, experience in this regard is slowly being accumulated model class by model class, and I think there is reason to be optimistic that convergence can be concluded with confidence for a very wide range of standard models with "default priors." This brings up the second major area of difficulty, namely, the specification of noninformative prior distributions. Years of sophisticated theoretical struggle with concepts such as finite additivity have yielded relatively little of practical value, and applied Bayesians continue to rely largely on well-known "flat priors." But here again, thinking of these priors as "defaults" may not be entirely bad, since the lack of established conventions is part of what has hindered Bayesians in their competition with traditional methods. Ongoing developments in default priors (e.g., in finding alternatives to the much-maligned $\text{Gamma}(\epsilon, \epsilon)$ prior for variance components) also seem promising.

Finally, we should note areas in which Bayesians have not too few solutions, but too many (thus leading to discord among competing camps). Here the most obvious area seems to be hypothesis testing and model choice. Gone are the days when all Bayesians agreed that Bayes factors were the only sensible solution: their reliance on proper priors and their nature as a (sometimes erratic) single number summary have greatly reduced the frequency of their appearance. Replacing them are a veritable alphabet soup of penalized likelihood criteria (AIC, BIC, DIC, NIC, TIC), conditional predictive summaries, posterior predictive loss criteria (Gelfand and Ghosh, 1998), and Bayesian p-values and posterior predictive checks (Gelman, Meng, and Stern, 1996). It seems that a lot more experience must be built up before any of these alternatives emerges as a "default" approach. Second, rather than developing richer and richer models for our data, it would likely behoove us to work harder to understand the models we have already got. Liu and Hodges (2003) recently offered a surprising characterization of bimodality in the balanced one-way random effects model, a setting most Bayesians would likely regard as well understood. A better understanding of the identifiability of variance components in hierarchical longitudinal and spatial models would certainly help us in both prior selection and MCMC algorithm design and implementation. A final (and long-standing) area of Bayesian discord might be the extent to which mere posterior summarization is sufficient for a standard analysis, and how many situations require more (e.g., full-blown decision theory, or reference to underlying casual structure).

25.3 Where we're going

Where will the field of applied Bayesian statistics be in 10 or 20 years? Legendary Bayesian thinker Dennis Lindley once famously predicted that the 21st century would be a "Bayesian century"; that prediction now looks somewhat optimistic, and in 1994 Prof. Lindley himself put off the century's arrival date to 2020 (Smith, 1995, p. 317). My own thinking is somewhere between that of Prof. Lindley and

Prof. Efron, in that, while applied Bayes will continue to gain prominence, ultimately the statistical and broader scientific communities will do as they have always done, retaining and using whatever classical *or* Bayesian ideas that deliver sensible answers for a reasonable investment (in terms of both human and computer time). To this end, the current and long-awaited development by SAS of an MCMC-based Bayes procedure will certainly win converts that an academic-style piece of freeware like WinBUGS never could.

In the biostatistical application areas with which I am most familiar, Bayesian methods have undoubtedly made a significant impact, and in many cases seem poised to fundamentally change the way scientific evidence is accumulated, analyzed, and reported. In clinical trials, the FDA Center for Devices and Radiological Health has been recommending Bayesian designs and stopping rules for ten years now, as a way of utilizing the large historical databases often available in device settings while simultaneously limiting the sample sizes needed to understand the safety and efficacy of new devices. As discomfort with the high costs of drug trials (in terms of both dollars and patient lives) continues to rise, the FDA Center for Drug Evaluation and Research is now also showing interest in hierarchical modeling and design, following the approach initially popularized by Thall, Simon, and Estey (1995). In spatial epidemiology, Bayesian methods for combining evidence across similar but distinct units are particularly helpful, since the random effects distribution can be tailored to the spatial pattern anticipated, regardless of whether the data are observed at point (say, latitude/longitude) or areal (say, county) level. The hierarchical framework is also very natural for resolving misalignment between two spatially oriented variables, or for handling data that are multivariate, spatiotemporally indexed, or feature a wide variety of other complexities (Banerjee, Carlin, and Gelfand, 2004). These are just two areas in my own work where sensible answers simply would not be available were we forced to rely solely on traditional methods.

While the MCMC-Bayes revolution will no doubt have significant repercussions on industry and government, the impact on academia will be especially interesting to observe. In a way, academic Bayesians have slit their own throats with MCMC, since an analysis that would have been *JASA*-worthy 15 years ago might now cause an associate editor only to yawn and say, "That's just another straightforward application of the Gibbs sampler, isn't it?" The prominence of Bayesian articles in the mainstream journals has led to more and more Bayesians on influential editorial boards. But the simultaneous maturation of the field and change of emphasis from convenient mathematical closed forms to complex numerical solutions implies that we may need to update our definition of "a good paper." The next 20 years will likely see applied Bayesian research spending less time on traditional statistical modeling and more time on integrating Bayesian thinking into the many substantive areas that have yet to benefit from it. Certainly, I think the days when a "good paper" featured a new method, some impressive (but often inapplicable) asymptotic theory, a simulation study or two, and an illustration with a toy data set ought to be behind us. Not only must real applications motivate our theory but they must also do it in a way that answers our nonstatistical colleagues' questions, rather than "shoehorning" them into models and paradigms we have sitting on the shelf.

The future of applied Bayes in statistical education is also interesting to ponder. The MCMC-Bayes revolution notwithstanding, the number of undergraduate Bayesian courses remains small; as mentioned above, most educators still seem to view the area as a secondary topic to be learned only after one has learned "the basics" of traditional statistical inference. One can certainly teach t-statistics and corresponding table lookups to students without any calculus background; they will not really understand what is going on, but should still be able to handle data sets that do not deviate from the standard designs. With its reliance on explicit modeling, most Bayesian courses and textbooks have historically admitted only students with at least a rudimentary understanding of distribution theory, a much higher threshold. But as more "point and click" MCMC-Bayes software appears, perhaps Bayesians will be able to reach the younger and less mathematically sophisticated audiences of Statistics 101? I do think this will happen, and indeed to have the impact on future scientific research that we desire, we must reach beyond the undergraduates of Bayes-dominated departments like Duke and Carnegie Mellon and start all undergraduates' training in statistical literacy with Bayesian thinking, since these are the very people who will one day be the scientists analyzing important questions in their own fields.

In closing, let me say that the future of applied Bayesian statistics appears bright, with the articles in this book providing particularly compelling evidence. I look forward to writing an even more upbeat assessment of the field for the book in honor of Don Rubin's 120th birthday.

26

Efficient EM-type algorithms for fitting spectral lines in high-energy astrophysics

David A. van Dyk and Taeyoung Park[1]

26.1 Application-specific statistical methods

In recent years, a progressive new trend has been growing in applied statistics: It is becoming ever more popular to build application-specific models that are designed to account for the hierarchical and latent structures inherent in any particular data generation mechanism. Such multilevel models have long been advocated on theoretical grounds, but the development of methodological and computational tools for statistical analysis has now begun to bring such model fitting into routine practice. In this chapter, we discuss one such application, the use of highly structured models to analyze spectral and spatial data obtained with modern high-resolution telescopes that are designed to study the high-energy end of the electromagnetic spectrum (e.g., X-rays and Gamma-rays). In particular, we consider the high-resolution data that is available from the space-based *Chandra X-ray Observatory*.

[1]Department of Statistics, University of California, Irvine and Department of Statistics, Harvard University, Cambridge, Mass. The authors gratefully acknowledge funding for this project partially provided by NSF grant DMS-01-04129 and by NASA Contract NAS8-39073 (Chandra X-ray Center). This chapter summarizes one thread of the work of the California–Harvard Astrostatistics Collaboration (www.ics.uci.edu/~dvd/astrostat.html). In addition to the authors, active participants include A. Connors, D. Esch, P. Freeman, H. Kang, V. L. Kashyap, X. L. Meng, A. Siemiginowska, E. Sourlas, Y. Yu, and A. Zezas.

Applied Bayesian Modeling and Causal Inference from Incomplete-Data Perspectives.
Edited by A. Gelman and X-L. Meng   © 2004 John Wiley & Sons, Ltd   ISBN: 0-470-09043-X

Launched in 1999 by the space shuttle *Columbia*, *Chandra* provides a new class of high precision instrumentation that allows for much more precise imaging of distant X-ray sources. The first author has been working on developing methods for handling this data since before *Chandra* was launched; there are a number of citations listed below that fill in many details of what is presented here and discuss related topics in astrostatistics.

X-ray telescopes such as *Chandra* can map nearby stars with active magnetic fields, the remnants of exploding stars, areas of star formation, regions near the event horizon of a black hole, very distant but very turbulent galaxies, or even the glowing gas embedding a cosmic cluster of galaxies. The production of X-ray emission requires temperatures of millions of degrees and indicates the release of stored energy such as that in very strong magnetic fields, extreme gravity, explosive nuclear forces, or shock waves in hot plasma. Thus, X-ray observations give astrophysicists a window into the physical processes involved in these turbulent regions of the universe, which is unavailable from observations of visible light. Because of the complexity of the sources themselves as well as the data collection process, however, unlocking this window requires sophisticated statistical modeling and analysis. For example, the recorded X-rays are a mixture of X-rays from a number of physical processes within the source. The X-rays are also subject to the so-called effective area, a nonignorable stochastic censoring process: The probability that an X-ray is observed depends on its energy, one of the variables of primary interest. The energy and the originating sky coordinates of each X-ray are observed with error and X-ray observations are subject to background contamination. (More background on the relevant astrophysics and instrumentation appears in van Dyk et al. (2004)). To handle these various factors we generally adopt a Bayesian perspective and construct highly structured multileveled models. Sophisticated computational tools such as EM-type algorithms and MCMC samplers are required for model fitting.

In this chapter, we describe a particular applied question that has come up in our work in astrophysics. Namely, we describe computational methods for fitting narrow emission lines in high-energy spectral analysis. Spectral analysis aims to describe the distribution of the energy of photons emitted from a particular source; here we focus on a high-energy interval of energies, the X-ray band. An emission line is a narrow range of energy with excess electromagnetic emission, relative to nearby intervals of energy. Such emission lines appear as sharp jumps in the distribution over a narrow range of energies. Emission lines are formed when electrons of energized ions fall down to lower energy shells and the excess energy is emitted in the form of a photon. Because of the distinct quantum differences between the energies of the electron shells of a particular ion, photons are emitted with one of a number of particular energies. Thus, we observe excess electromagnetic emission at these energies. The emission lines can be used to identify the ions and thus the composition of the source. The redshift of the emission lines can be used to compute the relative velocity of and the distance to the source. For these reasons, the precise fitting of emission lines is of key interest to astrophysicists.

This chapter is organized into four sections. In Section 26.2, we describe the statistical issues involved with the *Chandra's* data-generating mechanism and introduce a highly structured model that accounts for this mechanism. Our model is formulated in terms of several levels of missing data, which are critical in our formulation of the necessary computational techniques. The specific problem that we address in this chapter is introduced in Section 26.3, where we discuss the computational challenges that are involved with fitting narrow emission lines in high-energy spectral analysis. In Section 26.4, we discuss model checking techniques based on the posterior-predictive distribution.

26.2 The Chandra X-ray observatory

Data are collected on each X-ray photon that arrives at one of the detectors on board *Chandra*; the time of arrival, the two dimensional sky coordinates, and the energy are all recorded. Because of instrumental constraints, each of these four variables is discrete. Thus, the data can be complied into a four-way table of photon counts with margins corresponding to time, energy, and the two sky coordinates. Spectral analysis focuses on the one-way energy margin and image analysis models the two-way marginal table of sky coordinates. In this chapter, we focus on spectral analysis; see van Dyk and Hans (2002), Esch (2003), Esch, Connors, Karovska, and van Dyk (2004), and van Dyk et al. (2004) for discussion of image analysis of *Chandra* data.

A spectral model aims to describe the distribution of the energy of photons emitted from an astronomical source. This distribution can be formulated as a finite mixture model, in which the photon count in each energy bin is modeled as the sum of several independent Poisson random variables. A simplified form of this model might consist of a continuum term and an emission line. These terms represent two physical processes in the source; the continuum emission is a smooth function across a wide range of energies, while the emission line is highly focused at a particular energy. Thus, we might model the expected Poisson count in energy bin j as

$$\Lambda_j(\theta) = \Delta_j f(\theta^C, E_j) + \lambda p_j(\mu, \sigma), \quad \text{for } j = 1, \ldots, J \quad (26.1)$$

where Δ_j is the width of bin j, $f(\theta^C, E_j)$ is the expected counts per unit energy due to the continuum term at energy E_j, θ^C is the set of free parameters in the continuum model, λ is the expected counts due to the emission line, and $p_j(\mu, \sigma)$ is the proportion of an emission line centered at μ and with width σ that falls into bin j. A Gaussian or Lorentzian density function is often used to model the emission line, in which case σ might represent the standard deviation or some other measure of variability. There are a number of standard forms for the continuum term; here we use a power law, $f(\theta^C, E_j) = \alpha E_j^{-\beta}$, with $\theta^C = (\alpha, \beta)$.

While the model in (26.1) is of primary scientific interest, a more complex model is needed to address the data distortion introduced by instrumental effects and other aspects of the data collection procedure. For example, photons have a certain probability of being *absorbed* by interstellar or intergalactic media. Since this probability depends on the energy of the photon, the missing-data mechanism is nonignorable (Rubin, 1976a). A similar effect occurs inside the detector; rather than being reflected onto the detector, some photons are reflected away from or pass right through the mirror. The likelihood of this occurring again depends on the energy of the photon; this effect is known as the *effective area* of the detector. Even for photons that are recorded, their energy may be recorded with error; given the energy of the photon, there is a multinomial distribution that characterizes the likely energy that is recorded by the instrument. (In practice, the number of cells in these multinomial distributions is different from J; we index the cells that correspond to the observed data with $l = 1, \ldots, L$.) To account for these processes along with *background contamination*, (26.1) is modified via

$$\Xi_l(\theta) = \sum_{j=1}^{J} M_{lj} \Lambda_j(\theta) d_j u(\theta^A, E_j) + \lambda_l^B \qquad (26.2)$$

where M_{lj} is the probability that a photon with true energy in bin j is recorded in the multinomial cell l, d_j is the effective area of bin j, $u(\theta^A, E_j)$ is the probability that a photon with energy E_j is *not* absorbed, and λ_l^B is a Poisson intensity of the background counts in channel l. The multinomial distributions and effective area are presumed known from calibration. The absorption probability is parameterized using a smooth function, see van Dyk and Hans (2002) for details. Background contamination is quantified using a second observation from an area of black space near the source of interest, where all counts are assumed to be due to background contamination. More details, more general forms, and applications of this model can be found in van Dyk, Connors, Kashyap, and Siemiginowska (2001) and van Dyk and Kang (2003).

This data generation process can be described in terms of a number of steps and intermediate unobservable quantities. Each step starts with the output from the previous step and updates it in some possibly stochastic fashion. We begin with the energies of the continuum photons and the energies of the emission line photons. In the first step, these energies are mixed together. Next, a Bernoulli random variable is generated for each photon, with the probability of success depending on the energy of the photon. If this random variable comes up positive, the photon is observed; otherwise the photon is lost to absorption or the effective area of the instrument. In another step, error is added to the remaining photon energies via the conditional multinomial distributions. Finally, the data is contaminated with Poisson background counts.

This formulation of the data generation process leads naturally to a multilevel model that formalizes each of these intermediate quantities as missing data. Given the layers of missing data, the model falls into a sequence of simple standard models. For example, we might use a loglinear model for the Poisson counts from the continuum, or a binomial regression to account for absorption (van Dyk,

Connors, Kashyap, and Siemiginowska, 2001; van Dyk and Hans, 2002). Likewise, given the parameters for each of the stochastic steps, it is a simple application of the Bayes theorem to compute the conditional distribution of each of the layers of missing data. Thus, from a computational point of view, such tools as the EM or expectation/maximization algorithm (Dempster, Laird, and Rubin, 1977), the Data Augmentation algorithm (Tanner and Wong, 1987), the Gibbs sampler (e.g., Gelfand and Smith, 1990; Smith and Roberts, 1993), and other Markov chain Monte Carlo (MCMC) methods are ideally suited to highly structured models of this sort; see van Dyk (2003). The modular structure of these algorithms fits hand in glove with the hierarchical structure of our models. This allows us to divide a complex model-fitting task into a sequence of much easier tasks. The modular structure also allows us to take advantage of well-known algorithms that exist for fitting certain components of our model. For example, using the EM algorithm to handle a blurring matrix and background contamination in Poisson image analysis is a well-known (and often rediscovered) technique (Fessler and Hero, 1994; Lange and Carson, 1984; Lucy, 1974; Meng and van Dyk, 1997; Richardson, 1972; Shepp and Vardi, 1982). Even though this standby image reconstruction algorithm is unable to handle the richness of our highly structured model, we utilize it and its stochastic generalization as a step in our mode-finding and posterior-sampling algorithms.

In this short chapter, we only present one model that we hope illustrates the complexity of the data generation process, the models of this process, the algorithms required to fit the models, and the required inference and model-checking techniques. We emphasize, however, that the multilevel structure in the data generation process is inherent to the complex scientific processes studied and the instruments used in high-energy astrophysics. Thus, the missing-data framework, the related computational techniques, and methods for Bayesian inference and model checking have many waiting applications in high-energy astrophysics.

26.3   Fitting narrow emission lines

In this section, we outline some of the difficulties involved in fitting the location of a narrow emission line in (26.1). Our proposed solutions including EM-type algorithms, MCMC samplers, and data-analysis techniques along with detailed examples can be found in Park and van Dyk (2004) and Park, Siemiginowska, and van Dyk (2004).

When a Gaussian density function is used to model the emission line in the simplified spectral model given in (26.1), the standard EM algorithm renders straightforward calculation of the maximum likelihood estimate (MLE) or the posterior mode; to streamline our discussion, we focus on maximum likelihood estimation in this section. We construct a multilevel missing-data structure that accounts for background contamination, the effective area of the instrument, photon absorption, the blurring of photon energies, and the mixture of continuum and emission line photons, see van Dyk, Connors, Kashyap, and Siemiginowska (2001). For clarity, we consider an *ideal instrument* that produces counts that are a mixture of continuum and emission line photons, but these counts are not subject to the data distortion processes described in Section 26.2. Accounting for the various forms of data distortion causes no conceptual difficulty, but obscures the ideas involved with fitting an emission line. The counts from an ideal instrument are one of the levels of missing data in our formulation of the model that does account for data distortion; we call these counts the *ideal counts*. Notationally, we write $Y_j^{\text{ideal}} = Y_j^C + Y_j^L$ for each j, where Y_j^{ideal}, Y_j^C, and Y_j^L are the total ideal counts, the counts due to the continuum, and the counts due to the emission line in bin j respectively. Given the ideal counts, it is easy to construct an EM algorithm to fit (26.1); the missing data are the ideal counts split into continuum counts and emission line counts, that is, $\{(Y_j^C, Y_j^L), j = 1, \ldots, J\}$. Since the augmented-data log-likelihood is linear in these counts, the E-step simply computes the conditional expectation of the missing data. Because given the ideal counts, the photon counts due to the emission line in each bin follow binomial distributions, the conditional expectation of each is simply the total (ideal) photon counts times the relative magnitude of the emission line intensity and the combined continuum and emission line intensities at that bin. Specifically, given the current iterate of the model parameters, $\theta^{(t)} = (\theta^{C(t)}, \lambda^{(t)}, \mu^{(t)}, \sigma^{(t)})$, the E-step is given by

E-step: Compute $E[Y_j^L | \theta^{(t)}, Y_j^{\text{ideal}}]$ for each bin $j = 1, \ldots, J$, that is,

$$\widehat{Y}_j^L \equiv E[Y_j^L | \theta^{(t)}, Y_j^{\text{ideal}}] = Y_j^{\text{ideal}} \frac{\lambda^{(t)} p_j(\mu^{(t)}, \sigma^{(t)})}{\Delta_j f(\theta^{C(t)}, E_j) + \lambda^{(t)} p_j(\mu^{(t)}, \sigma^{(t)})}. \quad (26.3)$$

Next, the M-step of EM completes the update of the emission line location by computing the weighted average of the bin energies using the photon counts due to the emission line at every bin as the weights. In particular, the M-step updates the emission line location using

M-step: Compute $\mu^{(t+1)} = \sum_{j=1}^{J} E_j \widehat{Y}_j^L / \sum_{j=1}^{J} \widehat{Y}_j^L$,

where E_j is the mean energy in bin j. Generally the model includes other unknown parameters such as the continuum parameters and the emission line intensity, which are also updated in the M-step. Iteration between the E-step and the M-step forms what we call the "standard EM algorithm" for maximum likelihood estimation.

Given physical constraints on emission lines, it is often appropriate to replace the Gaussian line profile with a delta function. In this case, however, the standard EM algorithm breaks down. A delta function is a limiting case of a Gaussian density that results when the Gaussian variance goes to zero. Since the data are binned, the success probability of the binomial random variable in the E-step is zero for all of the bins except the one containing the previous iterate of the line location. That is, all of the photon counts attributed to the emission line are in one bin. Thus, the M-step necessarily returns the next iterate of the line location that is the same as the previous iterate. Since the EM algorithm begins with an arbitrarily specified value of a parameter, the algorithm will converge to the mean energy of the bin closest to the starting value in one iteration; the standard EM algorithm does not return the maximum likelihood estimate in this case.

To avoid this difficulty, we can update the line location at each iteration by maximizing the observed-data log-likelihood conditional on the other parameters in the model. To accomplish this, we simply compute the conditional observed-data log-likelihood at each possible value of the line location; because of the binning of the data, possible line locations within each bin are indistinguishable, and thus we are left with a finite number of possible distinguishable line locations. That is, this strategy updates the other model parameters by maximizing the augmented-data log-likelihood conditional on the line location using an EM iteration, and then updates the line location given the other model parameters without a missing-data formulation; this is an example of the ECME or expectation/conditional maximization either algorithm (Liu and Rubin, 1994). This algorithm allows groups of parameters to be updated by maximizing *either* the augmented-data log-likelihood or the observed-data log-likelihood while conditioning on the other parameters. The ECME algorithm is especially easy to formulate in this case because the conditional independence between the line location and the other parameters given the augmented data means that the E-step and conditional M-steps (CM-STEPS) for the other parameters are the same as in the standard EM algorithm. A difficulty with the ECME algorithm when used with real data that are subject to the data distortion processes described in Section 26.2, however, is that each iteration of the algorithm is computationally expensive, requiring the computation of the observed-data log-likelihood at each possible line location. Each evaluation involves computing, (26.2) which is time consuming because of the large dimension of the blurring matrix, M; this difficulty persists even when sparse matrix techniques are implemented.

As an alternative to ECME, we consider an AECM or alternating expectation/conditional maximization algorithm (Meng and van Dyk, 1997) that is computationally less expensive per iteration in this case. The AECM algorithm is so named because it allows the missing-data formulation to alternate for different groups of parameters. In terms of its use of missing data, the AECM algorithm finds middle ground between the EM and ECME algorithms. The AECM algorithm offers a more general formation than the ECME algorithm in that the CM-steps of AECM may conditionally maximize not only the observed-data log-likelihood or the conditional expectation of the augmented-data log-likelihood but also the conditional expectation of a *partially* augmented-data log-likelihood. That is, a portion of the missing data may be used to formulate some of the CM-steps in AECM. Thus, in our example, the ECME algorithm uses no missing data to formulate the CM-step for the emission line location, the EM algorithm uses all of the missing data, and the AECM algorithm allows us to formulate the CM-step using part of the missing data. In particular, we construct an augmented-data log-likelihood using the ideal counts as missing data, but do not separate the ideal counts into continuum and emission line counts. To update the line location in the AECM algorithm, we maximize this augmented-data log-likelihood conditional on the other model parameters. As with the ECME algorithm, the CM-step evaluates the augmented-data log-likelihood at each possible value of the line location, while fixing the other parameters at their current iteration. Because this evaluation does not involve the high-dimensional blurring matrix, each iteration is much quicker than those of the ECME algorithm.

The computational advantage of AECM, however, comes at a price: For some starting values, AECM exhibits the same pathological convergence as the EM algorithm, that is, the AECM algorithm can also get stuck at a point near its starting value and thus never reach a mode of the likelihood. In order to combine the stability of ECME and the speed of AECM, we propose a *Rotation* algorithm. In the Rotation algorithm, we run one ECME iteration followed by a number of AECM iterations, and repeat this procedure, rotating between ECME and AECM until convergence. For clarity, we refer to a rotation algorithm that runs m AECM iterations per ECME iteration as a Rotation(m) algorithm. In our experience, the rotation algorithms not only find the same mode as ECME for any starting value, but also outperform ECME in terms of required computation time.

To illustrate the application of our spectral model and the various EM-type algorithms, we use a *Chandra* observation of the high redshift quasar PG1634+706. Quasars are the most distant distinct detectable objects in the universe and their study has important consequences for cosmological theory. In particular, by measuring the location of the emission line of a quasar and accounting for the expansion of the universe, we can estimate the distance of the quasar from Earth. Thus, accurate fitting of emission line locations is central to the substantive scientific questions. We modeled this data using a power law continuum with the absorption model of Morrison and McCammmon (1983) to account for absorption due to the interstellar and intergalactic media, and a power law continuum for background contamination. The model was fitted via maximum likelihood using ECME, AECM, and the Rotation(1) and Rotation(9) algorithms. Figure 26.1(a) shows the fixed values of the AECM runs for each of the 51 equally spaced starting values between 1.0 keV and 6.0 keV. If a point in this plot does not lie on a horizontal line near 2.885 keV, it indicates that the AECM iteration is fixed at a point other than the mode of the likelihood, that is, AECM did not attain the maximum likelihood as shown in Figure 26.1(b). The middle two panels of Figure 26.1 plot the log-likelihood against computation time and against iteration number using the ECME, Rotation(1), and Rotation(9) algorithms; all three algorithms were started at 4.9 keV. These plots illustrate that the use of more AECM iterations in the Rotation algorithm can make the algorithm converge to a mode significantly more quickly and that the increase of the log-likelihood per iteration is about the same in these algorithms. The bottom two panels of Figure 26.1 compare the ECME, Rotation(1), and Rotation(9) algorithms in terms of total required computation time and the number of iterations required for convergence for 51 different starting values equally spaced between 1.0 keV and 6.0 keV. Figure 26.1(e) illustrates that the Rotation(9) algorithm converged most quickly among these algorithms with every starting value. In particular, the Rotation(1) and the Rotation(9) algorithms

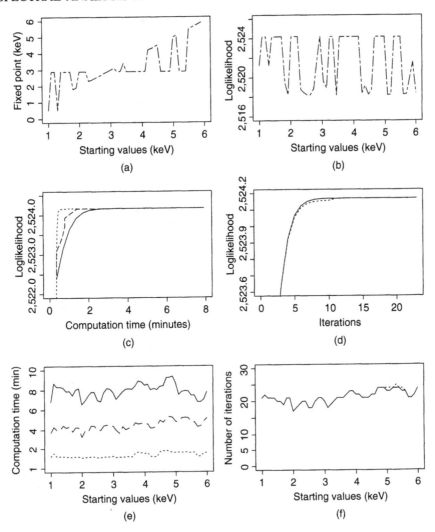

Figure 26.1   Comparison of the ECME, AECM, Rotation(1), and Rotation(9) algorithms. (a) and (b) illustrate that AECM can get stuck at a point other than a mode, thereby never reaching a mode of the likelihood; the AECM algorithm is represented by a dashed-dotted line. (c) and (d) illustrate the behavior of the log-likelihood evaluated at the iterates as a function of computation time and the iteration number using the ECME, Rotation(1), and Rotation(9) algorithms. The ECME, Rotation(1), and Rotation(9) algorithms are represented by solid, dashed, and dotted lines respectively. (e) and (f) compare the ECME and Rotation(1) and (9) algorithms in terms of computation time and the number of iterations required for convergence using 51 equally spaced starting values. The Rotation(9) algorithm is the quickest to converge among all of the algorithms considered for every starting value.

required only 1/2 and 1/10 of the computation time required by ECME respectively. While the algorithms all attain the same mode, 2.885 keV, for every starting value, Figure 26.1(f) indicates that the number of iterations required for convergence is almost identical for all of the algorithms with this data set; for a few starting values, the Rotation(9) algorithm takes more iterations than the other algorithms. Thus, in this example, the computation time for each AECM iteration is trivial relative to that of ECME, and when AECM is combined with ECME the average gain per iteration is equal to that of ECME alone.

26.4   Model checking and model selection

Residual plots and posterior-predictive methods (Gelman and Meng, 1996; Gelman, Meng, and Stern, 1996; Meng, 1994b; Rubin, 1981a, 1984) can be employed to check our spectral model specification. Both methods aim to check the self-consistency of the model, that is, the ability of the fitted model to predict the data to which the model was fit. The methods illustrated in this section were suggested for the spectral model by van Dyk and Kang (2003).

We consider the same model for Quasar PG1634+706 as discussed in Section 26.3 except that we compare three models for the emission line:

Model 0: There is no emission line.

Model 1: There is an emission line with fixed location in the spectrum.

Model 2: There is an emission line with unknown location.

The top two panels of Figure 26.2 compare the observed data with the fitted models under Models 0 and 1 in the first and second column respectively. The expected count per channel, $\Xi_l(\hat{\theta})$, is represented by a solid line and the predictive errors by dotted lines; $\hat{\theta}$ is the maximum likelihood estimate. The errors are computed using two standard deviations under the sampling model conditioning on $\hat{\theta}$; thus, these errors are based on a Gaussian approximation and do not account for the posterior variability of θ. The middle two panels of Figure 26.2 are mean subtracted versions of the first two panels, that is, these panels are residual plots. To better account for the Poisson nature of the data and the posterior variability in θ, we can compute residual errors using the posterior-predictive distribution. These plots appear as the final two panels in Figure 26.2; the jagged nature of the posterior-predictive residual errors is due to our Monte Carlo evaluation of this distribution. The advantage of the posterior-predictive errors is evident for the low counts in the high-energy tail of the spectra as shown in the residual plots of Figure 26.2. Comparing the two columns in Figure 26.2 near 2.885 keV also provides evidence for the inclusion of the emission line.

Posterior-predictive p-values can be used to compare the three models and, thus, to quantify the evidence in the data for the emission line. We base our comparisons

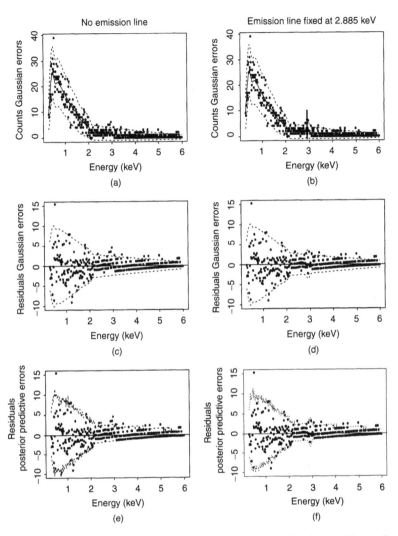

Figure 26.2   Model diagnostic plots. (a) and (b) show the data with predictive errors based on a Gaussian approximation; (c) and (d) show the residuals with errors based on a Gaussian approximation; and (e) and (f) show the residuals with errors based on the posterior predictive distribution. The two columns of the figure correspond to Models 0 and 1 respectively. The excess counts near 2.885 keV are apparent in the top two panels, thereby indicating evidence for the inclusion of the emission line in the model; the location of the emission line is represented by a vertical line in (b).

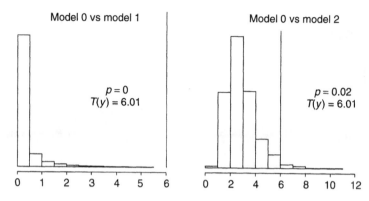

Figure 26.3   The posterior-predictive check. In each of the two histograms, the observed likelihood ratio test statistic (the vertical line) is compared with the posterior-predictive distribution of the test statistic under Model 0.

on the likelihood ratio test statistic,

$$T_i(y_{\mathrm{rep}}) = \log \left\{ \frac{\sup_{\theta \in \Theta_i} L(\theta|y_{\mathrm{rep}})}{\sup_{\theta \in \Theta_0} L(\theta|y_{\mathrm{rep}})} \right\}, \quad i = 1, 2,$$

where Θ_0, Θ_1, and Θ_2 represent the parameter spaces under Models 0, 1, and 2 respectively, and y_{rep} is a replicate data set. We can generate a sample from the posterior-predictive distribution of $T_i(y_{\mathrm{rep}})$ under Model 0; we use the EM-type algorithms described above to compute $T_i(y_{\mathrm{rep}})$. Histograms of $T_1(y_{\mathrm{rep}})$ and $T_2(y_{\mathrm{rep}})$ appear in Figure 26.3. Comparing these distributions with the observed values of the test statistics yields the posterior-predictive p-values in Figure 26.3. There is strong evidence for the presence of the emission line in the spectrum. Thus, Models 1 and 2 are preferable to Model 0.

27

Improved predictions of lynx trappings using a biological model

Cavan Reilly and Angelique Zeringue[1]

27.1 Introduction

Often statistics is viewed, and taught, as a series of procedures. In this view, methods are developed on the basis of some hypothesized data structure. The perspective that there are fixed data structures that can be treated as a whole misses the fascinating specificity of real-world problems. The field of time series prediction provides an excellent example of a well-defined data structure with a well-defined problem. In short, we assume we have a real-valued stochastic process that depends on time and our goal is to predict values of this process at some point in the future. If we assume the process is stationary, then there are representation theorems that provide us with a parameterized representation of any such series. Hence, to predict the series, we fit one of these parameterized forms and extrapolate. There are other classes of stochastic processes that have been developed to deal with nonstationary series, and while none of these has the same status as autoregressive moving averages, the same strategy is advocated: find a suitable parametric form from a class and estimate the parameters.

This general approach to statistics is often not the best approach to data analysis. As an example, we will consider prediction of the often-analyzed series of Canadian

[1]Division of Biostatistics, School of Public Health, University of Minnesota, Minneapolis, Minn.

Applied Bayesian Modeling and Causal Inference from Incomplete-Data Perspectives.
Edited by A. Gelman and X.-L. Meng © 2004 John Wiley & Sons, Ltd ISBN: 0-470-09043-X

lynx trapped in the Mackenzie River area from 1821 to 1934 (Elton and Nicholson, 1942). We will develop a model using just the first 80 years, and then use this model to predict the series for the next 34 years. These data have been analyzed dozens of times see (Tong, 1990 for a review), often by methods that have no basis in population biology. For example, several early analyses fit a sine curve to the population over time and cleaned up the remaining lack of fit with an autoregression (Bulmer, 1974; Campbell and Walker, 1977). But why would a sine curve describe the dynamics of the lynx population? Clearly the lynx population fluctuates, but sine curves, or even finite linear combinations of such curves, are certainly not the only periodic functions. Perhaps such a model even provides good predictions, but could we do better using knowledge of the biology involved?

Our statistical model of the lynx series should be based on the biological context. This means that the model should attempt to describe fluctuations in the series in terms of the source of the fluctuations. As mentioned above, most approaches to statistical models of the lynx series have modeled the series as having fluctuations that are attributable to some form of autocorrelation in the series without attempting to understand why there would be such autocorrelation. The approach presented here assumes that these fluctuations are due to fluctuations in the primary food source of the lynx, namely, the snowshoe hare. The problem with this approach is that there is no data on the hare population for this period; hence we will need to impute the hare population, at least implicitly.

To understand the basis of the model developed below, we first note an important fact about the Canadian lynx. The Canadian lynx is an unusual predator in terms of its diet. This predator relies almost exclusively on a diet of snowshoe hare. When the hare become scarce in a region, the lynx will either move to other regions or slowly starve to death rather than switch their food source (McCord and Cardoza, 1982; Keith, 1990; Poole, 1994; Slough and Mowat, 1996; Brand and Keith, 1979). Other similar predators, such as the bobcat, will change their diet according to what food sources are available. Hence, our statistical model should attribute the source of fluctuations in the lynx population to fluctuations in the size of the hare population.

27.2 The current best model

There have been many attempts to model the lynx series: indeed, this series is considered a benchmark by many who work in nonlinear time series analysis. A rather comprehensive treatment of methods existing up to 1990 can be found in Tong (1990). As mentioned in the introduction, the first attempts at modeling this series combined autoregressions with sine curves. In 1980, Tong and Lim published a paper in which they used a self exciting threshold autoregression (SETAR) to model the lynx series. They had noticed that the series increased at a different rate than it declined, hence sine curves were inappropriate. SETAR models can display this behavior. Basically this model fits a different autoregression to the upswings and the downturns in the population. For model selection issues, they employed

Akaike's information criterion. Many other models have been fit to this data with varying degrees of success. Almost all of these models have been based on some proposed form of autocorrelation in the series. In reviews of various treatments, Lim (1987) and Lai (1996) both rated Tong's SETAR model to be the best in overall fit.

27.3   Biological models for predator prey systems

The most fundamental model of the interaction of a predator species with a prey species is provided by the Lotka–Volterra equations. These equations assume that the number of hare would increase exponentially in the absence of predation and the number of lynx would decay exponentially in the absence of hare. In addition, when there are lynx present in the system, the hare population will decrease exponentially at a rate depending on the population of lynx, and similarly the population of lynx will increase exponentially at a rate depending on the hare population. If $u_1(t) = $ the number of lynx at time t, and $u_2(t) = $ the number of snowshoe hare at time t, then this simple framework implies the following set of differential equations that describe the dynamics of the interaction between these two species

$$\frac{du_1}{dt} = -\alpha_1 u_1 + \beta_1 u_1 u_2$$

$$\frac{du_2}{dt} = \alpha_2 u_2 - \beta_2 u_1 u_2,$$

where α_j, β_j for $j = 1, 2$ are positive parameters.

From a biological perspective, this model has the obvious shortcoming that it does not consider the effect of other predators on the population of snowshoe hare. That is, to have a model that represents the interaction of species in this habitat, we should have more terms in the second equation of the form $-\beta_j u_j u_2$ for $j = 3, \ldots, J$, where $J - 1$ is the number of predators that consume snowshoe hare. Indeed, one can imagine a system of equations where there is an equation for each predator and an equation for each prey that describes which animals consume each other in a habitat. What makes the equation for the lynx unique is that it only depends on the hare population. To take advantage of this property of the lynx equation, we suppose there are two types of snowshoe hare: those that ultimately are consumed by lynx and those that are not. We can split the equation for the total hare population into two equations: one of the two equations will govern the dynamics of the population of hare that are consumed by lynx and one equation for all the other hare. The first of these equations will not depend on the population of any other predator and will be exactly of the form of the second equation above. These two equations will be related, but we assume that the effect of competition between hares is negligible compared to the effect of birth and death on the population. Such an assumption is a basic tenet of the Lotka-Volterra equations. Hence the effect of other predators is just that now in the basic Lotka-Volterra equations presented above, $u_2(t)$ = the number of snowshoe hare alive at time t that are ultimately consumed by lynx. Of course, we cannot measure the number of hare today that will eventually be consumed by lynx, but it is nonetheless a well-defined concept. Actually, just determining the number of hare in a given habitat is a hard problem.

Another biological shortcoming of this model is the assumption that in the absence of predators, the snowshoe hare population will increase without bound. Clearly this is not realistic, as ultimately the food source of the hare will become depleted. To remedy this shortcoming, other terms are often added to the right side of the equations that include powers of the population of the species on the left side of the equation so that this behavior is ruled out. Rather than taking this route, we think of the system of equations as a useful model only when conditions are such that neither species dies out. That these conditions are applicable to the lynx/hare system over the last several hundred years, and that therefore this model is appropriate for the lynx/hare system, is obvious from the continued survival of both species.

A mathematical aspect of this model that has led some to conclude that it is not useful as a model in practice is that these equations are not structurally stable: small changes in the parameter values can lead to radical changes in the behavior of solutions. This has led some to abandon these equations or modify them to obtain a system that is better behaved. While this instability does make model fitting difficult, we can still use this set of equations to estimate parameters and make predictions, as we demonstrate in what follows. We do not think this structural instability makes the model unrealistic, as the world is full of phenomena that are quite sensitive to parameters.

27.4   Some statistical models based on the Lotka-Volterra system

Our first statistical model is based on the Lotka–Volterra system presented above. We observe the number of lynx trapped each year, $y(t)$ for 80 years. Although the number of lynx and hare can only take integer values, we model these quantities by real valued processes, as in the biological models presented above. We suppose that the expected proportion of lynx trapped each year is some constant proportion of the total number of lynx residing in the region, so that

$$y(t) = \alpha_0' u_1(t)\delta(t),$$

where $\delta(t), t = 1, \ldots, 80$ is a sequence of unit mean iid random variables that are independent of $u_1(t)$. For the purposes of conducting inference, we further assume these are lognormally distributed errors. The resulting model has 8 parameters: $\alpha_1, \alpha_2, \beta_1, \beta_2, u_1(1), u_2(1), \alpha_0'$, and σ, the standard deviation of the lognormally distributed errors.

This model is poorly identified; hence, we turned to the scientific literature in an attempt to construct informative priors. There are several methods that have been suggested for estimating parameters in the system. For example, one can construct an artificial habitat for hare so that no predation takes place. Observations on the hare population in such a setting could provide estimates of the birth rate of hare. But even in such situations, it is not clear that the birth rate is what it would be if there were lynx present. In any event, we can then assume that the birth rate of hare that ultimately get consumed by lynx is the same as the overall hare birthrate and obtain an informative prior for the birth rate parameter α_2. Other methods have been used to estimate the birthrate of hare, such as counting the mean number of young surviving. Similar techniques have been used to estimate the death rate of lynx (Poole, 1994; Slough and Mowat, 1996; Brand and Keith, 1979).

Unfortunately, we found that unless we used prior distributions with smaller standard deviations than the prior information really indicates, the posterior is too diffuse, as we describe below in the section on posterior simulation. For this model, the model parameters and the predictions themselves diverged as the Metropolis algorithm proceeded. Despite this, the predictions of the model at the best local mode we could find were very good, but we are reluctant to recommend the use of such predictions in general.

A simple reparameterization leads to a model with six parameters, and the resulting model behaves much better. This reparameterization can be thought of as just changing the units of the system. By letting $\theta_1(t) = \log(\beta_2 u_1(t))$ and $\theta_2(t) = \log(\beta_1 u_2(t))$ we obtain the system,

$$\log(y(t)) = \alpha_0 + \theta_1(t) + \epsilon(t)$$

$$\frac{d\theta_1}{dt} = e^{\theta_2} - \alpha_1$$

$$\frac{d\theta_2}{dt} = \alpha_2 - e^{\theta_1},$$

where $\epsilon(t)$ for $t = 1, \ldots, 80$ is a sequence of independent normal measurement errors. We then have six parameters in the model $(\theta_1(1), \theta_2(1), \alpha_0, \alpha_1, \alpha_2,$ and $\sigma)$. Unfortunately, although it is not immediately transparent, these six parameters are not identifiable.

To understand the nature of the identifiability problem here, we need to consider the trajectories of the system. The system has a non-hyperbolic fixed point at $(\theta_1 = \log \alpha_2, \theta_2 = \log \alpha_1)$. If we take the ratio of the equations that define the system, we obtain the differential equation

$$\frac{d\theta_1}{d\theta_2} = \frac{e_2^\theta - \alpha_1}{\alpha_2 - e_1^\theta},$$

which can be solved to yield an equation that describes the trajectories of the system in phase space

$$\alpha_2 \theta_1(t) - e^{\theta_1(t)} + \alpha_1 \theta_2(t) - e^{\theta_2(t)} = \alpha_2 \theta_1(1) - e^{\theta_1(1)} + \alpha_1 \theta_2(1) - e^{\theta_2(1)}.$$

If we define $f(x) = \alpha_2 x - e^x$, then f is concave and has a unique maximum at $\log \alpha_2$, hence provided $\alpha_2 \theta_1(1) - e^{\theta_1(1)} + \alpha_1 \theta_2(1) - e^{\theta_2(1)} + e^{\theta_2(t)} - \alpha_1 \theta_2(t) < \alpha_2(\log \alpha_2 - 1)$, there are two distinct solutions to the previous equation, one less than $\log \alpha_2$ and one greater than $\log \alpha_2$. We can repeat this argument using a condition on $\theta_1(t)$ too, hence the set of trajectories implied by the model is a collection of closed curves. Moreover, we can see from the equation that for trajectories near the fixed point, these curves will be approximately ellipses. For the lynx data, given this parameterization, the data supports the trajectory being very close to the fixed point for the θ_2 dimension, hence an elliptical trajectory with respect to that dimension. But if the trajectory is an ellipse and we only have data related to the θ_1 axis, then any translation of the trajectory along the θ_2 axis will yield the same fit to the data. When we attempted to find the posterior mode or generate samples from the posterior, we noticed that α_2 and $\theta_2(1)$ always moved together—this is what we expect given the elliptical trajectories. Given this identifiability problem, we simply fix $\theta_2(1)$, the rescaled initial number of hare that are ultimately consumed by lynx, at some arbitrary value and use noninformative priors for the other parameters in the model. In general, fixing $\theta_2(1)$ may reduce the set of possible trajectories, but this does not appear to be the case for this data set. Also, by fixing $\theta_2(1)$, we clearly cannot interpret α_1, but α_2 is still interpretable. The resulting model has five parameters that we estimate from the data.

Prior information on the system

There have been a large number of field studies aimed at understanding the population dynamics of lynx and hare. None of these have generated long time series of the sort on which we will base our predictions. Instead, these studies typically observe the numbers of animals over a short time period. Of the facts that these studies have identified, a consistent observation has been that the lynx population reaches its peak 1 to 2 years after the hare population reaches its peak. That is, once the hare population starts to decline, the lynx population follows suit. The Lotka–Volterra system has the property that periodic solutions have a fixed period, hence we use a prior distribution on the system that states that the difference in time between the two peaks is 1.5 years with a standard deviation of 0.25. When we discuss computing the posterior at a location in parameter space we will make clear how one can use this prior information.

27.5 Computational aspects of posterior inference

Given the structure of our model, computation is quite difficult. Note that we have no data on the number of hare at any point in time. The point of using the Lotka–Volterra system is to have a functional form for the number of lynx over time that is consistent with models from population biology. Although we think the formulation of the system in terms of the number of lynx and hare is quite intuitive, one can take the hare out of the system and obtain a second order differential equation for the lynx dynamics. Since we ultimately solve the system numerically, we end up converting back to two first-order equations in any event.

Computing the posterior at a location in parameter space

Since there is no explicit solution to the system of equations presented above, computation of the likelihood is not straightforward. We compute the log-likelihood at a point in parameter space $(\theta_1(1), \theta_2(1), \alpha_0, \alpha_1, \alpha_2, \sigma)$ by first computing the contribution to the log-likelihood of the first observation $y(1)$. Since $\log y(1) \sim N(\alpha_0 + \theta_1(1), \sigma^2)$ this term is straightforward. To compute the contribution of $y(2)$ to the log-likelihood, we first numerically integrate the system forward in time one step to obtain $\theta_1(2)$ and $\theta_2(2)$, then we use $\log y(2) \sim N(\alpha_0 + \theta_1(2), \sigma^2)$ to determine the contribution of the second time point to the log-likelihood. Note that $\theta_1(2)$ will be a function of α_1 and α_2. If we iterate this process, we can compute the log-likelihood for all of the data in this fashion. Finally, given that we have computed the log-likelihood we simply add the terms from the log-prior to obtain the log-posterior.

To perform the numerical integration, we use the fourth-order Runge–Kutta method (for implementation see Press et al., 1992). In order to use a prior distribution on the distance between the peaks of the series, we need to modify the basic procedure outlined above. As described above, we will only have the values of the solution to the system of differential equations at integer values. While this is adequate for computing the log-likelihood, we actually need the values of the solution for times between the integer valued times in order to determine at what time the peak of each series occurs. To this end, we integrate the system forward in time and save the solution each tenth of a year. Then we examine the value of the solutions over this finer time scale in order to determine when the peaks occur in each series. From the time of the peaks of the two series, it is easy to get the distance between the peaks implied by the set of parameter values $(\theta_1(1), \theta_2(1), \alpha_0, \alpha_1, \alpha_2)$. We then use this distance between the peaks in the term for the log-prior. Since the distance between the peaks is the same for all peaks, we can save some computational time by only integrating over this fine scale for the first pair of peaks.

Finding posterior modes

Although our posterior is only five-dimensional, finding posterior modes is quite difficult since the posterior is computed by numerically solving a system of differential equations. We found that using the simulated annealing algorithm for optimization of functions with continuous arguments presented in Press et al. (1992) allowed us to find posterior modes with some success.

Since the use of that algorithm is not at all standardized, we briefly indicate how we were able to successfully use the method. The simulated annealing algorithm of Press et al. is a stochastic mode-finding algorithm based on the downhill simplex method combined with a Metropolis-type algorithm. This algorithm has three parameters whose values greatly influence the utility of the approach: the initial computational temperature, the number of iterations at each temperature, and the percentage the computational temperature should decrease when lowered. We found that using an initial computational temperature of 1 that gets lowered every 500 iterations by 90% was useful for finding local modes here. Choice of the initial computational temperature has, in our experience, been the most important parameter when using this algorithm. One should monitor the best solution as the temperature is decreased. If the initial temperature is selected too high, then these best solutions tend not to be as good as the initial value. If this value is selected too low, then the algorithm usually converges quickly to a local mode.

Simulating from the posterior distribution

Since we can compute the log-posterior as described above, we can use the Metropolis algorithm to draw simulations from the posterior distribution. While we are actually only concerned with predictions based on the posterior mode, we used the Metropolis algorithm as a check on the propriety of the posterior distribution. We used the general strategy outlined in Gelman, Carlin, Stern, and Rubin (2003): a multivariate normal jumping distribution with an estimated covariance matrix that is scaled so that 30 to 40% of the jumps are accepted. Since we were not able to successfully compute the numerical derivatives of the log-posterior with adequate accuracy, we ran the chain for several thousand iterations to obtain an estimate of the covariance matrix, then used this estimate in the next run of the chain. It was by using the Metropolis algorithm with multiple chains that never converged that we were able to conclude that the model with six parameters and noninformative prior distributions did not give a proper posterior distribution. Similarly, when we used priors constructed from the literature, as previously mentioned, the chains still did not mix adequately to declare convergence of the chains (using Gelman and Rubin's $\sqrt{\hat{R}}$). As sometimes happens, although the posterior is mathematically proper when we use informative priors, if these priors are not adequately informative, the posterior can numerically behave as if it is not proper.

27.6 Posterior predictive checks and model expansion

While the model performs quite well in terms of prediction, if we perform diagnostic checks just using the first 80 years of data and our fitted model, we discover an important discrepancy between the model and the data. In Figure 27.1, we see a graph of the residuals at the posterior mode and a graph of the mean of

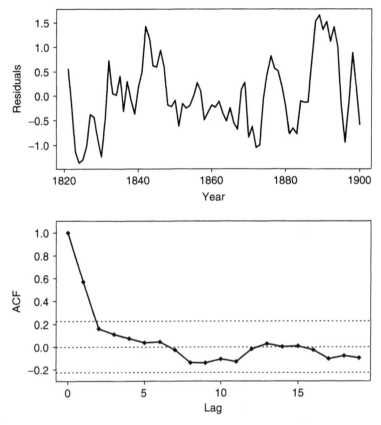

Figure 27.1   The residuals at the posterior mode and the mean of the posterior distribution of the residuals when there is no autoregressive component. There is autocorrelation at one lag.

the posterior distribution of the autocorrelation function of the residuals. We do not need to compute the posterior predictive distribution of the residuals in this example even though we are doing a posterior predictive check because we simply have iid Gaussian noise added to a functional form; hence, we know how large the autocorrelation function should be if there is really no autocorrelation. There is evidently substantial autocorrelation at lag one. This is not surprising given that there is an extensive literature indicating the presence of autocorrelation in this series, and here we see how posterior predictive checks can automatically detect such deviations from iid errors. There are basically two potential sources for this autocorrelation: the model dynamics are inadequate or the equation relating the dynamics to the measurements is incorrect. Since the model dynamics are based on the biological background, we expand our model to consider more realistic models for the way the number of lynx trappings relate to the number of lynx. In particular, the assumption that the proportion of lynx trapped is constant over time seems questionable. We would expect that the effort of trappers to capture lynx is a function of the demand for lynx pelts. As lynx pelts are luxury items, the demand would be greatly affected by fluctuations in the business cycle. To model this effect, we suppose that the measurement errors are a realization from an autoregression. To determine the order of the autoregression, we fit the smallest number of terms to this autoregression so that there is no autocorrelation in the posterior predictive residuals. This exercise led us to conclude that a first-order autoregression (with parameter ϕ) is adequate to describe the deviation from iid errors. In Figure 27.2, we see the residuals at the mode and the mean of the posterior distribution of the residuals. In Figure 27.3, we see the fitted curve and the predictions for the lynx and the scaled hare population (scaled to fit on the graph). In particular, notice the asymmetry of the rise and decline in the populations over time.

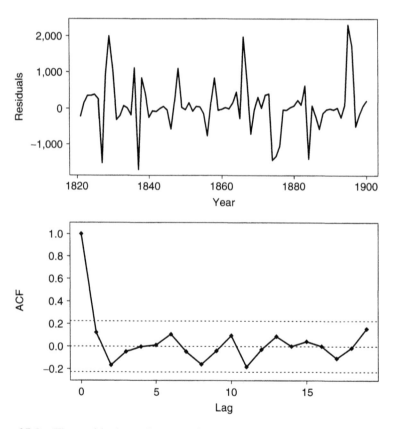

Figure 27.2   The residuals at the posterior mode and the mean of the posterior distribution of the residuals when there is a first-order autoregressive component. There is no evidence for autocorrelation.

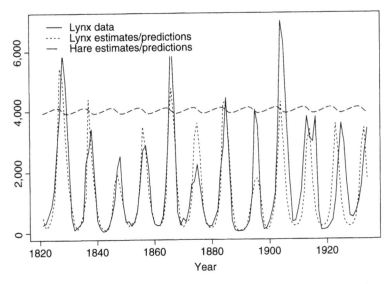

Figure 27.3   The lynx series, and the fitted values for the lynx and the hare. (The hare are scaled to fit in the graph.) The model is only fit to the first 80 years: the fitted values beyond year 1900 represent predictions.

27.7   Prediction with the posterior mode

Of course, without some regularity on the log-posterior we can never be sure that we have really found the global optimum. After running the simulated annealing algorithm for many iterations with many restarts from new locations in parameter space, we eventually became convinced that the best of the modes we had identified is the global optimum. Then we used this global optimum to make predictions. To obtain the predictions, we use the parameter values we found at the optimum and integrate the system forward in time starting from year 80 (the parameter values are $\alpha_0 = 14.4309$, $\alpha_1 = 804.209$, $\alpha_2 = 0.0006318$, $\theta_1(1) = -8.2474$, $\theta_2(1) = 6.6888$, $\sigma = 0.7151$, $\phi = 0.7431$). Although there are perhaps better ways to quantify the quality of a set of correlated predictions, we use the root-mean-square error of the predictions to quantify the quality of the predictions. For the above model, this quantity is 1,481.6. As noted above, perhaps the most widely supported model for this series is Tong's SETAR model. Tong fit his model to the entire series of 134 observations and using some model fit criteria, he eventually arrived at a 14-parameter model. To compare Tong's model to the model proposed here, we used Tong's parameter estimates (obtained from the entire series) and with his model made predictions starting from year 80 for the rest of the series. Strictly speaking, we should compare the predictions from our model to the predictions from a SETAR model fit to only the first 80 years of the series. In any event, the root-mean-squared error from Tong's model is 1,599.3; hence our model is better in terms of prediction even though Tong got to use more data and his model has more than twice as parameters. While the model developed here generates accurate predictions, there are some large discrepancies between the fitted curve and data (e.g., around 1865). A better fit could be obtained by allowing noise in the system, that is, use a system of stochastic differential equations. Such a model would be more realistic as we would expect stochastic disturbances (e.g., the weather) to impact animal populations.

27.8   Discussion

We have shown here how using models based on the science at hand, when combined with state-of-the-art statistical methods, can greatly improve our long-term predictive ability. Similar phenomena are known to exist in prediction of economic time series, but in that case it is usually accepted that nonstructural models, such as time series models, can outperform structural models (those based on economic theory) in the short term. We have also illustrated that nonlinear dynamical models can be of use in applications, and are not useless pieces of theory from textbooks. The numerical challenges of such model fitting are not to be underestimated, but they are not insurmountable.

28

Record linkage using finite mixture models

Michael D. Larsen[1]

28.1   Introduction to record linkage

A goal of record linkage is to identify pairs of records (a, b), a from file A and b from file B, that correspond to the same person or entity. If there are no unique codes that identify the matching pairs of records, then links can be designated by comparing variables contained in the two files. In US census operations, social security number (SSN) is not collected, but first and last name, street address and house number, and other information are recorded. Often a great deal of work, including name and address parsing and standardization, is required to prepare files for comparison. If unique SSN's were recorded accurately for all individuals in both files, then the linkage task would be greatly simplified.

At the US Bureau of the Census, record linkage is an important step in under-count estimation and coverage evaluation. In order to evaluate the 1990 census, the Bureau of the Census conducted a post-enumeration survey (PES). The PES database was matched to census records. The number of individuals counted in both the census and the PES and the numbers counted in one but not in the other canvas, under an assumption of independence between enumerations, yields an overall estimate of the population. The actual estimation procedure is much more complicated in its details, but the idea is essentially the same. The 1990 PES is discussed in articles in volume 88 of *Journal of the American Statistical Association*

[1]Department of Statistics and CSSM, Iowa State University, Ames, Iowa.

Applied Bayesian Modeling and Causal Inference from Incomplete-Data Perspectives.
Edited by A. Gelman and X-L. Meng   © 2004 John Wiley & Sons, Ltd   ISBN: 0-470-09043-X

(1993) and in volume 9 of *Statistical Science* (1994). See also Elliott and Little (2000). Census 2000 included a post-enumeration survey as part of its Accuracy and Coverage Evaluation (ACE; see, e.g., Hogan, 2000). Record linkage methods also are used at census for unduplication and address matching.

Clerical review of records is time consuming and costly. Computers can quickly score the level of agreement on recorded information between records in the two files, and, given a decision procedure, designate pairs as links or nonlinks. The method of scoring agreements and disagreements, how to combine evidence from several fields of information, and setting cutoff values are practical and theoretical problems.

Bill Winkler has made many advances in record linkage applications at the US Bureau of the Census, Statistical Research Division (Winkler, 1994, 1995, and references therein). Tom Belin while in graduate school at Harvard studied census record linkage problems with Bill and Don, leading to Belin (1993) and Belin and Rubin (1995). The author followed in this line of work and with Don produced Larsen and Rubin (2001). It is characteristic of Don's approach to scientific problems that models are used to learn from data. Don saw the potential advantage of modeling record linkage data using latent class models and, more generally, mixture models to fit important structures in the data. This chapter presents some results that have come out of that approach. It also describes ideas for incorporating experience with similar record linkage operations into methods and analyses and for accounting for sources of uncertainty in results.

Section 28.2 discusses some theory of record linkage. Section 28.3 presents latent class and mixture models and describes estimation procedures. Section 28.4 summarizes the application of methods to census test data sets. Sections 28.5 and 28.6 focus on extensions, regression analysis of linked files and hierarchical models, that are being developed. Section 28.7 provides summary comments.

28.2 Record linkage

Assume the files A and B do not contain duplicates and that K comparisons can be made between information in pair of records (a, b). Comparisons in the census context are made using name, address, and demographic information. Let $\{v_k(a), k = 1, \ldots, K\}$ be K pieces of information in file A on person a and $\{w_k(b), k = 1, \ldots, K\}$ be the information in file B on person b. The comparison of information on pair (a, b) yields a comparison vector $\gamma(a, b) = \{\gamma_k(a, b), k = 1, \ldots, K\}$, where $\gamma_k(a, b) = 1$ if $v_k(a) = w_k(b)$ and 0 otherwise. If $a \in A$ and $b \in B$ were produced by the same person, then it would be expected that $\gamma(a, b)$ would contain more "1" entries than if they were produced by different people. Illustrations are given in Winkler (1995).

Not all pairs are compared since such a procedure would create many pairs and most of them would not be plausible as matches (e.g., a and b live in different geographical areas, have different first letters of last name, etc.). The records in the two files are divided into S "blocks" and comparisons are made between records within blocks. Let the records in block s in file A be denoted $A(s)$ and in file B, $B(s)$. It is assumed here that blocking does not create errors in matching, that is, that records involved in matches are blocked together.

Fellegi and Sunter (1969) proposed decomposing the space $A \times B$ into sets M, the matches, and U, the unmatched or nonmatching pairs. They showed that, given probabilities $\Pr(\gamma|M)$ and $\Pr(\gamma|U)$ for all comparison vectors γ, if pairs with $\Lambda = \Pr(\gamma|M)/\Pr(\gamma|U)$ above a cutoff are designated links and pairs with the ratio below a second cutoff are designated nonlinks, then the number of undeclared pairs is minimized at the error levels corresponding to the two cutoff levels.

In practice, there have been several methods developed for estimating the probabilities involved in the ratio or for producing the ratio itself. In many applications, the log ratio $\log \Lambda$, also called the *weight*, is used. The components $\Pr(\gamma_k|M)/\Pr(\gamma_k|U)$ are sometimes taken to have a multiplicative (log-additive) effect: $\log \Lambda = \sum_{k=1}^{K} (\log \Pr(\gamma_k|M) - \log \Pr(\gamma_k|U))$. Some applications sum specified values for agreement or disagreement on each field of information, rather than using logs of estimated probabilities.

Belin and Rubin (1995), see also Belin et al. (2004) in this volume, used training data to estimate transformations to normalize the distributions of weights for matches and for nonmatches. They then fit mixtures of transformed normals to data from new record linkage operations. The transformed normal mixture models yield estimated error rates and cutoff values. Winkler (1995), Larsen and Rubin (2001), and references therein fit mixture models directly to the comparison vector data.

28.3 Mixture models

Consider the observed data to be the patterns of agreements for the pairs of records: $\{\gamma(a, b), a \in A(s), b \in B(s), s = 1, \ldots, S\}$. Before clerical review determines match status, the data can be viewed as arising from a mixture of comparison vectors from matches and nonmatches. The probability of observing pattern γ is

$$\Pr(\gamma) = \Pr(\gamma|M)p_M + \Pr(\gamma|U)p_U, \tag{28.1}$$

where $\Pr(\gamma|M)$ and $\Pr(\gamma|U)$ are the probabilities of the pattern among the matches and nonmatches respectively, p_M is the probability that a pair is a match, and $p_U = 1 - p_M$. The observed-data likelihood is a product of (28.1) over all pairs (a, b), where the pair (a, b) determines a value of $\gamma = \gamma(a, b)$.

The conditional-independence model specifies that the conditional probability of pattern γ is the product of the conditional probabilities for agreeing or disagreeing on the K fields:

$$\Pr(\gamma|C) = \prod_{k=1}^{K} \Pr(\gamma_k|C)^{\gamma_k} (1 - \Pr(\gamma_k|C))^{1-\gamma_k},$$

with $C \in \{M, U\}$. Loglinear models can be used to incorporate interactions between fields of information within classes.

Let $z(a, b) = 1$ if pair (a, b) is a match and 0 otherwise. The vector of indicators **z** is unobserved before clerical review. If matches are unique and $z(a, b) = 1$, then $z(a, b')$ and $z(a', b) = 0$ for $a' \neq a$ and $b' \neq b$. Taking this into account greatly complicates the structure of the complete data; rather than consider a pair (a, b), it would be necessary to consider all pairs in a block simultaneously. In practice, pairs are considered individually (Fellegi and Sunter, 1969; Winkler, 1994; Larsen and Rubin, 2001) and a linear sum assignment algorithm (Jaro, 1989; Winkler, 1994) is used to force one-to-one matching after parameter estimation.

The probability that $z(a, b)$ equals 1 given the agreement pattern is

$$\Pr(M|\gamma) = p_M \Pr(\gamma|M)/\Pr(\gamma).$$

If **z** were observed and pairs considered individually, the complete-data likelihood would be a product over pairs (a, b) of

$$\left[\Pr(\gamma|M) \, p_M\right]^z \left[\Pr(\gamma|U) \, p_U\right]^{1-z}.$$

If there are duplicate records for a person in either file, then the files could be linked to themselves to identify and remove the duplicates before they are linked together.

Maximum likelihood estimation

A convenient method of finding maximum likelihood estimates (MLE) of mixture model parameters is to treat the unobserved indicators **z** as missing data and use the EM (Dempster, Laird, and Rubin, 1977) or ECM (Meng and Rubin, 1993) algorithms for estimation. The EM algorithm is used for the conditional-independence model and for models using loglinear models with direct estimates of parameters. The ECM algorithm accommodates models that would involve iterative proportional fitting or other indirect estimation. More complex models have been used to model discrete mixtures (e.g., Becker and Yang, 1998) and in record linkage (Winkler, 1995; Larsen and Rubin, 2001, and references therein).

In the case of the conditional-independence model, given current estimates of parameter values, the E-step is completed by computing $\Pr(z(a, b) = 1|\gamma(a, b))$, the expected value of $z(a, b)$, for all pairs (a, b), which is the same for each unique comparison vector γ. Call the values $Ez(a, b)^{(t+1)}$. The M-step is completed by calculating maximum likelihood estimates of parameters with entries in **z** held at their current expectations. The estimate of p_M is the sum of $Ez(a, b)^{(t+1)}$ over pairs (a, b) divided by n, the total number of pairs. The estimate of $\Pr(\gamma_k(a, b) = 1|M)$ is the sum of $Ez(a, b)^{(t+1)}$ over pairs (a, b) for which $\gamma_k(a, b)$ equals 1 divided by the sum of $Ez(a, b)^{(t+1)}$ over all pairs. The estimate of $\Pr(\gamma_k(a, b) = 1|U)$ is the sum of $1 - Ez(a, b)^{(t+1)}$ over pairs (a, b) for which $\gamma_k(a, b)$ equals 1 divided by the sum of $1 - Ez(a, b)^{(t+1)}$ over all pairs. The algorithm iterates between the E- and M-steps until the observed-data likelihood converges to a maximum.

Bayesian estimation

Bayesian analysis multiplies the likelihood by a prior distribution on the parameters to reflect uncertainty about parameter values. Prior distributions for the record linkage parameters described above can be relatively noninformative, but still reflect knowledge about record linkage in similar applications. Census operations, often involving more than 100,000 record pairs, have been conducted in several locations over the last couple of decades.

Logically, the probability of agreeing on a comparison should be higher among matches than among nonmatches. Relatively weak prior distributions for the latent class model can be specified using beta distributions to reflect this *a priori* belief and the fact that the percent of pairs that are matches is restricted by the relative sizes of the files. Alternatively, data from a past record linkage operation that has been reviewed by clerks could be used to form a prior distribution. The proportion of pairs that are matches could be taken as the mean of the distribution for p_M. In the latent class model, the mean probability of agreeing on a field of information for matches and nonmatches could be set equal to the values observed previously. Belin and Rubin (1995) and Larsen and Rubin (2001) observed that parameter estimates vary substantially by record linkage location. The variability in the prior distributions should be high enough to allow the current data to strongly influence the posterior distribution. The degree of blocking also will affect the types of comparison patterns observed in a record linkage operation.

Posterior distributions can be simulated by sampling from alternating conditional distributions in steps analogous to those of EM and ECM. Given values for parameters after iteration step t, each variable $z(a, b)$ is drawn from a Bernoulli distribution with parameter equal to $\Pr(z(a, b) = 1 | \gamma(a, b))$. Call the values $z(a, b)^{t+1}$. Given the drawn values of components of $\mathbf{z}$, the parameters are drawn from current conditional distributions.

If the prior distribution on the proportion p_M is a Beta(α_M, β_M) distribution, then given $\mathbf{z}^{(t+1)}$ p_M has a beta distribution with parameters $\alpha_M + \sum z(a, b)^{t+1}$ and $\beta_M + \sum (1 - z(a, b))^{t+1}$. Here summation is over pairs (a, b). In the latent class case, an independent Beta$(\alpha_{Mk}, \beta_{Mk})$ prior distribution on $\Pr(\gamma_k(a, b) = 1 | M)$ produces a conditional

$$\text{Beta}\left(\alpha_{Mk} + \sum z_{ab}^{(t+1)} \gamma_k(a, b), \beta_{Mk} + \sum z_{ab}^{(t+1)} (1 - \gamma_k(a, b))\right)$$

distribution, independently for $k = 1, \ldots, K$, where $z_{ab} = z(a, b)$. An independent Beta$(\alpha_{kU}, \beta_{kU})$ prior distribution on $\Pr(\gamma_k(a, b) = 1 | U)$ yields a

$$\text{Beta}\left(\alpha_{Uk} + \sum (1 - z_{ab})^{(t+1)} \gamma_k(a, b), \beta_{Uk} + \sum (1 - z_{ab})^{(t+1)} (1 - \gamma_k(a, b))\right)$$

distribution, independently for $k = 1, \ldots, K$. The algorithm cycles between drawing values of parameters and values of missing indicator variables until drawn values are being sampled from the posterior distribution.

In the case of models with interactions within classes that do not have direct maximum likelihood estimates, simulation of parameters given current values of indicators $z(a, b)$ involves simulation analogous to iterative proportional fitting. The methods of Gelman et al. (2003, Section 16.8) can be used within each class given the indicators $z(a, b)$. Schafer (1997, Chapter 8) and Larsen (1994, 1996), when studying with Don Rubin, explored simulation methods for these general models.

The end product of the Bayesian sampling algorithm is several sets of values for mixture model parameters and several sets of imputed links. Each set of parameter values could be used to calculate probabilities that pairs are matches, $\Pr(z(a, b) = 1|\gamma(a, b))$. Each set of imputed links is a set of drawn values for **z** that identify which pairs (a, b) are designated as links at a particular iteration of the algorithm.

28.4 Application

Larsen and Rubin (2001) developed record linkage methodology using mixture models by addressing four practical concerns and applied the methods to five census data sets. First, they selected a mixture model from a set of candidate models by comparing probabilities estimated via maximum likelihood under the models to empirical probabilities obtained from data.

Second, an initial fit based on the selected model was used to sort the pairs of records according to their estimated probabilities of matching. Pairs with high probabilities are designated links, whereas those with low probabilities are designated nonlinks.

Third, a few cases were selected using the model for immediate review by clerks. It is common when using probabilistic record linkage to examine some record pairs to check that the models are accurately separating matches from nonmatches. Record linkage, when there are many nonmatching and few matching pairs, is most effective when it finds the matches. When comparisons between records yield mostly agreements (disagreements), there are mostly matches (nonmatches) and not much mixing of the two groups. At intermediate levels of agreement, which correspond to intermediate probabilities of matching, the groups are less well separated. It is possible then, based on estimated probabilities, to select some pairs to review that are harder to classify.

Fourth, the model is refit using both the clerically classified cases reviewed in step three and the remaining unclassified cases. The procedure iterates through identifying records, clerical review, and refitting as time and clerical resources allow, or until little additional improvement is needed or made. The procedure was stopped when two consecutive clerical review steps found few new matches.

Five data sets, described in Larsen and Rubin (2001), were created from the 1988 and 1990 Census and Post-Enumeration Survey (PES) operations at separate sites. The choice of a reasonable initial model along with clerical review and updating provided some robustness to the procedure, which worked well on these data

Data Set	Pairs Reviewed	Declared Link Match	nonM.	Undeclared Match	nonM.	Declared Nonlink Match	nonM.
D88a	10,000	10,074	0	961	1,471	57	103,742
D88b	10,000	6,762	3	34	112	82	49,780
D90a	4,000	3,286	8	221	182	89	33,541
D90b	3,200	2,836	3	573	367	79	34,937
D90c	2,400	1,203	1	0	0	58	37,952

Table 28.1 Number of matches and nonmatches ("nonM.") designated as links and nonlinks and undeclared at 0.001 estimated false-match and false-nonmatch rates for each data set after some clerical review and model updating.

sets. Table 28.1 presents some of the results from Larsen and Rubin (2001). Most of the pairs (the total number of pairs at each site is the sum of the last six columns) are correctly classified after reviewing approximately ten percent of the file.

Experiments on forming prior distributions were conducted with these data sets. Results reported in Larsen and Rubin (2001) used maximum likelihood estimation. The census data sets ranged in size from 37,327 to 116,305 pairs of records and used ten ($K = 10$) fields of information, yielding $2^{10} = 1,024$ possible agreement/disagreement vectors. Using fairly weak prior information, as described in Section 28.2, estimates of parameters did not differ substantially from the maximum likelihood estimates. The designations of pairs as links and nonlinks, using posterior means as parameter values in probability calculations, also did not change appreciably, sometimes not at all.

One modification to the procedure that was implemented using the Bayesian results was to select the cutoff values for specified error rates based on the 10th percentile of the simulated error distribution. That is, based on a set of simulated parameters one can determine how many links should be declared at a specified error level. In these applications across simulations, the 10th percentile of the number of links was lower than the mean number of links. This procedure was more conservative in terms of sending more pairs to clerical review and making fewer classification errors. That is, the Bayesian approach allowed a different, more conservative procedure to be implemented. As more pairs were clerically reviewed and the fit of models updated, differences among procedures using the 10th percentile, the median, and the mean decreased.

When stronger prior information was used, as would be expected, the parameter estimates changed. In most cases, the performance in terms of identifying matches and nonmatches was approximately the same. When the prior distribution based on the data set from a rural site with poor quality address information (site D90c in Larsen and Rubin (2001) was given substantial weight, however, the number of misclassifications increased for the other sites. The differences in site characteristics then can influence the performance of model-based record linkage methods.

Two extensions have been studied and are described briefly below. Section 28.5 concerns regression analysis of files created through record linkage when there is possible matching error. Section 28.6 presents ideas for a hierarchical record linkage model that would accommodate sites with varying characteristics.

28.5   Analysis of linked files

Consider the following regression model

$$y_i = x_i'\beta + \epsilon_i, \quad i = 1, \ldots, n,$$

where $x_i = (x_{i1}, \cdots, x_{ip})'$ is a vector of p known covariates, $E(\epsilon_i) = 0$, $var(\epsilon_i) = \sigma^2$, and $cov(\epsilon_i, \epsilon_j) = 0$ for $i \neq j, i, j = 1, \ldots n$. Suppose that the response (Y) is in file B, the covariates (X) are in file A, the two files are linked imperfectly, and there is at most one link in the other file for each record. The true pairs (x_i, y_i) are not observable. Instead, one observes $u_i's$ that may or may not correspond to x_i.

Scheuren and Winkler (1993) proposed the following model for u_i's:

$$u_i = \begin{cases} y_i & \text{with probability } q_{ii} \\ y_j & \text{with probability } q_{ij} \text{ for } i \neq j, \end{cases}$$

where $\sum_{j=1}^{n} q_{ij} = 1, i, j = 1, \ldots, n (i \neq j)$.

A naive estimator of β would be

$$\hat{\beta}_N = (X'X)^{-1}X'U,$$

where $X = (x_1', \ldots, x_n')'$ and $U = (u_1, \ldots, u_n)'$. This estimator is biased due to the imperfect linkage of response and predictor variables.

An improved estimator that utilized the estimated probabilities of being links for adjustment of regression results was presented by Scheuren and Winkler (1993). An iterative procedure to correct for outliers was presented by Scheuren and Winkler (1997). Lahiri and Larsen (2000, 2004) developed an alternative estimator of β and its standard error, which they studied using simulation. Table 28.2 presents some simulation results from Lahiri and Larsen (2004), which show an improvement in performance using their procedure.

One difficulty encountered in the work of Lahiri and Larsen (2004) concerned the expression of uncertainty due to record linkage parameter estimation on adjusted regression results. When the linkage probabilities were treated as fixed values for the purposes of regression adjustment, confidence intervals based on the estimates and standard errors were too short to cover the simulation slope values. An improvement in results was achieved by using the Bayesian approach for record linkage. For each set of simulated record linkage parameter values, the regression adjustment estimates and standard errors can be calculated. Several sets of

Method	Sum of Absolute Deviations $\sum\lvert\hat{\beta}-\beta\rvert$	Sum of Squared Errors $\sum(\hat{\beta}-\beta)^2$
Naive regression	21.14	2.24
Robust regression	17.45	1.54
Scheuren–Winkler adjustment	15.53	1.43
Lahiri–Larsen adjustment	11.54	0.84

Table 28.2   Regression results based on 400 simulated data sets using four estimators.

estimates and their standard errors can be combined using formulas for multiple imputation inference (Rubin, 1987b). This approach incorporates more variability into regression error estimates and produces improved coverage.

28.6   Bayesian hierarchical record linkage

The algorithms presented previously do not enforce one-to-one linking. That is, for an individual a in file A, there is no constraint in the model that forces the probability that a has a match to be 1. Several records in file B could agree closely with person a and have high probabilities of matching person a. The algorithms also do not explicitly incorporate all constraints due to blocking. Specifically, if pairs within a particular block tend to agree much more than is typical, the sum over pairs of the probabilities could add to more than the possible number of matches. Also, the parameter estimates produced overall might be very different than would be produced for subsets of record pairs, and the estimate of the probability of agreeing on a comparison among matches could be lower than the estimate among nonmatches.

Larsen (1999, 2002) discusses extensions of current models and algorithms that could address these concerns. If it is assumed that there are no duplicates in files A and B, then there should be at most one unique match for each record in each file. That is, if $z(a, b) = 1$ if pair (a, b) is a match and zero otherwise, then $\sum_{a \in \text{block } s} z(a, b) \leq 1$ and $\sum_{b \in \text{block } s} z(a, b) \leq 1$. Logical parameter restrictions can be expressed as $\Pr(\gamma_k\lvert M) \geq \Pr(\gamma_k\lvert U), k = 1, \ldots, K$.

A hierarchical model could allow each area or block to adapt to its own config- uration of agreement/disagreement patterns, but limit the variability by relating the parameters across sites or blocks to one another. Within block s, the (independent) prior distributions on parameters could be

$$\Pr(\gamma_k = 1\lvert M, s) \sim \text{Beta}(\alpha_{sMk}, \beta_{sMk})$$

$$\Pr(\gamma_k = 1\lvert U, s) \sim \text{Beta}(\alpha_{sUk}, \beta_{sUk})$$

$$\text{and } p_{Ms} \sim \text{Beta}(\alpha_s, \beta_s).$$

Across blocks the parameters of these distributions could be related to one another through the following (independent) distributions:

$$\theta_{sMk} = \text{logit}\left(\frac{\alpha_{sMk}}{\alpha_{sMk} + \beta_{sMk}}\right) \sim N(\mu_{\theta Mk}, \sigma^2_{\theta Mk}),$$

$$\theta_{sUk} = \text{logit}\left(\frac{\alpha_{sUk}}{\alpha_{sUk} + \beta_{sUk}}\right) \sim N(\mu_{\theta Uk}, \sigma^2_{\theta Uk}),$$

$$\tau_{sMk} = \log(\alpha_{sMk} + \beta_{sMk}) \sim N(\mu_{\tau Mk}, \sigma^2_{\tau Mk}),$$

$$\tau_{sUk} = \log(\alpha_{sUk} + \beta_{sUk}) \sim N(\mu_{\tau Uk}, \sigma^2_{\tau Uk}),$$

$$\theta_s = \text{logit}\left(\frac{\alpha_s}{\alpha_s + \beta_s}\right) \sim N(\mu_\theta, \sigma^2_\theta), \quad \text{and} \quad \tau_s = \log(\alpha_s + \beta_s) \sim N(\mu_\tau, \sigma^2_\tau),$$

for $s = 1, \ldots, S$ indexing blocks and $k = 1, \ldots, K$ indexing comparison fields.

The simulation of the posterior distribution is more involved than in Section 28.2 because of the second level of the hierarchy and the number of parameters. If the prior distribution for the parameters across the blocks is chosen as above, then the simulation of the posterior distribution can be accomplished using the Metropolis–Hastings algorithm within a Gibbs sampling sequence. Implementation of this algorithm is being studied.

Bayesian record linkage has been studied also by Fortini, Liseo, Nuccitelli, and Scanu (2001) and Fortini, Nuccitelli, Liseo, and Scanu (2002). These authors avoid placing prior distributions on parameters by averaging over them analytically.

28.7   Summary

The use of latent class and mixture models has proved to be a productive application of models to a complicated applied problem. Some common practices, for example, reviewing a subset of records, can be incorporated into the estimation, model selection, and analysis procedures. It has been possible to build on Bill Winkler and Don Rubin's proposal for latent class and mixture modeling of record linkage comparison data to address additional concerns. Further work is being conducted on methods of analysis for linked files and for implementing hierarchical models of record linkage. The topic of automated record linkage has increasing relevance to research and society as administrative and survey information increases in detail and availability (Winkler, 2003). It will also play a role in counterterrorism efforts (Gomatam and Larsen, 2004).

From interaction with Don Rubin, the author has observed the value of using models as tools to learn about data, the importance of trying to understand and model sources of variability, and the benefit and pleasure of interacting with subject area experts.

29

Identifying likely duplicates by record linkage in a survey of prostitutes

Thomas R. Belin, Hemant Ishwaran, Naihua Duan, Sandra H. Berry and David E. Kanouse[1]

29.1 Concern about duplicates in an anonymous survey

The Los Angeles Women's Health Risk Study (LAWHRS) was a survey of female street prostitutes in Los Angeles County that aimed to provide insight into the evolution of the AIDS epidemic in the early 1990s (Kanouse et al., 1999). Goals of the study included estimating the size of the female street-prostitute population in Los Angeles, determining seroprevalence of the HIV virus among female street prostitutes, measuring the prevalence of sexual and drug-related risk behaviors associated with HIV transmission, measuring the frequency of condom use and other preventive behaviors, and relating HIV status to behavior patterns and prostitute characteristics.

The LAWHRS was designed as a probability sample of areas of Los Angeles, times of day, and days of the week (Duan, Kanouse, and Berry, 1992). The

[1]Department of Biostatistics, University of California, Los Angeles, Calif., Cleveland Clinic Foundation, Cleveland, Ohio, and RAND Corporation, Santa Monica, Calif.

Applied Bayesian Modeling and Causal Inference from Incomplete-Data Perspectives.
Edited by A. Gelman and X-L. Meng © 2004 John Wiley & Sons, Ltd ISBN: 0-470-09043-X

area frame was assembled from the police, health officials, and study consultants including former prostitutes, and special procedures were developed for field staff (interviewers, drivers, and phlebotomists) to go through sampled areas beginning at randomly selected start points, to approach women for interviews in a systematic fashion until agreement was obtained from a woman in the area, and to obtain informed consent (Kanouse et al., 1999). Interviews typically took roughly 45 min, and participants were asked to provide a blood sample to be tested for exposure to HIV, syphilis, and hepatitis B. Women were paid $25 for participation. Blood-test results were not immediate, but women could obtain test results by calling RAND to arrange an appointment. Test results were stored and retrieved using a "distinguishing" code constructed at the time of interview by stringing together a set of responses to seemingly innocuous questions such as, "What is the first letter of your mother's maiden name?" We allude to the code as "distinguishing" rather than "identifying" because it succeeded in distinguishing between individuals in the study without allowing anyone to associate the information with a person in the way that a name or social security number would allow.

After eligibility for the study was established through a question about trading sex for money or drugs in the previous year, a screening question sought to avoid duplicate interviews by asking, "Have you been interviewed already by the Los Angeles Women's Health Risk Study?" An informed consent procedure was administered to respondents who acknowledged eligibility and who stated that they had not been interviewed previously. As part of the protocol, participants were told that they would not be asked to disclose their name, address, or other information that could be used to identify them personally. The informed consent form was signed by the interviewer, who certified that the respondent had reviewed all of the points on the form.

The present chapter describes an approach that was developed to address concerns that arose over possible duplicate interviews. The payment for participation was judged to be large enough that it might provide motivation for individual prostitutes to participate more than once. But because the interviews were anonymous, duplicate interviews could not be identified in a straightforward way. Some insights were possible based on the distinguishing codes associated with individual prostitutes for retrieving blood-test results. Although such codes would not enable personal identification in the manner of a name or social security number, they enabled the research team to assess whether individuals had participated more than once, at least to the extent that participants would answer the questions the same way in successive interviews.

The study produced 998 completed interviews, representing roughly 61% of the 1,629 women who were approached for screening. On the basis of the distinguishing codes, there were 55 individuals who were "known" to be interviewed more than once: 50 individuals had duplicate identifying codes in the database, and 5 individuals had codes that were observed in triplicate. However, distinguishing codes were not available for approximately 23% of the interviews (Ishwaran, Berry, Duan, and Kanouse, 1991). Further, it was suspected that some subjects might have misled interviewers when answering questions to be used for identification purposes. This gave rise to concern that there were additional undetected duplicate interviews in the database, carrying the potential to bias estimates of HIV prevalence and other outcomes of interest.

The present chapter summarizes collaborative work building on methods for calibrating error rates in record linkage settings using a mixture-model framework (Belin and Rubin, 1995). The effort strengthened the foundation of the LAWHRS by providing evidence that the study had indeed succeeded in interviewing a large number of different street prostitutes. In this chapter, we review relevant frameworks for record linkage that relate to the problem of identifying duplicates, after which we describe the procedure used to identify duplicates in the LAWHRS and offer comments for future applications.

29.2 General frameworks for record linkage

The problem of identifying duplicate records in the LAWHRS is framed here as a problem of record linkage, which generally refers to a technique for identifying individual records in one or more databases that correspond to the same person. Early theoretical work on record linkage (e.g., Newcombe, Kennedy, Axford, and James, 1959; Fellegi and Sunter, 1969) gave rise to strategies for bringing together candidate matched pairs of records. These authors recognized that there were inherent uncertainties in automated procedures requiring investigators to develop tolerances for false linkages, or false matches. Fellegi and Sunter (1969) outlined a procedure for estimating false-match rates that made use of the estimated probabilities of agreement on components within individual records that provided the basis for their procedure to assign weights characterizing closeness of agreement between record pairs. For example, typographical and transcribing errors might result in gender agreeing 99% of the time between pairs of records referring to the same person, while chance agreement would suggest that gender might agree 50% of the time between pairs of records from different people. Assuming independence of agreement across fields of information within records, one can estimate probabilities of false match as a function of the agreement weights, which can then be used to establish a cutoff between record pairs treated as matches and record pairs treated as nonmatches. Belin (1993) shows that the performance of record-linkage procedures can depend critically on decisions about where to set such a cutoff. But as noted in Belin and Rubin (1995) and Larsen and Rubin (2001), the Fellegi–Sunter approach can founder on violations of the independence assumption, giving rise to inaccurate estimates of false-match rates. (For example, in census applications considered in those articles, agreement on first name would clearly not be independent of agreement on gender.)

Belin and Rubin (1995) finesse the independence assumptions in the Fellegi–Sunter framework by tapping outside information, namely previously processed databases that have already been reviewed for accuracy by teams of matching clerks. These databases, with matching-clerk determinations taken as a gold standard, offer information regarding the distribution of agreement weights for true-matched pairs and the distribution of agreement weights for false-matched pairs. The key innovation of Belin and Rubin (1995) involved viewing the agreement weights in a database not yet reviewed by matching clerks as arising from a mixture of weights for true matches and weights for false matches. In this context, it is not essential that the agreement weights be derived from a procedure such as that of Fellegi and Sunter (1969); more crucial is that the procedures used to develop agreement weights are exchangeable across applications. Two-component mixture models could then be fit in a current database by making use of informative priors derived from previously processed data. Other elements of the estimation strategy include the EM algorithm (Dempster, Laird, and Rubin, 1977) to obtain posterior modes for mixture-model parameters, the SEM algorithm to obtain asymptotic standard errors (Meng and Rubin, 1991), and multiple imputation to average over uncertainty about appropriate normalizing transformations (Rubin, 1987b).

Another strategy for calibrating error rates in record linkage is described by Larsen and Rubin (2001), who extend the Fellegi–Sunter approach in a more flexible framework that allows dependence of agreement among fields of information in records. The mixture-model idea is still central, as the set of all pairs of records is partitioned on the basis of latent indicators into separate classes, but the models in this context are mixtures for discrete data (reflecting agreement or disagreement on each of the several characteristics). In the census application that served as a motivating example, three-class mixtures were explored, where the fitted models tended to divide pairs into same-household matches, same-household nonmatches, and different-household nonmatches. Instead of assuming the existence of a large database that has already been processed, Larsen and Rubin (2001) propose to achieve accurate calibration of false-match rates by fitting a mixture model, selecting subsets of the original record pairs to be reviewed for accuracy of matching determinations, and iterating the model-fitting and review process until only a small proportion of record pairs reviewed in successive review cycles appear to be matches.

29.3   Estimating probabilities of duplication in the Los Angeles Women's Health Risk Study

In the LAWHRS, the availability of 50 known duplicates and 5 known triplicates based on information in the "distinguishing codes" presented an opportunity to use the Belin and Rubin (1995) framework to assess the extent of additional duplication in the database. Specifically, the known duplicates and triplicates would provide a "training" data set where, for a given metric summarizing closeness of agreement between answers in the balance of the interview, the training data would provide information on the distribution of the agreement metric between duplicate interviews on the same person as well as information on the distribution of the agreement metric between interviews on different people. We elaborate by summarizing a procedure for choosing a distance metric, discussing the estimation of mixture components, and describing findings from the LAWHRS.

Choosing a distance metric

There were 14 questionnaire items that were used to construct the individual distinguishing codes in the LAWHRS. Because these items were not available on all individuals, it was necessary to use other questionnaire items to develop a distance metric to summarize closeness of agreement between records. Ishwaran, Berry, Duan, and Kanouse (1991) list 107 questionnaire items that were available for inclusion in a distance metric. While it would have been possible to use all items in the metric, it was presumed that some items would contribute substantially to the ability to distinguish records while other items would not. To avoid including items in the distance metric that were largely adding noise to the assessment, it was decided to include items in the distance metric only if there was evidence that they would contribute to the ability to distinguish individuals.

This problem was conceptualized using a testing framework to decide whether a pair of records represents two different individuals or two records from the same individual. Suppose the database has n records. Let questionnaire items be indexed by i, where $i = 1, 2, \ldots, 107$, let record pairs be indexed by j, where $j = 1, 2, \ldots, n(n-1)/2$, and let $\delta_{i,j}$ represent the indicator function comparing record pair j on question i, with the result equal to 1 if there is agreement and equal to 0 if there is disagreement. The set of record pairs can be partitioned into the set T of truly matched (or duplicate) pairs and the set F of false-matched pairs. If we let p_{0i} represent the probability of agreement on question i between two different individuals and p_{1i} represent the probability of agreement on question i in two interviews with the same individual, we can write

$$\delta_{i,j} \sim \begin{cases} \text{Bernoulli}(p_{0i}) & \text{if } j \in F \\ \text{Bernoulli}(p_{1i}) & \text{if } j \in T. \end{cases}$$

The training data set provides information that can be used to estimate p_{0i} and p_{1i}. Specifically, we can let

$$\hat{p}_{0i} = \frac{\sum_{j \in F} \delta_{i,j}}{|F|}$$

and

$$\hat{p}_{1i} = \frac{\sum_{j \in T} \delta_{i,j}}{|T|},$$

where $|\cdot|$ represents the cardinality of a set. We assume that the 50 record pairs with duplicate distinguishing codes and the 5 record triples with triplicate distinguishing codes in the training set refer to 55 different individuals (e.g., we assume that a person did not respond twice with one set of answers to the distinguishing-code questions and twice more with a different set of answers to the distinguishing-code questions). Then, except for the triplicates, we should have near independence among the $\delta_{i,j}$ for $j \in T$. Therefore,

$$|T|\hat{p}_{1i} \sim \text{Bin}(|T|, p_{1i}),$$

approximately. However, there may be quite a bit of dependence among the $\delta_{i,j}$ when $j \epsilon F$. For instance, the first record from duplicate pair A will be compared against both records of a different duplicate pair B, and the second record from duplicate pair A will also be compared against both records from duplicate pair B. These four comparisons are apt to yield the same indicator values if the duplicates are well matched. Ignoring triplicates, we should expect

$$|F|\hat{p}_{0i} \sim 4\,\text{Bin}(|F/4|, p_{0i}),$$

at least approximately.

For items where $p_{0i} = p_{1i}$, which would not be useful for distinguishing true-matched and false-matched pairs, we would have

$$\hat{p}_{1i} - \hat{p}_{0i} \sim N(0, s^2)$$

where

$$s = \sqrt{\frac{\hat{p}_{1i}(1 - \hat{p}_{1i})}{|T|} + \frac{4\hat{p}_{0i}(1 - \hat{p}_{0i})}{|F|}}$$

based on using the fact that $\hat{p}_{1i}$ is independent of $\hat{p}_{0i}$ and applying a normal approximation.

These results were used to motivate a decision rule to include question i in the metric summarizing closeness of agreement if

$$\hat{p}_{1i} > \hat{p}_{0i} + 3.1\,s,$$

which corresponds to an event in the upper 0.1 percentile of the normal reference distribution. This decision rule suggested that 55 of the original 107 items would be included in the distance metric.

The second part of the algorithm involved assigning a weight to a chosen question. If question i was deemed suitable, then the weight for this question was calculated as

$$w_i = \frac{1}{2}\left(\frac{p_{1i}}{p_{0i}} + \frac{1 - p_{0i}}{1 - p_{1i}}\right).$$

The w_i's ranged from values of 1.54 to 28.40. Although not formally equivalent to the weighting procedure outlined in Fellegi and Sunter (1969), which involves logarithms of the ratios of conditional probabilities of agreement given T and F, this weighting scheme has the property of assigning high weights to questions that have a high probability of agreement under the alternative hypothesis as well as to those questions that have a high probability of disagreement under the null hypothesis. Agreement weights Y_j were calculated by summing the w_i values across all questionnaire items represented in record pair j and rescaling so that the maximum agreement weight would equal 100.

Figure 29.1 displays the distribution of agreement weights for the true- and false-matched pairs. There is some overlap, but it is clear that the metric provides a strong basis for distinguishing true and false matches.

In line with the findings of Belin (1993), we anticipated that even *ad hoc* approaches to assigning weights would perform reasonably well. For purposes of comparison, using the same questionnaire items as were included in the distance metric described above, unit weights were assigned for agreement on field i. As expected, this metric capturing a count of the number of fields of agreement between records produced well-separated components, but the original weighting

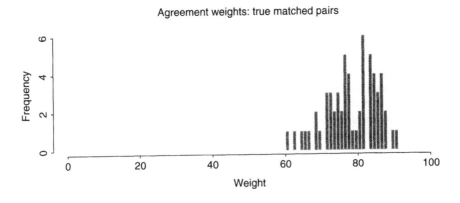

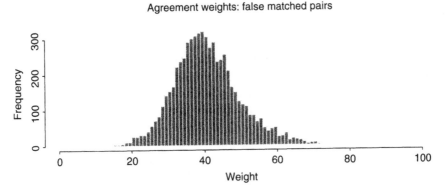

Figure 29.1   Histograms of agreement weights for (a) true-matched pairs and (b) false-matched pairs, showing clear separation but some overlap.

scheme appeared to produce better separation in the region of overlap between the two components.

Estimating mixture components

Formally, the problem of determining the extent of duplication in the LAWHRS can be framed as a mixture problem, where the distribution of all distances can be partitioned into two components, one characterizing the distribution of distances associated with true-matched pairs and one characterizing the distribution of distances associated with false-matched pairs. The problem is complicated by the presence of many pairs being associated with each individual record. For example, among the 115 records in the training data set (comprised 50 duplicate pairs and 5 triplicate sets), one could construct 6,555 record pairs, of which 65 are true matches (the 50 duplicates plus 3 matched pairs for each triplicate), and the remaining 6,490 are false matches. The problem is further complicated by the fact that the size of the training set is small compared to the size of the set being investigated for duplicates, where there were nearly 500,000 pairs to be considered. While the distribution of weights for true matches might be fairly comparable between the training and target databases, the greater number of possible pairs in the target database raised the possibility of a different distribution of weights for false matches between the training and target databases.

A first-pass approach sought to gauge the extent of the duplication by identifying best candidate matches for each record using the newly developed closeness-of-agreement metric and then reviewing candidate matches manually. The largest weight associated with a false-matched pair in the training data set was 80.3 on the 100-point scale, so in this first pass, it was decided to consider all pairs with weights above 80.3 to be duplicates. Individual distinguishing codes were also assessed to judge whether some of the cases might be part of triples or quadruples. This process yielded 25 new suspected doubles, 7 new suspected triples (4 of which were duplicates in the training set), and 2 new quadruples (1 of which was a triplicate in the training set). According to this tally, overall there were 2 quadruples, 11 triples (7 newly suspected plus 4 of the 5 triples in the training data, with the other triple in the training data now looking like a quadruple), and 71 duplicates (25 newly suspected plus 46 of the 50 duplicates in the training data, the other 4 duplicates now appearing to be parts of triples). The implied number of duplicate interviews was 3 for each of the quadruples, 2 for each of the triples, and 1 for each of the duplicates, or 99 overall. This represented roughly 10% of the 998 completed interviews.

An alternate approach to assessing the extent of contamination of the set of interviews through duplication was based on a probability model described in Belin and Rubin (1995). Working with the weights for the best candidate matches, the model assumes that the weights for true matches and weights for false matches are each normally distributed after application of two-parameter Box–Cox transformations, with distinct transformations for each component to address possibly different skewness in the weight distributions. The transformations are indexed by a power parameter γ and a scaling parameter g corresponding to the geometric mean of the observations in the following way:

$$\psi(w; \gamma, g) = \begin{cases} \frac{w^\gamma - 1}{\gamma(g)^{\gamma - 1}} & \text{if } \gamma \neq 0 \\ g \log(w) & \text{if } \gamma = 0. \end{cases}$$

The two transformations are estimated from the training sample by the use of a grid search of the likelihood. To facilitate identification of the mixture distribution in the target sample, the training data are also used to provide an estimate of the ratio of the variances of the component distributions on the transformed scales. Subsequently, a mixture model is fit to the weights for best candidate matches from the target sample. An EM algorithm is available to obtain estimates of the component means and the unconstrained component variance parameter, after which posterior probabilities of duplication are available as ratios of transformed-normal component densities (Belin and Rubin, 1995). Using this approach, the aggregate probability of false match (duplication) was estimated to be 14.9%. This departed somewhat from the assessment from the manual procedure, partly due to the impact of cases where the record pair was not clearly a duplicate. But the broader conclusion of investigators from both approaches was that duplication, while a concern that merited attention, was not at a level that would completely undermine findings from the very demanding fieldwork.

Results from LAWHRS

Estimates for the street-prostitute workforce were obtained by combining sampling weights from the LAWHRS with estimated probabilities of duplication to downweight the impact of potential duplicates. Key findings from the LAWHRS are summarized in Berry, Kanouse, Duan, and Lillard (1992) and Kanouse et al. (1992). Survey participants saw a mean of 30.2 clients per week. Vaginal sex without a condom occurred in 12% of most recent transactions, and oral sex without a condom occurred in 21% of most recent transactions, with some transactions involving both. In 30% of transactions, the client requested a condom, with a condom being used in 97% of those cases. Meanwhile, in 11% of transactions the client requested that a condom not be used, with condoms not being used in roughly half of those transactions. While practical considerations involving supervision of phlebotomy delayed the initiation of blood testing, blood draws were available for over half of the sample. Laboratory analysis suggested seropositive rates of 2.5% for HIV-1 antibodies, 33% for hepatitis B surface antibodies, and 34% for past or present syphilis infection. The emerging profile suggests that both street prostitutes and their clients are at a substantial risk for sexually transmitted diseases including HIV and that the amount of risk assumed is an outcome of a negotiation process.

29.4   Discussion

The application of Belin and Rubin's mixture-model technique for identifying duplicate interviews was successful on multiple levels in the context of this interesting applied context. First, the availability of the method provided a probabilistic framework for incorporating evidence about duplication, which was desirable in a context where great effort had been expended to obtain a probability sample. The results, which appeared consistent with a manual approach that had face validity, helped to instill confidence not only in the method itself but also, on account of the nonthreatening magnitude of the estimated duplication, in the findings of the study as a whole.

Further confidence in the method derived from an anecdote that the investigators alluded to as "the search for the three-faced Eve." Feedback from field workers had identified a case as a likely triplicate, although in the clerical process, linkage was lost between the record of the third interview and two other records that could be classified as duplicates using the distinguishing codes. Although the investigators were prepared to have project staff review the hundreds of hard copies of interviews to try to find the lurking triplicate, the weighting scheme suggested a candidate match that was readily identified as the third member of the triple without requiring such extensive manual effort. This process added a measure of face validity to the methodology.

An anonymous reviewer noted the possibility of comparing characteristics of duplicates and nonduplicates to assess potential systematic relationships. That is, to build on terminology from Rubin (1976a), one could consider whether records were "duplicated completely at random," with no systematic differences between duplicates and nonduplicates, or were "duplicated at random", allowing the possibility that duplicates and nonduplicates may differ on covariates. One could further consider defining weighting classes based on covariate data and using weighting adjustments to assess whether certain quantities of interest, such as AIDS prevalence, are disproportionately affected by duplication.

A final comment concerns implications of this methodology in disclosure-avoidance problems. Survey organizations routinely offer pledges of confidentiality to survey participants, yet there is often considerable interest in having data files from censuses and annual surveys available for public use. The concern raised by the record-linkage methods described here is that another avenue might become available for identifying individuals in public-use data files by aggregating information on seemingly innocuous characteristics. Many individuals in the United States would be uniquely identifiable given a set of, say, 50 pieces of covariate information. As a hypothetical, suppose that a data user knew 50 pieces of covariate information on an individual from public sources (e.g., white, male, age 55, city of residence, etc.) and suppose that a public-use data file supplies records of personal income along with all of the same 50 items except city of residence. In such a setting, record-linkage techniques might be used by adversarial individuals to try to break confidentiality in public-use files. The implied challenge may require not just that imputation be used for disclosure avoidance, as suggested in Rubin (1993) and Raghunathan, Reiter, and Rubin (2003) but also that the imputation procedures scramble covariate information across individuals rather than just drawing entire individual records using hot-deck or approximate Bayesian bootstrap procedures. Raghunathan, Reiter, and Rubin (2003) recognize that choices between model-based and resampling-based imputation procedures involve trade-offs affecting both precision and protection against disclosure. Guarding against record-linkage technology implies another layer of challenge in disclosure-avoidance problems. The idea of joining seemingly distinct statistical frameworks into a unified whole, which paid off in assessing the extent of duplication in the Los Angeles prostitute survey, might also be important to a successful disclosure-avoidance strategy.

30

Applying structural equation models with incomplete data

Hal S. Stern and Yoonsook Jeon[1]

Structural equation models (SEM) are commonly used by social scientists to draw inferences about phenomena of interest. The models consist of systems of equations relating latent variables of scientific interest and the various observations that are designed to measure the latent variables. It is not uncommon for individuals in a study to be missing some of the measured variables. Many applied researchers in the past dealt with such incomplete data by ignoring any cases with missing values. This is certainly inefficient as it can lead to the loss of one-third or more of a typical study population and may be misleading if those with missing values differ from those with complete cases in any systematic way. A number of likelihood-based or least-squares approaches to accommodating incomplete cases have been developed (see, e.g., Little and Rubin, 2002). Approaches based on maximum likelihood estimation rely on asymptotic theory. In this chapter, the Bayesian approach to posterior inference for structural equation models with incomplete data is described and illustrated with data from a study of adolescent development. Structural equation models are introduced in Section 30.1, with an emphasis on the confirmatory factor or measurement model. The Bayesian approach to inference for SEM with incomplete data including relevant notation and algorithms is described in Section 30.2. Section 30.3 applies the methodology to data from the Iowa Youth and Family Project concerning psychological development of adolescents.

[1]Department of Statistics, University of California, Irvine, Calif., and Citi Cards, New York. The authors thank K. Conger and M. Rueter for the IYFP data, and two referees and the editors for their helpful comments. The authors research was partially supported by grants from the National Science Foundation, National Institute of Mental Health, and Iowa State University.

Applied Bayesian Modeling and Causal Inference from Incomplete-Data Perspectives.
Edited by A. Gelman and X-L. Meng © 2004 John Wiley & Sons, Ltd ISBN: 0-470-09043-X

30.1   Structural equation models

Latent variable model

The general SEM consists of two pieces: the latent variable model and the confirmatory factor model. The latent variable model describes relationships among a set of latent variables. The variables typically represent theoretical constructs, like self-esteem, that are difficult to measure. The model distinguishes between endogenous and exogenous latent variables. Endogenous variables are determined by other variables within the model and exogenous variables are determined outside of the model. The mathematical representation for the latent variable model is

$$\eta_i = B\eta_i + \Gamma\xi_i + \zeta_i$$

$$\xi_i \sim N(0, \Phi) \tag{30.1}$$

$$\zeta_i \sim N(0, \Psi),$$

for $i = 1, \ldots, n$, where η_i is an $m \times 1$ vector of endogenous latent variables, ξ_i is an $r \times 1$ vector of exogenous latent variables, ζ_i is an $m \times 1$ vector of error variables, which are assumed to be uncorrelated with the exogenous variables, B is an $m \times m$ coefficient matrix for the endogenous latent variables (with zero elements on the diagonal), Γ is an $m \times r$ coefficient matrix for the exogenous latent variables, Φ is an $r \times r$ diagonal variance–covariance matrix of the exogenous latent variables, and Ψ is an $m \times m$ diagonal variance–covariance matrix of the error variables. It is assumed that $(I - B)$ is nonsingular (Bollen, 1989). The errors and exogenous variables are assumed independent across individuals.

Confirmatory factor model

The confirmatory factor model represents the link between observed variables Y and latent variables Z. The observed variables are often indicators or scales designed to measure the underlying latent variables. The mathematical representation of the confirmatory factor model is

$$Y_i = \mu + \Lambda Z_i + \epsilon_i$$

$$\epsilon_i \sim N(0, \Sigma) \tag{30.2}$$

$$Z_i \sim N(0, R),$$

for $i = 1, \ldots, n$, where Y_i is a $p \times 1$ vector of observed variables, μ is a $p \times 1$ mean vector, Z_i is a $q \times 1$ vector of latent variables, Λ is a $p \times q$ coefficient matrix, ϵ_i is a $p \times 1$ error vector, which is uncorrelated with the latent variable Z_i, R is a $q \times q$ variance–covariance matrix of the latent variables, and Σ is the

$p \times p$ variance–covariance matrix of the errors of measurement. The error vectors and latent variables are assumed independent across individuals. With complete data, the model is often expressed in terms of centered data (so that $\mu = 0$) but the mean vector is necessary for our incomplete data setting.

Combining the models

The latent variable and confirmatory factor models are combined in a single model by taking $Z_i = (\xi_i^T, \eta_i^T)^T$ with $q = m + r$. Then the parameters of the latent variable model B, Γ, Φ, Ψ represent a particular structural hypothesis about the nature of the covariance matrix R of the latent variables in the confirmatory factor model. In particular, we have

$$R = \begin{pmatrix} \Phi & \Phi\Gamma^T((I-B)^{-1})^T \\ (I-B)^{-1}\Gamma\Phi & (I-B)^{-1}(\Gamma\Phi\Gamma^T + \Psi)((I-B)^{-1})^T \end{pmatrix} \tag{30.3}$$

Identification

A model is said to be identified if it is possible to identify each parameter from the elements of the population variance–covariance matrix. The identification issue is relevant to both the confirmatory factor model and the latent variable model. Identification of the confirmatory factor model requires that the parameters Λ, R, and Σ be identified from the $p(p+1)/2$-element variance–covariance matrix of Y. It is common to identify the model by introducing restrictions, for example, setting some elements of Λ and Σ to zero. Common restrictions include the assumption of uncorrelated errors of measurement (off-diagonal elements of Σ equal to zero) and/or the assumption that each observed variable measures a single latent variable (only a single nonzero λ_{ij} coefficient in each row of Λ). In addition, identification of the confirmatory factor model requires a "scaling" of the latent variables. This is required because the contribution of the latent variables to the observed variance is $\Lambda R \Lambda^T$ and it is always possible to find an infinite number of equivalent (Λ, R) pairs. There are two ways to produce a scale for the latent variables: set the variances of the latent variables to one (i.e., make R a correlation matrix) or take each latent variable's variance to be equal to the variance of one of the observed variables (i.e., set a single λ_{ij} in each row equal to 1).

Identification of the latent variable model implies further that the parameters in the latent variable model can be distinguished. Much has been written about identification for the full structural equation model (including confirmatory factor and latent variable components), for example, see Bollen (1989). This chapter assumes that the latent variable model is recursive, which guarantees that the full structural equation model is identifiable if the confirmatory factor model is identifiable. A latent variable model is said to be recursive if there are no reciprocal causation or feedback relationships among the latent variables, and the error in one equation is not correlated with the errors in the other equations. Mathematically, a recursive model has lower triangular B matrix and diagonal Φ matrix.

30.2 Bayesian inference for structural equation models

Inference for structural equation models is often done without relying on formal probability models. Parameters are estimated to minimize some measure of distance between the observed variance matrix and the population variance matrix implied by the model (see, for example, Bollen, 1989). Though it is possible to accommodate missing data with such approaches (Yuan and Bentler, 1996), it is generally more straightforward to accommodate missing data with an underlying probability model and that is the approach of this chapter.

The confirmatory factor model with complete data

Under the assumed normal distributions, the marginal distribution of the observed data is $Y_i \overset{iid}{\sim} N(\mu, \Lambda R \Lambda^T + \Sigma), i = 1, 2, \ldots, n$. Assuming $\mathbf{Y} = (Y_1, \ldots, Y_n)$ is completely observed, maximum likelihood (ML) estimates and associated inference can be obtained directly from the Gaussian likelihood. Throughout this chapter we focus on Bayesian inference. The Bayesian approach to inference incorporates a prior distribution for the parameters and provides inferences based on the posterior distribution $p(\mu, \Lambda, R, \Sigma | Y) \propto L(\mu, \Lambda, R, \Sigma | Y) p(\mu, \Lambda, R, \Sigma)$. Maximum likelihood inference relies on large sample theory, whereas the Bayesian approach does not. We discuss the choice of prior distributions a bit later in this section.

For computational convenience, and to facilitate our approach to incomplete data, we introduce the augmented complete-data likelihood incorporating the latent variables $\mathbf{Z} = (Z_1, \ldots, Z_n)$. The joint distribution of Y_i and Z_i is easily obtained by multiplying the marginal distribution of Z_i and the conditional distribution of Y_i given Z_i. This yields

$$\begin{pmatrix} Y_i \\ Z_i \end{pmatrix} \overset{iid}{\sim} N \left(\begin{pmatrix} \mu \\ 0 \end{pmatrix} \begin{pmatrix} \Lambda R \Lambda^T + \Sigma, & \Lambda R \\ R \Lambda^T & R \end{pmatrix} \right) \tag{30.4}$$

for $i = 1, \ldots, n$. Rubin and Thayer (1982, 1983) describe the use of the EM algorithm for obtaining maximum likelihood estimates in the confirmatory factor model by treating the latent variables as missing data in the augmented model. Bayesian inference can be carried out in much the same way. We obtain inferences from the posterior distribution $p(\mu, \Lambda, R, \Sigma, \mathbf{Z} | Y)$ and then marginalize over Z. This approach also provides the flexibility for reporting inferences about the latent variables (factor scores) if desired.

To identify the confirmatory factor model, we assume that R is a variance–covariance matrix and that a single element in each row of Λ is equal to one. This identifies the scale of each latent variable as the scale of one of its indicators.

Model specification with missing data

If there are missing values, then we write $Y_i = (Y_{obs,i}, Y_{mis,i}), i = 1, \ldots, n$, and let $\mathbf{Y} = (Y_{obs}, Y_{mis})$ represent the entire data set. Throughout we assume that missing values are missing at random (MAR) in the sense of Rubin (1976a). This means that we can draw inferences based on the observed-data likelihood without modeling the missing-data mechanism. Though the data itself does not provide enough information to determine whether MAR is a plausible assumption, it is common to initially assume MAR and then assess the sensitivity of the inferences to this assumption. One strategy for data analysis with missing values is multiple imputation (see, for example, Rubin, 1987), wherein multiple complete data sets are constructed using a probabilistic imputation model to fill-in the missing values. The other common strategy, which is followed here, is to carry out likelihood-based or Bayesian inference by averaging over the missing values. Collins, Schafer, and Kam (2001) compare these two strategies in the context of SEM.

Finkbeiner (1979) carries out a likelihood-based analysis with missing data by averaging over the missing values. Jeon (1998) instead uses the augmented model likelihood that includes $\mathbf{Z}$ and extends the EM approach of Rubin and Thayer (1982) to accommodate the unintentionally missing data Y_{mis} along with the intentionally missing (i.e., latent) $\mathbf{Z}$. We use a Bayesian approach based on the augmented model. The augmented model can be written as

$$\begin{pmatrix} Y_{obs,i} \\ Y_{mis,i} \\ Z_i \end{pmatrix} \overset{ind}{\sim} N \left(\begin{pmatrix} \mu_{obs,i} \\ \mu_{mis,i} \\ 0 \end{pmatrix}, \Omega = \begin{pmatrix} \Omega_{oo,i} & \Omega_{om,i} & \Omega_{oz,i} \\ \Omega_{mo,i} & \Omega_{mm,i} & \Omega_{mz,i} \\ \Omega_{zo,i} & \Omega_{zm,i} & \Omega_{zz,i} \end{pmatrix} \right) \quad (30.5)$$

for $i = 1, \ldots, n$. In this notation, we use m or mis to denote elements corresponding to missing values, o or obs to denote elements corresponding to observed values, and z to denote elements corresponding to latent variables. It follows from earlier results that

$$\Omega = \begin{pmatrix} (\Lambda R \Lambda^T + \Sigma)_{oo,i} & (\Lambda R \Lambda^T + \Sigma)_{om,i} & (\Lambda R)_{oz,i} \\ (\Lambda R \Lambda^T + \Sigma)_{om,i} & (\Lambda R \Lambda^T + \Sigma)_{mm,i} & (\Lambda R)_{mz,i} \\ (R\Lambda^T)_{zo,i} & (R\Lambda^T)_{zm,i} & R \end{pmatrix}.$$

The joint posterior distribution of the model parameters, the missing data, and the latent variables is

$$p(\mu, \Lambda, R, \Sigma, Z, Y_{mis}|Y_{obs}) \propto \left(\prod_{i=1}^{n} p(Y_{obs,i}, Y_{mis,i}, Z_i|\mu, \Lambda, \Sigma, R) \right)$$

$$\times p(\mu, \Lambda, \Sigma, R). \quad (30.6)$$

Notation for carrying out the Bayesian analysis

The posterior distribution, both with complete data and without, is difficult to study analytically. Instead, we study the posterior distribution by using Markov chain Monte Carlo to generate samples from it and then computing numerical summaries from the simulated values. A key issue in applying the Bayesian method to the analysis of data using SEM is that the variance–covariance matrix Σ and the coefficient matrix Λ typically have special structure that must be taken into account.

In this chapter, we restrict attention to the common case in which Σ is either diagonal (the observed variables $\mathbf{Y}$ are independent given the latent variables $\mathbf{Z}$) or block diagonal (conditional correlations allowed but only within subsets of the vector $\mathbf{Y}$). We can combine the two cases since the diagonal matrix can be viewed as a block diagonal matrix with each block consisting of a single element. Let the number of blocks be m and denote these by $\Sigma_1, \ldots, \Sigma_m$. The number of elements in block k is p_k. The model for $\mathbf{Y}$ given $\mathbf{Z}$ can be written as a series of independent models defined on the blocks. This requires some fairly detailed notation that is provided next.

Let $Y_{i,k}^*$ be the $p_k \times 1$ subvector of Y_i corresponding to the kth block, $Y_{i,k}^* = (y_{i,k(1)}, y_{i,k(2)}, \ldots, y_{i,k(p_k)})^T$, where $y_{i,k(j)}$ is the observation corresponding to the jth variable in the kth block for unit i. The notation $k(j)$ allows us to account for the possibility that the elements of Y may need to be reordered to create the block diagonal form. Define μ_k^* to be the corresponding $p_k \times 1$ subvector of μ, Λ_k^* to be the corresponding $p_k \times q$ submatrix of Λ with $k(j)$th row $\Lambda_{k(j)}^T = (\lambda_{k(j),1}, \lambda_{k(j),2}, \ldots, \lambda_{k(j),q})$, and $\epsilon_{i,k}^*$ to be the corresponding $p_k \times 1$ subvector of ϵ. The model within the kth block can be written as follows:

$$Y_{i,k}^* = \mu_k^* + \Lambda_k^* Z_i + \epsilon_{i,k}^*, \quad i = 1, \ldots, n.$$

We next accommodate the fact that for a confirmatory factor model some elements of Λ_k^* are assumed *a priori* to be equal to zero. We reorder the elements in each row of Λ_k^* so that the zero elements are together and make the corresponding change to Z_i, writing $Z_{i,k(j)} = (Z_{i,1,k(j)}^T, Z_{i,0,k(j)}^T)^T$ and $\Lambda_{k(j)}^T = \left(\Lambda_{1,k(j)}^T, \Lambda_{0,k(j)}^T\right) = \left(\Lambda_{1,k(j)}^T, 0\right)$, where $Z_{i,1,k(j)}$ is the $b_{k(j)} \times 1$ vector corresponding to the latent variables with nonzero coefficients, $Z_{i,0,k(j)}$ is the $(q - b_{k(j)}) \times 1$ vector corresponding to the latent variables with zero coefficients, $\Lambda_{1,k(j)}^T$ is the row vector of elements to be estimated in $\Lambda_{k(j)}^T$, $\Lambda_{0,k(j)}^T$ is the row vector of *a priori* zero elements. Then the model for an individual variable, say the jth variable, within the kth block can be written as

$$y_{i,k(j)} = \mu_{k(j)} + \Lambda_{1,k(j)} Z_{i,1,k(j)} + \epsilon_{i,k(j)}. \tag{30.7}$$

If we treat the Z_i as known (they will be available at the appropriate stage of the MCMC algorithm developed below), then we can combine the individual variable models (30.7) into a multivariate regression model for block k (i.e., a regression model with multiple responses),

$$Y_{i,k}^* = \mu_k^* + Z_{i,k}^* \Lambda_k^{**} + \epsilon_{i,k}^*, \quad i = 1, 2, \ldots, n, \tag{30.8}$$

where Λ_k^{**} is the $\sum_{j=1}^{p_k} b_{k(j)} \times 1$ vector obtained by concatenating the $\Lambda_{1,k(j)}$ vectors for each of the block's variables into a single column, and $Z_{i,k}^*$ is the $p_k \times \sum_{j=1}^{p_k} b_{k(j)}$ matrix with jth row equal to $Z_{i,1,k(j)}^T$ in the columns corresponding to the relevant elements of Λ_k^{**} and zero elsewhere.

Prior distributions

We choose noninformative flat prior distributions for location parameters and regression coefficients and conjugate prior distribution for the variance components,

$$p(\mu) \propto 1;$$

$$p(\lambda|\Sigma) \propto 1, \text{ where } \lambda \text{ is any element of } \Lambda(\text{not fixed at 0 or 1});$$

$$\Sigma_j \sim \text{inverse-Wishart}(\nu_j, S_j), \quad j = 1, \ldots, m;$$

$$R \sim \text{inverse-Wishart}(\nu_{m+1}, S_{m+1}).$$

The inverse-Wishart distribution is a conjugate prior distribution for a multivariate normal variance matrix. A limitation of the inverse-Wishart is that it assumes a single degrees-of-freedom parameter, thus assuming equal information about each variable. In the present case, this should not be a problem if the blocks contain few variables. More generally alternative prior distributions can be used for the variance–covariance matrices, as described for example in Gelman, Carlin, Stern, and Rubin (2003). The flat prior distributions for μ and for the elements of Λ are not proper distributions, but the resulting posterior distribution is a proper distribution.

Gibbs sampling for structural equation models

Gibbs sampling (Geman and Geman, 1984; Gelfand and Smith, 1990) can be used for simulating from the posterior distribution because each of the full conditional distributions (the posterior distribution of one subvector of parameters or missing data given all of the others) are known distributional forms. We now describe the Gibbs sampling algorithm for the confirmatory factor model, including the possibility of missing data, by specifying the full conditional distributions:

Full conditional distribution of μ
Given all of the model parameters and conditioning on the augmented complete data containing $\mathbf{Y}$ and $\mathbf{Z}$, the distribution of μ is normal,

$$\mu|\mathbf{Y}, \mathbf{Z}, \Lambda, \Sigma, R \sim \text{N}(\mu_\mu, V_\mu),$$

where $\mu_\mu = \overline{Y} - \Lambda\overline{Z}$, $V_\mu = \frac{1}{n}\Sigma$, $\overline{Y}$ is the sample mean of the observed or indicator variables and $\overline{Z}$ is the sample mean of the factor scores.

Full conditional distribution of Z_i

The conditional distribution of Z_i, the latent variables for the ith observation, can be obtained from its joint normal distribution with Y_i given in (30.4),

$$Z_i \mid Y_i, \mu, \Lambda, \Sigma, R \sim N(\mu_{z,i}, V_z), \quad i = 1, 2, \ldots, n,$$

where $\mu_{z,i} = R\Lambda^T(\Lambda R \Lambda^T + \Sigma)^{-1}(Y_i - \mu)$, and $V_z = R - R\Lambda^T(\Lambda R \Lambda^T + \Sigma)^{-1}\Lambda R$. Draws for different individuals are independent.

Full conditional distribution of $Y_{\mathrm{mis},i}$

In a similar manner, the conditional distribution of the missing data is obtained from the joint normal distribution of $(Y_{\mathrm{obs},i}, Y_{\mathrm{mis},i}, Z_i)$ given in (30.5),

$$Y_{\mathrm{mis},i} \mid Y_{\mathrm{obs},i}, Z_i, \mu, \Lambda, \Sigma, R \sim N(\mu_{Y,i}, V_{Y,i}), \quad i = 1, 2, \ldots, n,$$

where

$$\mu_{Y,i} = \mu_{\mathrm{mis},i} + \begin{pmatrix} \Omega_{mz,i} & \Omega_{mo,i} \end{pmatrix} \begin{pmatrix} \Omega_{zz,i} & \Omega_{zo,i} \\ \Omega_{oz,i} & \Omega_{oo,i} \end{pmatrix}^{-1} \begin{pmatrix} Z_i \\ Y_{\mathrm{obs},i} - \mu_{\mathrm{obs},i} \end{pmatrix},$$

$$V_{Y,i} = \Omega_{mm,i} - \begin{pmatrix} \Omega_{mz,i} & \Omega_{mo,i} \end{pmatrix} \begin{pmatrix} \Omega_{zz,i} & \Omega_{zo,i} \\ \Omega_{oz,i} & \Omega_{oo,i} \end{pmatrix}^{-1} \begin{pmatrix} \Omega_{zm,i} \\ \Omega_{om,i} \end{pmatrix}.$$

The draws for different individuals are independent.

Full conditional distribution of R

Given the latent variables, the posterior distribution of the unstructured variance–covariance matrix R is straightforward,

$$R \mid Y, \mu, \Lambda, \Sigma, Z \sim \text{inverse-Wishart}\left(n + v_{m+1}, \left(\textstyle\sum_{i=1}^{n} Z_i Z_i^T + S_{m+1}\right)^{-1}\right).$$

Full conditional distributions of Λ and Σ

The conditional posterior distributions for Λ and Σ are specified separately for each block using the representation in (30.8). For the kth block, we obtain

$$\Lambda_k^{**} \mid Z_{i,k}^*, Y_{i,k}^*, \mu_k^*, R, \Sigma_k \overset{ind}{\sim} N\left(\hat{\Lambda}_k^{**}, \left(\sum_{i=1}^{n} (Z_{i,k}^*)^T \Sigma_k^{-1} Z_{i,k}^*\right)^{-1}\right),$$

where $\hat{\Lambda}_k^{**} = \left(\sum_{i=1}^{n} (Z_{i,k}^*)^T \Sigma_k^{-1} Z_{i,k}^*\right)^{-1} \sum_{i=1}^{n} (Z_{i,k}^*)^T \Sigma_k^{-1}(Y_{i,k}^* - \mu_k^*)$, and

$$\Sigma_k \mid Y_{i,k}^*, \Lambda_k^{**}, Z_{i,k}^*, \mu_k^*, R \sim \text{inverse-Wishart}\,(n + v_k, W_k)$$

where $W_k = \left(\sum_{i=1}^{n} (Y_{i,k}^* - \mu_k^* - Z_{i,k}^* \Lambda_k^{**})(Y_{i,k}^* - \mu_k^* - Z_{i,k}^* \Lambda_k^{**})^T + S_k\right)^{-1}$.

Beyond the confirmatory factor model

The Gibbs sampling algorithm we have described allows us to obtain simulations from the posterior distribution of the parameters for the confirmatory factor model. Because we have assumed a recursive latent variable model with fully lower triangular B, it turns out that there is a one-to-one correspondence between the parameters (B, Γ, Φ, Ψ) of the latent variable model and the elements of the variance–covariance matrix of the latent variables R. Then posterior inferences for the latent variable parameters can be can be obtained directly from the posterior distribution of R by referring to the definition (30.3) in Section 30.1.

An alternative is to put prior distributions on Φ, Ψ, Γ, B and work directly with the joint posterior distribution (30.6) after making the transformation from R to the latent variable parameters. For models where there is no longer a one-to-one relationship between R and the elements of the latent variable model, this alternative would be the only way to proceed.

30.3   Iowa Youth and Families Project example

The data set and the model

Data from the Iowa Youth and Families Project (IYFP), a longitudinal study concerning the welfare of rural families that was carried out at the Institute for Social and Behavioral Research at Iowa State University, provides an example for illustrating our approach. Each of 451 rural Iowa families was interviewed four times during the period 1989 to 1992. One particular analysis by researchers Rand Conger and Martha Reuter concerned the development of adolescent problem solving behavior. There are 11 indicators in that analysis corresponding to 4 latent variables in their conceptual model. The first latent variable, "warmth, communication, and listening (1989)," is measured by 3 indicators (denoted Y_1, Y_2, and Y_3). The second latent variable, "adolescent problem solving behavior (1990)," is measured by 3 indicators (Y_4, Y_5, and Y_6). The third latent variable, "adolescent cynical, contemptuous attitude (1991)," is measured by 2 indicators (Y_7, Y_8). The fourth latent variable, "adolescent problem solving behavior (1992)," is measured by 3 indicators (Y_9, Y_{10}, and Y_{11}). The scientists expected a positive association between positive family communication in the first year (the first latent variable) and adolescent problem solving ability in the final year (the fourth latent variable). They also expected a positive association between problem solving ability in years 2 and 4 but a negative association between having a cynical attitude in year 3 and problem solving in year 4. We might expect the indicators for the second latent variable and the fourth latent variable to be correlated because the same items are used to measure the same latent variable at two different times. For this study, only 295 families out of the 451 had a completely observed $\mathbf{Y}$ vector; a complete case analysis loses 35% of the sample. Causes of missing data included unanswered questions and unusable videotape (which was used to define some indicators). The investigators were content to begin with the MAR assumption.

The matrix Λ relates the observable variable $\mathbf{Y}$ to the four latent variables. As described earlier, we identify the matrix R by selecting one indicator to define the scale of the latent variables. For example, Y_1 defines the scale of Z_1,

$$\Lambda^T = \begin{pmatrix} 1 & \lambda_2 & \lambda_3 & 0 & 0 & 0 & 0 & 0 & 0 & 0 & 0 \\ 0 & 0 & 0 & 1 & \lambda_5 & \lambda_6 & 0 & 0 & 0 & 0 & 0 \\ 0 & 0 & 0 & 0 & 0 & 0 & 1 & \lambda_8 & 0 & 0 & 0 \\ 0 & 0 & 0 & 0 & 0 & 0 & 0 & 0 & 1 & \lambda_{10} & \lambda_{11} \end{pmatrix}.$$

The conditional variance–covariance matrix of the indicator variables given the latent variables, Σ, can be written as a block diagonal matrix by reordering the rows and columns so that there are eight blocks: $\{Y_1\}, \{Y_2\}, \{Y_3\}, \{Y_7\}, \{Y_8\}, \{Y_4, Y_9\}, \{Y_5, Y_{10}\}, \{Y_6, Y_{11}\}$. The bivariate blocks reflect the expected correlation between those indicators measured both at time 2 and 4. Let θ_i denote the diagonal element corresponding to variable Y_i and also introduce $\theta_{12}, \theta_{13}, \theta_{14}$ as the off-diagonal elements for the two-variable blocks.

The latent variable model relates the four latent variables. Owing to the longitudinal nature of the study, it is natural to think of Z_1 as exogenous (ξ in the notation of Section 30.1) and the other latent variables as endogenous ($(Z_2, Z_3, Z_4) = (\eta_2, \eta_3, \eta_4)$ in the notation of Section 30.1). Then we can write the latent variable model for this example in the notation (30.2) as

$$\begin{pmatrix} z_{i,2} \\ z_{i,3} \\ z_{i,4} \end{pmatrix} = \begin{pmatrix} 0 & 0 & 0 \\ \beta_{32} & 0 & 0 \\ \beta_{42} & \beta_{43} & 0 \end{pmatrix} \begin{pmatrix} z_{i,2} \\ z_{i,3} \\ z_{i,4} \end{pmatrix} + \begin{pmatrix} \gamma_2 \\ \gamma_3 \\ \gamma_4 \end{pmatrix} z_{i,1} + \begin{pmatrix} \zeta_{i,2} \\ \zeta_{i,3} \\ \zeta_{i,4} \end{pmatrix},$$

with Φ a scalar variance parameter of the exogenous variable (Z_1) and Ψ a 3×3 diagonal variance matrix of the endogenous variables (Z_2, Z_3, Z_4). We have assumed a recursive latent variable model (lower triangular matrix B and diagonal matrix Ψ). There are 10 free parameters (3 elements of B, 3 elements of Γ, 1 element of Φ and 3 elements of Ψ), which can be identified from the 10 unique elements of R.

Bayesian inference

Prior distributions are chosen in accord with the earlier discussion. Specifically, we choose flat prior distributions for μ and the elements of Λ and

$$\theta_j \sim \text{scaled inverse-}\chi^2(0.01, 1), \ j = 1, 2, 3, 7, 8;$$

$$\Sigma_{(l,m)} \sim \text{inverse-Wishart}(2, I_2), \ (l, m) = (4, 9), (5, 10), (6, 11);$$

$$R \sim \text{inverse-Wishart}(4, I_4),$$

where I_2 and I_4 are 2×2 and 4×4 identity matrices respectively, and

$$\Sigma_{(4,9)} = \begin{pmatrix} \theta_4 & \theta_{12} \\ \theta_{12} & \theta_9 \end{pmatrix}, \ \Sigma_{(5,10)} = \begin{pmatrix} \theta_5 & \theta_{13} \\ \theta_{13} & \theta_{10} \end{pmatrix}, \ \Sigma_{(6,11)} = \begin{pmatrix} \theta_6 & \theta_{14} \\ \theta_{14} & \theta_{11} \end{pmatrix}.$$

The MCMC algorithm is described in Section 30.2. We carried out 5,000 iterations for each of 5 independently chosen starting points. The Gelman and Rubin (1992) potential scale reduction factor based on the last halves of the 5 chains are less than 1.2 so we take these simulations as representative of the posterior distribution. Numerical summaries of the posterior distribution are provided in Table 30.1. The left side of the table gives the estimated posterior medians and 95% central posterior intervals for the elements of R.

Because the parameters in our recursive latent variable model are completely determined by the elements of R, we do not need any additional simulation effort to draw inferences about them. We take each posterior draw of R and solve to obtain a posterior sample for the latent variable parameters. The right side of Table 30.1 shows summaries of the posterior distribution for the latent variable model parameters. The results confirm the researchers' hypotheses in that the posterior distribution for β_{42}, the association between problem solving ability in years 2 and 4, is concentrated on positive numbers, while β_{43}, the association between having a cynical attitude in year 3 and problem solving in year 4, is primarily concentrated on negative values.

Assessing fit

An essential part of a data analysis using SEM is to assess the quality of the fit of the model. The Bayesian approach easily allows model checking via the posterior predictive approach of Rubin (1984) and Gelman, Meng, and Stern (1996). We

Param	Posterior Median	95% Posterior Interval	Param	Posterior Median	95% Posterior Interval
$R_{(1,1)}$	8.96	(6.72, 11.74)	ϕ_1	8.96	(6.72, 11.74)
$R_{(1,2)}$	0.23	(0.08, 0.41)	ψ_2	0.12	(0.02, 0.19)
$R_{(1,3)}$	−1.30	(−2.17, −0.60)	ψ_3	1.11	(0.56, 2.04)
$R_{(1,4)}$	0.40	(0.04, 0.78)	ψ_4	0.69	(0.44, 0.98)
$R_{(2,2)}$	0.15	(0.09, 0.22)	γ_2	0.03	(0.01, 0.04)
$R_{(2,3)}$	−0.14	(−0.26, −0.05)	γ_3	−0.11	(−0.19, −0.03)
$R_{(2,4)}$	0.13	(0.06, 0.20)	γ_4	0.01	(−0.05, 0.06)
$R_{(3,3)}$	1.91	(0.91, 3.36)	β_{32}	−0.88	(−2.08, −0.21)
$R_{(3,4)}$	−0.32	(−0.56, −0.11)	β_{42}	0.82	(0.19, 1.64)
$R_{(4,4)}$	0.89	(0.59, 1.21)	β_{43}	−0.12	(−0.34, 0.05)

Table 30.1 Numerical summaries of the posterior distribution of the confirmatory factor parameter R and the associated latent variable model parameters.

define one or more test statistics $T(\mathbf{Y})$ or discrepancies $T(\mathbf{Y}; \Upsilon)$ (the latter may depend on the unknown parameters that are generically denoted by Υ here) and compare the observed value (or observed distribution in the case of a discrepancy) with an appropriate reference distribution. The reference distribution is obtained by generating replicate data, say $\mathbf{Y}^{\mathrm{rep}}$ from its posterior predictive distribution; this is done by simulating Υ from its posterior distribution and then $\mathbf{Y}^{\mathrm{rep}}$ from the distribution for $\mathbf{Y}$ given Υ. The replicate data $\mathbf{Y}^{\mathrm{rep}}$ is a plausible value for what we might expect if the study were repeated with the same (but currently unknown) parameter value. We can choose T to be a traditional SEM measure of fit, for example, $T(\mathbf{Y}; \Upsilon) = tr(S(\Lambda R \Lambda^T + \Sigma)^{-1})$, a measure of the distance between the observed sample covariance matrix S and the model-implied covariance matrix. It is not a problem that the definition of T depends on missing values because we have included the missing values in our simulation algorithm. For each posterior draw Υ, we simulate a replicated data set $\mathbf{Y}^{\mathrm{rep}}$ and then calculate $T(\mathbf{Y}; \Upsilon)$ and $T(\mathbf{Y}^{\mathrm{rep}}; \Upsilon)$. The two sets of values can be displayed in a scatterplot and/or summarized by a tail-area probability, the probability that $T(\mathbf{Y}^{\mathrm{rep}}; \Upsilon) \geq T(\mathbf{Y}; \Upsilon)$. In the IYFP example, the points are well scattered around a 45-degree line indicating no obvious lack of fit. A more thorough assessment would consider other diagnostic measures as well.

30.4   Summary and discussion

In this article, we explore statistical inference for structural equation models with missing data using a Bayesian approach. Methods are developed for confirmatory factor models and linked to a particular class of latent variable models. A computational algorithm is developed for generating samples from the posterior distribution of the model parameters. In addition, we discuss how the fit of the model is assessed using posterior predictive model checks. The influence of Rubin's work can be seen in the incorporation of latent variables as missing data following Rubin and Thayer (1982), the general approach to inference for missing data (Little and Rubin, 2002), the Gelman and Rubin (1992) developments regarding MCMC, and the application of posterior predictive model checks (Rubin, 1984).

31

Perceptual scaling

Ying Nian Wu, Cheng-En Guo, and Song Chun Zhu[1]

31.1 Introduction

Vision as statistical learning and inference

Vision can be posed as a statistical learning and inference problem. As an over-simplified account, let W be a description of the outside scene in terms of "what is where," let I be the retina image, and let $p(W, I)$ be the joint distribution of W and I.[2] Then visual learning is to learn $p(W, I)$ from training data, and visual perception is to infer W from I based on $p(W|I)$.

There are two major schools on visual learning and perception. One school is operation oriented and learns the inferential process defined by $p(W|I)$ directly, often in the form of an explicit transformation $W \approx F(I)$. This scheme is mostly used in supervised learning, where W is the object category, and is given in training data. The other school is representation oriented and learns the generative process $p(W)$ and $p(I|W)$ explicitly, then perception is to invert the generative process

[1]Department of Statistics and Computer Science, University of California, Los Angeles, Calif. We thank the two editors for advice on presentation. The work is partially supported by NSF grant IIS-0222967.

[2]In a philosophically more rigorous formulation, we may assume the existence of an underlying world, which is a functional. When this functional acts on the physical equipment, it gives what we call "W." When this functional acts on the retina cells, it gives what we call "I." A distribution over this "world functional" leads to the joint distribution of W and I. See for example, Mumford and Gidas (2001).

Applied Bayesian Modeling and Causal Inference from Incomplete-Data Perspectives.
Edited by A. Gelman and X-L. Meng © 2004 John Wiley & Sons, Ltd ISBN: 0-470-09043-X

by maximizing or sampling $p(W|I) \propto p(W)p(I|W)$. This scheme is Bayesian in nature, the prior distribution $p(W)$ may also be accounted for by a regularization term such as smoothness or sparsity. This scheme is often used in unsupervised learning where W is not available in training data.

In the literature, there are a number of statistical theories proposed for vision. In representation-oriented school, Grenander (1993) and Mumford (1994) proposed pattern theory as a paradigm for vision (see also Geman and Geman, 1984; Amit, Grenander, and Piccioni, 1991; Grenander and Miller, 1994; Geman, Potter, and Chi, 2002), for important contributions that are related to pattern theory). Olshausen and Field (1996) proposed the sparsity principle as a general strategy employed by primitive visual cortex, and use it to learn linear bases from natural images, and these bases are considered mathematical models for simple visual cells (see also Bell and Sejnowski, 1997), on independent component analysis for learning edge filters from natural images). The sparsity principle was also investigated by Candes and Donoho (1999) in the framework of harmonic analysis on wavelets and curvelets. Zhu, Wu, and Mumford (1997) and Wu, Zhu, and Liu (2000) proposed a class of Markov random field models (Besag, 1974; Cressie, 1993) for textures, and studied the minimax entropy principle and the equivalence of ensembles for feature statistics based on linear filters. In the operation-oriented school, contributions were made by Amit and Geman (1997), and Blanchard and D. Geman (2003), who stressed the importance of computing efficiency in visual perception. Tu and Zhu (2002) proposed data-driven Markov chain Monte Carlo (MCMC) for integrating operation-oriented methods into representation-oriented schemes.

As evidenced by the above theories, to understand visual learning and perceptual inference, it is crucial to identify fundamental visual phenomena and understand the underlying statistical principles. The proposed work is to study a ubiquitous visual phenomenon that we call *perceptual scaling*.

Perceptual scaling

The left column of Figure 31.1 displays three images of an ivy wall taken at three different distances. For the image at near distance, we perceive individual leaves, including their edges and shapes. For the image at far distance, however, we only perceive a collective foliage impression without discerning individual structures. While the near-distance image looks regular and simple, with sparse structures, the far-distance image appears random and complex, with rich details. Why does the same pattern result in different perceptions at different distances? Can we find a mathematical theory to formally explain this perceptual transition over scale?

This transition from sparse structures to collective textures is ubiquitous in outdoor scenes, and we call such transition *perceptual scaling*. For instance, the images of branches and twigs in the right column of Figure 31.1 also exhibit such a scaling effect. Currently, it is still unclear whether this transition is a continuous

Figure 31.1 Perceptual scaling: transition from sparse structures to collective textures over distance.

one or a quantum jump. It is likely that there exists a small gray area where both structure interpretation and texture interpretation are equally plausible.

Perceptual scaling typically presents itself in a single image of a static natural scene, because objects and patterns can appear at a wide variety of distances and depths from the viewer. See Figure 31.2 for two examples, where the leaves and branches give us different impressions at different scales. Thus, a mathematical theory that accounts for this scaling effect is crucial for a visual system to successfully interpret virtually any natural scenes.

As another example of perceptual scaling, in Figure 31.3, the left image gives us vivid 3D impression of shapes, whereas the right image only gives us an overall impression of roughness.

Figure 31.2   Perceptual scaling: the same patterns can appear at different scales in a single image.

Perceptual scaling also manifests itself in motion scenes (e.g., Doretto, Chiuso, Wu, and Soatto, 2003). For instance, when we look at sea surface, we perceive the shapes of big waves and we can trace their motions, whereas for the large number of small ripples, their shapes are not perceptible and their motions are not trackable.

There have been many interesting theories on the issue of scaling in the literature, such as scale space theory (e.g., Lindeberg, 1994), multiresolution analysis (Mallat, 1989), fractals (Mandelbrot, 1982), spectrum and simple statistics of natural images (Ruderman and Bialek, 1994; Mumford and Gidas, 2001; Chi, 2001; Simoncelli and Olshausen, 2001). However, none of these theories are concerned with the effect of image scaling on our perception of particular patterns such as those in Figure 31.1.

Given the fact that visual perception is a statistical inference problem, and complexity and randomness must be studied in a statistical framework, we argue that perceptual scaling is a statistical phenomenon. In particular, our approach relies heavily on the concept of entropy. This concept has its root in statistical mechanics, and can be understood as counting (in log-scale) the size of certain equivalence class (or ensemble). It also plays a central role in information theory, where it counts the average number of bits for coding the signal.

In this chapter, we prove two scaling laws in vision: if we get farther from a visual pattern, then (1) the resulting retina image becomes less sparse, and (2) the underlying pattern becomes less perceptible. The two scaling laws have interesting implications in the possible strategy employed by visual cortex, and reveal the connection between wavelet sparse coding and Markov random fields.

Figure 31.3   Perceptual scaling: from 3D shapes to texture impression of roughness.

31.2   Sparsity and minimax entropy

Wavelets and Markov random fields

The simple neuron cells in the primitive visual cortex (called V1) are mathematically modeled by a set of localized, oriented, and elongated linear bases/filters,

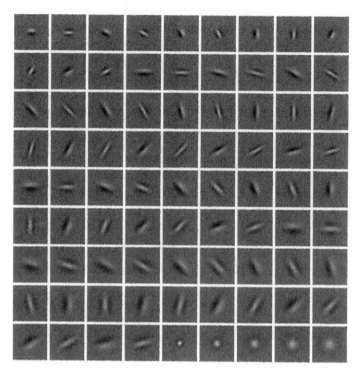

Figure 31.4   Linear bases/filters as mathematical model of V1 cells.

$\{B_{x,y,k}\}$, where (x, y) indexes the location, and k indexes the shape, such as orientation and scale. See Figure 31.4 for an illustration.

There are two major classes of representations for nature images; both involve the above local bases/filters.

Wavelets and sparse coding: This representation is generative (Lewicki and Olshausen, 1999)

$$c_{x,y,k} \sim p(c), \tag{31.1}$$

$$I = \sum c_{x,y,k} B_{x,y,k} + \epsilon, \tag{31.2}$$

where $c_{x,y,k}$ are coefficients for representing I in the form of (31.2), and ϵ is the residual error. The key principle is the sparsity principle (Olshausen and Field, 1996), where $\{B_{x,y,k}\}$ is assumed to be over-complete, that is, the number of bases exceeds the number of pixels, but for a typical image, only a small number of $c_{x,y,k}$ are significantly different from 0, that is, the prior distribution $p(c)$ in (31.1) is a long-tail distribution such as mixture of normals (Olshausen and Millman, 2000; see also George and McCulloch (1997), for independent but closely related work on Bayesian variable selection in regression). The sparsity assumption can also be expressed in a nonprobabilistic form by a regularization or penalty term (Candes and Donoho, 1999). If we treat $\{B_{x,y,k}\}$ as unknown parameters, then we can learn them from natural images (Olshausen and Field, 1996).

Markov random fields (MRFs) and feature statistics: For a homogeneous local image patch, which is still denoted by I for simplicity, we compute filter responses $r_{x,y,k} = \langle I, B_{x,y,k} \rangle$ for all the filters within this patch (Malik and Perona, 1990), and then for each type of filter k, we compute the histogram $H_k(I)$ by pooling $r_{x,y,k}$ over all (x, y) in this patch. The image patch is then represented by the set of histograms $H_k(I)$ (Heeger and Bergen, 1995; Portilla and Simoncelli, 2000). The basic idea is to consider the ensemble of images (Wu, Zhu, and Liu, 2000):

$$\Omega = \{I : H_k(I) = H_k(I^{\mathrm{obs}}), \forall k\}, \tag{31.3}$$

which collects all the images I that share the same histograms as the observed image I^{obs}. This ensemble is called *Julesz ensemble* by Wu, Zhu, and Liu (2000). One can model I as following the uniform distribution over the Julesz ensemble Ω according to the maximum entropy principle, where maximum entropy here means we are completely ignorant about I except that it is in a particular Julesz ensemble. This uniform distribution is equivalent to an MRF model or a Gibbs distribution (Wu, Zhu, and Liu, 2000),

$$f(I) = \frac{1}{Z} \exp\left\{ \sum_k \langle \lambda_k, H_k(I) \rangle \right\} = \frac{1}{Z} \exp\left\{ \sum_k \sum_{x,y} \lambda_k(\langle I, B_{x,y,k} \rangle) \right\},$$

$$\tag{31.4}$$

where λ_k is a vector of the same dimension as $H_k(I)$, so it can also be viewed as a one-dimensional step function over the bins of the histogram $H_k(I)$. Z is the normalizing constant that depends on $\{\lambda_k\}$. This model is called FRAME model (Filter, Random field, And Maximum Entropy) by Zhu, Wu, and Mumford (1997). If $\{H_k(I)\}$ are taken to be other statistics (e.g., moments instead of histograms), then the corresponding $\{\lambda_k()\}$ become other functions (e.g., polynomials instead of step functions). It is just a matter of parameterization.

The set of filters can be learned so that the volume of the Julesz ensemble Ω, that is, $|\Omega|$, or the entropy of the fitted MRF model $f(I)$ in (31.4), is minimum. This is the minimum entropy principle. Here minimum entropy means that we want to be as certain about I as possible, so we want the corresponding Julesz ensemble to be as small as possible. Or in other words, we want the most meaningful set of filters to describe I. Inferentially, one can estimate $\{B_{x,y,k}\}$ and λ_k in the FRAME model by maximum likelihood. Computationally, this can be accomplished by stochastic gradient algorithm.

Although both the sparsity principle and the minimum entropy principle are about representing the image with minimum complexity, the philosophies and the mathematical structures in wavelet model and the FRAME model are very different. Philosophically, the wavelet model is constructive, where I is deterministically constructed by superposition of local bases. The FRAME model is restrictive, where

I is defined stochastically by restricting histograms of filter responses. Mathematically, the $\{B_{x,y,k}\}$ in the wavelet model are bases, and the corresponding $c_{x,y,k}$ compete to explain I, so there is lateral inhibition among them, that is, if one base is active in explaining I, then it will inhibit other overlapping bases. The $\{B_{x,y,k}\}$ in the FRAME model are filters, and there is no lateral inhibition among the filter responses $r_{x,y,k}$.

It is worth mentioning that, if $\{B_{x,y,k}\}$ is complete, that is, the number of bases is the same as the number of pixels, then both models reduce to independent component analysis (Bell and Sejnowski, 1997). One may call the latter the "restructive" scheme, because it involves a one to one transformation between I and the coefficients $\{c_{x,y,k}\}$ or the responses $\{r_{x,y,k}\}$. The principle behind independent component analysis is the factorial coding principle, which is closely related to both sparsity principle and the minimum entropy principle.

(a)   (b)

(c)   (d)

Figure 31.5   Feature statistics. (a) and (c) are observed images. (b) and (d) are "reconstructed" by matching feature statistics.

Complexity regimes

The complexity behavior of the two models are also different.

Figure 31.5 shows two examples of feature statistics representation. (a) and (c) are observed images, and (b) and (d) are respectively the "reconstructed" images. However, the reconstruction is of a statistical nature: (b) and (d) are sampled from the respective Julesz ensembles Ω (31.3) by matching feature statistics. We can see that this representation is appropriate for random images such as image (a). It captures texture information, but does not do a good job in capturing salient structures.

Figure 31.6 shows two examples of sparse coding. (a) and (c) are observed images, (b) and (d) are images reconstructed by 300 bases. We used the matching

Figure 31.6   Sparse coding. (a) and (c) are observed images. (b) and (d) are respectively the reconstructed images using 300 bases.

Figure 31.7 From sparse coding to feature statistics. (a) Observed near-distance image. (b) Reconstructed by sparse coding with 1,000 bases. (c) Observed far-distance image. (d) "Reconstructed" by matching feature statistics.

pursuit algorithm of Mallat and Zhang (1993) to select the bases (in a manner very similar to forward stepwise regression). We can see that sparse coding is very effective for images with sparse structures, such as image (a). However, the texture information is not well represented.

To summarize, the wavelet sparse coding model is effective in low entropy regime where images have order and structures, such as the shape and geometry. We call this regime "sketchable." The FRAME model is effective in high entropy regime where images have less structures, such as stochastic texture. We call this regime "nonsketchable." The competition between these two models in terms of some model selection criterion such as minimum description length (e.g., Hansen and Yu, 2000). This competition may give us a threshold that tells us when we should stop using sparse coding representation and switch to feature statistics.

Figure 31.7 displays some preliminary results. (a) and (c) are images of an ivy wall at near-distance and far-distance respectively. (b) is reconstructed near-distance image using sparse coding representation with 1,000 bases selected by the matching pursuit algorithm. (d) is statistically reconstructed far-distance image using feature statistics representation by matching histograms of filter responses.

In a previous paper (Guo, Zhu, and Wu, 2003), we studied and experimented with a primal sketch model (the name comes from the book by Marr, 1982), where the image I is divided into sketchable part I_{sk} and nonsketchable part I_{nsk}. The model for I is $p(I) = p(I_{sk})p(I_{nsk}|I_{sk})$. I_{sk} is modeled by wavelet sparse coding. $p(I_{nsk}|I_{sk})$ is modeled by the FRAME model, with I_{sk} being the boundary conditions. Or in other words, I_{nsk} interpolates I_{sk} by matching local feature statistics.

See Figure 31.8 for an example, where (a) is the observed image; (b) depicts the sketch version of the image, where each base in representing I_{sk} is replaced by a small line segment (or a circle for center-surround base); (c) is the synthesized image, where the structures are reconstructed by sparse coding, and the textures are generated by matching feature statistics. See Figure 31.9 for two more examples.

The prior models for the spatial arrangements of local bases is a pairwise Gibbs point process model (see also Stoyan, Kendall, and Mecke, 1987; Wu, Zhu, and, Guo, 2002) that takes care of continuity, joints, and closures of the local bases. We call such a model the Gestalt field.

In the next two sections, we will prove two scaling laws that explain the transition from sparse structures to stochastic textures.

31.3   Complexity scaling law

Let I be the image of a pattern observed at a certain distance, and let us assume that I is generated by a physical process that can be summarized by a probability distribution $p(I)$. Let Λ be the lattice on which I is defined.

Definition 1 *Image Complexity, denoted by $\mathcal{H}(I)$, is defined as the entropy of $p(I)$, that is, $\mathcal{H}(I) = -\sum_I p(I) \log p(I)$. The **complexity rate** is defined as $\mathcal{H}(I)/|\Lambda|$.*

When we move away from a scene, the change of image involves both local smoothing and down-sampling. As a first step, we shall only study the effect of down-sampling, while ignoring the effect of local averaging. To simplify the situation even further, let us assume that we down-sample I by a factor of 2 along both vertical and horizontal axes. Then there are four down-sampled versions, and let us denote them by $I_-^{(k)}$, $k = 1, 2, 3, 4$, each defined on a down-sampled lattice Λ_-, so that $|\Lambda_-| = |\Lambda|/4$. See Figure 31.10 for an illustration.

Theorem 1 *Complexity Scaling Law.*

$$1) \quad \mathcal{H}(I_-^{(k)}) \leq \mathcal{H}(I), \ k = 1, \ldots, 4.$$

$$2) \quad \frac{1}{|\Lambda_-|} \sum_{k=1}^{4} \mathcal{H}(I_-^{(k)})/4 \geq \frac{1}{|\Lambda|}\mathcal{H}(I).$$

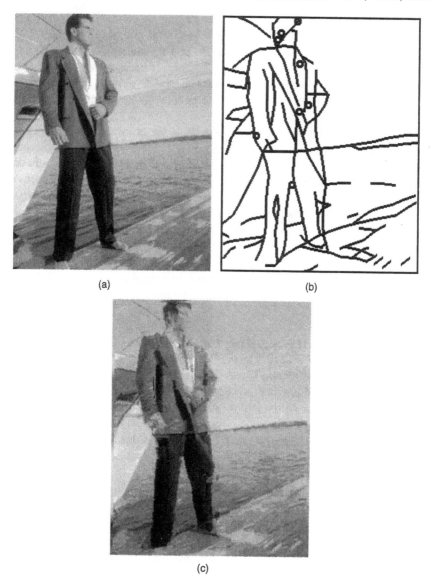

Figure 31.8   Primal sketch: (a) Observed image. (b) Image sketch with each base replaced by a line segment (or a circle). (c) Synthesized image.

Proof. 1) $p(I|I_-^{(k)}) = p(I)/p(I_-^{(k)})$ since $I_-^{(k)}$ is fully determined by I. Thus

$$\mathcal{H}(I) - \mathcal{H}(I_-^{(k)}) = E_I \left[-\log \frac{p(I)}{p(I_-^{(k)})} \right] = \mathcal{H}(I|I_-^{(k)}) \geq 0.$$

Figure 31.9   Primal sketch: (a) Observed image. (b) Synthesized image.

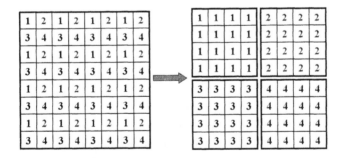

Figure 31.10   The four down-sampled versions of the original image.

2) Let $\mathcal{M}()$ denote mutual information,

$$\sum_{k=1}^{4} \mathcal{H}(I_{-}^{(k)}) - \mathcal{H}(I) = \mathrm{E}\left[\log \frac{p(I)}{\prod_k p(I_{-}^{(k)})} \right]$$

$$= \mathcal{M}(I_{-}^{(k)}, k = 1, 2, 3, 4) \geq 0.$$

One can also understand this result from the perspective of Komolgorov complexity. The shortest algorithmic coding length of I must be greater than or equal to the shortest coding length of any of the $I_-^{(k)}$, but must be smaller than or equal to the sum of the shortest coding lengths of the four $I_-^{(k)}$.

In Theorem 1, we only consider the effect of down-sampling, without considering the effect of local averaging. But from information theoretical perspective, the purpose of local averaging is to make the entropy of down-scaled I_- as close to the entropy of I as possible in order to maintain as much information as possible. As a result, the complexity rate of I_- will be even larger if we take into account the local smoothing effect.

This theorem tells us that if we down-sample an image, the image looks more random. This can be easily understood from real-life experience. For instance, for the ivy wall pattern in Figure 31.1, when we move farther away from it, we lose information, so the complexity is decreasing. But we see more leaves within the unit area of the visual field, so the complexity rate is increasing.

The complexity scaling law we have proved has far reaching implications on sparsity principle (Olshausen and Field, 1996). At near distance, the complexity rate is very low, so sparsity principle applies. But as the viewer moves farther from the underlying pattern, the complexity rate of the image will increase, so that there may not exist any sparse deterministic representation of the image, and the sparsity principle is violated. As a result, the visual system can only interpret the image by some summaries that cannot determine the image deterministically, and these summaries are feature statistics. This may explain the perceptual transition from sparse coding to feature statistics.

31.4 Perceptibility scaling law

The purpose of vision is to make inference about the outside world. Now, let us study the issue of perceptual transition in an inferential framework, under the slogan that "vision = inverse graphics."

Let W describe the outside world that produces the image I. Let us assume that both W and I are properly discretized, and that W is detailed enough to determine I uniquely, that is, $I = g(W)$, where the many to one function $g()$ can be thought of as a graphics process. For natural patterns such as foliage and grass, W is typically very complex, including detailed descriptions of all the leaves and strands of grass. Such visual complexity is a defining characteristic of natural scenes and is a key factor for visual realism in graphics and paintings.

Suppose W is generated by a physical process that can be summarized by a distribution $p(W)$ (we shall not engage in a philosophical discussion on whether there exists a true $p(W)$). Given $W \sim p(W)$, and $I = g(W)$, we have $p(W|I) = p(W, I)/p(I) = p(W)/p(I)$. $p(W, I) = p(W)$ because I is fully determined by W. This distribution defines an inversion of the graphics equation $I = g(W)$.

Definition 2 *Scene complexity, denoted by $\mathcal{H}(W)$, is defined as the entropy of $p(W)$.*

Definition 3 *Imperceptibility, denoted by $\mathcal{H}(W|I)$, is defined as the average conditional entropy of $p(W|I)$, that is, $\mathcal{H}(W|I) = -\sum_{W,I} p(W, I) \log p(W|I)$.*

Imperceptibility is defined as the average of the conditional entropy over the data I.

Theorem 2 *Let $W \sim p(W)$, and $I = g(W)$, then $\mathcal{H}(W|I) = \mathcal{H}(W) - \mathcal{H}(I)$. That is, **imperceptibility = scene complexity − image complexity**.*

This theorem can be easily understood from the fact that joint entropy = marginal entropy + average of conditional entropy. This fact is the key to the proofs of several theorems in this chapter.

The imperceptibility $\mathcal{H}(W|I)$ gives a general definition of "ill-posedness" of the inversion problem. Here the concept of imperceptibility only means the possibility of estimating W under a particular physics representation of W.

For an image I, its down-scaled version I_- can be obtained by local smoothing and down-sampling, and the process can be represented by a many to one reduction function $R()$, such that $I_- = R(I)$.

Theorem 3 *Perceptibility Scaling Law. For $W \sim p(W)$, $I = g(W)$, if $I_- = R(I)$ with $R()$ being any many to one reduction function, then $\mathcal{H}(W|I_-) \geq \mathcal{H}(W|I)$. That is, imperceptibility becomes larger with down-scaling.*

If $\mathcal{H}(W|I_-)$ is too large, we can only perceive some aspect of W, that is., $W_- = \rho(W)$, for some many to one reduction $\rho()$, such that $\mathcal{H}(W_-|I_-)$ is small. It is possible to find such a W_-, because of the following theorem.

Theorem 4 *For $W \sim p(W)$, $I = g(W)$, and $I_- = R(I)$, $W_- = \rho(W)$, we have $\mathcal{H}(W_-|I_-) \leq \mathcal{H}(W|I_-)$.*

Here W_- can be a coarser representation of W, where the scale of the elements in W_- may be larger than that of W. It is possible that there still exists a g_-, such that $I_- = g_-(W_-)$, but it is most likely that this is only approximately true. It is also likely that W_- may only correspond to some statistical property of I_-, or in other words, that $p(I_-|W_-)$ with a high entropy rate. That is, although W defines I deterministically via $I = g(W)$, W_- may only defines I_- statistically via a probability distribution $p(I_-|W_-)$. While W represents sparse structures, W_- may only represent collective textures.

This perceptibility scaling law provides a possible explanation to the perceptual transition from sparse structures to stochastic textures.

31.5   Texture = imperceptible structures

The visual cells in the primitive visual cortex V1 may correspond to various types of local descriptors for local structures appearing at different scales, locations, and orientations. Olshausen and Field (1996) proposed a sparsity principle as a V1 strategy. This principle holds that for a typical image, only a small number of local descriptors need to be selected to interpret the image. We argue that the sparsity principle only accounts for part of V1 representations and activities. This is because the number of local descriptors is much less than the number of all possible image patches. As a result, there are a lot of image patches that cannot be well represented by local descriptors, or there are no sparse representations for such image patches. Such image patches often correspond to patterns viewed at a far distance, so that both the complexity rate and the imperceptibility are high. These image patches cannot be accounted for by the sparsity principle. Then what are the possible representations for them?

One possible choice is to summarize them into feature statistics, that is, they are interpreted statistically as textures (or more precisely stochastic textures), instead of structures. Then what feature statistics should we use? The next theorem sheds light on this question.

Theorem 5 *For $F = F(I)$ be a set of feature statistics, (i) If $W \sim p(W)$, $I = g(W)$, then*

$$D(p(W|I)\|p(W|F)) = E_W \left[\log \frac{p(W|I)}{p(W|F)} \right]$$

$$= \mathcal{H}(W|F) - \mathcal{H}(W|I) = \mathcal{H}(I|F).$$

(ii) If $W \sim p(W)$ and $[I|W] \sim p(I|W)$, then

$$D(p(W|I)\|p(W|F)) = E_{W,I} \left[\log \frac{p(W|I)}{p(W|F)} \right]$$

$$= \mathcal{H}(W|F) - \mathcal{H}(W|I) = \mathcal{M}(W, I|F).$$

Here $D()$ denotes Kullback–Leibler divergence, and $\mathcal{M}()$ denotes mutual information.

Result (i) justifies the minimum entropy principle we discussed before. That is, to minimize $\mathcal{H}(W|F)$ over a set of possible $\{F()\}$, we need to minimize $\mathcal{H}(I|F)$. In result (ii), $\mathcal{M}(W, I|F)$ measures the sufficiency of F.

This theorem shows that in order to choose good feature statistics, we must have $p(W|F)$ to be close to $p(W|I)$. This makes us believe that F must be derived from some intermediate results in the computation of $p(W|I)$.

We propose the following strategy for primitive visual cortex. For each local patch around pixel (x, y), that is, $I_{x,y}$, there can be a number of local descriptors to describe it. Let $w_{x,y}$ index the possible local descriptor as well as its parameters. Then by fitting a local model, we compute $p(w_{x,y}|I_{x,y})$. This can be done efficiently in a parallel manner.

For those pixels (x, y) with very low $\mathcal{H}(p(w_{x,y}|I_{x,y}))$, we use sparse coding representation, that is, we select a small number of local descriptors to represent those pixels, while respecting our prior knowledge for the spatial arrangements of these local descriptors.

For those pixels (x, y) with very high imperceptibility $\mathcal{H}(p(w_{x,y}|I_{x,y}))$, the underlying structures cannot be unambiguously determined. As such, we abort the effort of committing a particular $w_{x,y}$. Instead, we pool the local posterior $p(w_{x,y}|I_{x,y})$ over (x, y) into texture statistics. That is, texture = pooling of imperceptible structures. This should be complimentary to the sparsity principle.

This complementary principle bridges deterministic structures and stochastic textures in a very elegant manner. It also has interesting implications on the two conjectures of Julesz on textures (Julesz, 1981), as well as the phenomenon of lateral inhibition in neuroscience.

For the wavelet sparse coding model $I = \sum c_{x,y,k} B_{x,y,k} + \epsilon$, the local model is $I_{x,y} = c_{x,y,k} B_{x,y,k} + \epsilon$. If the bases are not perceptible, we can pool local posterior over (x, y). One can show that the pooled statistics is very close to the histograms of filter responses. If we assume such feature statistics, then we are led to the Markov random field model (31.4). Thus, we establish an interesting link between wavelet sparse coding theory and Markov random field theory. We shall further investigate this connection, which should be interesting to both wavelet community and spatial statistics community.

31.6 Perceptibility and sparsity

The inferential concept of perceptibility also arises from the coding perspective. That is, we only assume $I \sim p(I)$, and W is an augmented variable purely for the purpose of coding I, via a model $W \sim f(W)$ and $I|W \sim f(I|W)$. In this coding scheme, for an image I, we first estimate W by a sample from the posterior distribution $f(W|I)$, then we code W by $f(W)$ with coding length $-\log f(W)$. After that, we code I by $f(I|W)$ with coding length $-\log f(I|W)$. So the average coding length is $-\mathrm{E}_p \left[\mathrm{E}_{f(W|I)}(\log f(W) + \log f(I|W)) \right]$.

Theorem 6 *The average coding length is* $E_p[\mathcal{H}(f(W|I))] + \mathrm{D}(p\|f) + \mathcal{H}(p)$. *That is,* **coding redundancy = imperceptibility + error**. *Here* $\mathcal{H}(f(W|I))$ *is the entropy of* $f(W|I)$ *conditional on* I, *and* $\mathrm{D}(p\|f)$ *is the Kullback–Leibler distance.*

The relationship between perceptibility and sparsity deserves more investigation. To make the idea more concrete, let us consider the sparse coding model $I = \sum c_{x,y,k} B_{x,y,k} + \epsilon$. If the image is very complex, then even the sparsest representation still has a large number of bases, so that sparsity principle is violated. One may ask, what is wrong with a nonsparse representation? This can be answered by perceptibility. That is, if the sparsest representation still has a large number of bases, then there can be a lot of representations that are only slightly less sparse, but can approximate I with equally small error ϵ. Or in other words, there can be a lot of "equivalent" representations, so that there is ambiguity as to which one to use. This ambiguity may be mathematically defined, and clearly it is closely related to imperceptibility. In wavelet sparse coding theory, this issue of ambiguity has not been studied. But it is clearly of fundamental importance to vision applications, because the representation is to be used in later stages of visual processing.

At the end of this chapter, the authors would like to acknowledge that the work presented here has close connections to Rubin's statistics in the following four aspects. First, we study the problem in the Bayesian framework. Second, the issue of unsupervised learning fits naturally into the EM framework. Third, the loss of information over distance is essentially a matter of missing data. Fourth, the perceptibility issue leads to the issue of defining the estimand.

References

Abadie, A., and Imbens, G. (2002). Simple and bias-corrected matching estimators for average treatment effects. *Econometrica*.

Abowd, J., and Woodcock, S. (2001). Disclosure limitation in longitudinal linked data. *Confidentiality, Disclosure and Data Access: Theory and Practical Applications for Statistical Agencies*. New York: North Holland.

Abramowitz, M., and Stegun, I. A., eds. (1964). *Handbook of Mathematical Functions with Formulas, Graphs, and Mathematical Tables*. Washington, D.C.: National Bureau of Standards.

Adams, A. S., Soumerai S. B., and Ross-Degnan, D. (2001). Use of antihypertensive drugs by medicare enrollees: does type of drug coverage matter? *Health Affairs* **20**, 276–286.

Adler, G. (1994). A profile of the Medicare Current Beneficiary Survey. *Health Care Financing Review* **15**, 153–163.

Agresti, A. (1984). *Analysis of Ordinal Categorical Data*. New York: Wiley.

Albert, J. H., and Chib, S. (1993). Bayesian analysis of binary and polychotomous response data. *Journal of the American Statistical Association* **88**, 669–679.

Amari, S. (1995). Information geometry of the EM and em algorithms for neural networks. *Neural Networks* **8**, 1379–1408.

Amit, Y., and Geman, D. (1997). Shape quantization and recognition with randomized trees. *Neural Computation* **9**, 1545–1588.

Amit, Y., Grenander, U., and Piccioni, M. (1991). Structural image restoration through deformable templates. *Journal of the American Statistical Association* **86**, 376–387.

Andersen, P. K., Klein, J. P., and Palacios, R. T. (1997). Estimation for variance in Cox regression model with shared gamma frailties. *Biometrics* **53**, 1475–84.

Anderson, M. J., and Fienberg, S. E. (1999). *Who Counts? The Politics of Census-Taking in Contemporary America*. New York: Russell Sage Foundation.

Anderson, T. W. (1958). *An Introduction to Multivariate Statistics*. New York: Wiley.

Applied Bayesian Modeling and Causal Inference from Incomplete-Data Perspectives.
Edited by A. Gelman and X-L. Meng © 2004 John Wiley & Sons, Ltd ISBN: 0-470-09043-X

Angrist, J., and Evans, W. (1998). Children and their parents' labor supply: evidence from exogenous variation in family size. *American Economic Review* **88**, 450–477.

Angrist, J., Graddy, K., and Imbens, G. W. (2000). The interpretation of instrumental variables estimators in simultaneous equations models with an application to the demand for fish. *Review of Economic Studies* **67**, 499–527.

Angrist, J., and Imbens, G. W. (1995). Two-stage least squares estimation of average causal effects in models with variable treatment intensity. *Journal of the American Statistical Association* **90**, 431–442.

Angrist, J., Imbens, G. W., and Rubin, D. B. (1996). Identification of causal effects using instrumental variables (with discussion). *Journal of the American Statistical Association* **91**, 444–472.

Arjas, E., and Gasbarra, D. (1994). Nonparametric Bayesian inference from right censored survival data, using the Gibbs sampler. *Statistica Sinica* **4**, 505–524.

Artz, M. B. (2002). Impact of generosity level of outpatient prescription drug coverage on prescription drug events and expenditure among older persons. *American Journal of Public Health* **92**, 1257–1263.

Ashenfelter, O. (1978). Estimating the effects of training programs on earnings. *Review of Economics and Statistics* **60**, 47–57.

Ashenfelter, O., and Card, D. (1985). Using the longitudinal structure of earnings to estimate the effect of training programs. *Review of Economics and Statistics* **67**, 648–660.

Athens, L. (1997). *Violent Criminal Acts and Actors Revisited*. Urbana: University of Illinois Press.

Avitzour, D. (1995). Stochastic simulation Bayesian approach to multitarget tracking. *IEE Proceedings Radar, Sonar and Navigation* **142**, 41–44.

Baker, S. G., and Lindeman, K. S. (1994). The paired availability design: a proposal for evaluating epidural analgesia during labor. *Statistics in Medicine* **13**, 2269–2278.

Banerjee, S., Carlin, B. P., and Gelfand, A. E. (2004). *Hierarchical Modeling and Analysis for Spatial Data*. London: CRC Press.

Barker, F. G., Chang, S. M., Gutin, P. H., Malec, M. K., McDemott, M. W., Prados, M. D., and Wilson C. B. (1998). Survival and functional status after resection of recurrent glioblastoma multiforme. *Neurosurgery* **42**, 709–720.

Barnard, J., Frangakis, C. E., Hill, J., and Rubin, D. B. (2003). A principal stratification approach to broken randomized experiments: a case study of school choice vouchers in New York City (with discussion). *Journal of the American Statistical Association* **98**, 299–323.

Barnard, J., McCulloch, R., and Meng, X. L. (2000). Modeling covariance matrices in terms of standard deviations and correlations, with application to shrinkage. *Statistica Sinica* **10**, 1281–1311.

Basu, S., Sen, A., and Bannerjee, M. (2003). Bayesian analysis of competing risks with partially masked cause of failure. *Applied Statistics* **52**, 77–93.

Bastos, F. I., and Strathdee, S. A. (2000). Evaluating effectiveness of syringe exchange programmes: current issues and future prospects. *Social Science and Medicine* **51**, 1771–1782.

Beaton, A., and Zwick, R. (1992). Overview of the national assessment of educational progress. *Journal of Educational Statistics* **17**, 95–109.

Becker, M. P., and Yang, I. (1998). Latent class marginal models for cross-classifications of counts. *Sociological Methodology* **28**, 293–325.

Belin, T. R. (1993). Evaluation of sources of variation in record linkage through a factorial experiment. *Survey Methodology* **19**, 13–29.

Belin, T. R., Ishwaran, H., Duan, N., Berry, S., and Kanouse, D. (2004). Identifying likely duplicates by record linkage in a survey of prostitutes. In this volume.

Belin, T. R., and Rubin, D. B. (1995). A method for calibrating false-match rates in record linkage. *Journal of the American Statistical Association* **90**, 694–707.

Bell, A., and Sejnowski, T. J. (1997). The 'independent components' of natural scenes are edge filters. *Vision Research* **37**, 3327–3338.

Bentler, P. M., and Tanaka, J. S. (1983). Problems with EM algorithms for ML factor analysis. *Psychometrika* **48**, 247–251.

Bergstralh, E. J., Kosanke, J. L., and Jacobsen, S. L. (1996). Software for optimal matching in observational studies. *Epidemiology* 7, 331–332. www.mayo.edu/hsr/sasmac.html.

Berkane, M., ed. (1997). *Latent Variable Modeling and Applications to Causality.* Lecture Notes in Statistics 120, New York: Springer-Verlag.

Berry, S. H., Kanouse, D. E., Duan, N., and Lillard, L. A. (1992). Risky and non-risky sexual transactions with clients in a Los Angeles probability sample of female street prostitutes. *VIII International Conference on AIDS/III STD World Congress*, Poster Abstracts, volume 2. Amsterdam: PoD 5604.

Bertsekas, D. P. (1991). *Linear Network Optimization*. Cambridge, Mass.: MIT Press.

Besag, J. (1974). Spatial interaction and the statistical analysis of lattice systems (with discussion). *Journal of the Royal Statistical Society B* **36**, 192–236.

Besag, J. (1986). On the statistical analysis of dirty pictures (with discussion). *Journal of the Royal Statistical Society B* **48**, 259–302.

Bethlehem, J. G., Keller, W. J., and Pannekoek, J. (1990). Disclosure control of microdata. *Journal of the American Statistical Association* **85**, 38–45.

Biscarat, J. C. (1994). Almost sure convergence of a class of stochastic algorithms. *Stochastic Processes and their Applications* **50**, 83–89.

Bishop, Y. M. M., Fienberg, S. E., and Holland, P. W. (1975). *Discrete Multivariate Analyses: Theory and Practice*. Cambridge, Mass.: MIT Press.

Blanchard, G., and Geman, D. (2003). Hierarchical testing designs for pattern recognition. Technical Report, Mathematical Science, Johns Hopkins University.

Blustein, J. (2000). Drug coverage and drug purchases by medicare beneficiaries with hypertension. *Health Affairs* **37**, 2265–2281.

Bollen, K. A. (1989). *Structural Equations with Latent Variables*. New York: Wiley.

Booth, J. G., and Hobert, J. P. (1999). Maximizing generalized linear mixed model likelihoods with an automated Monte Carlo EM algorithm. *Journal of the Royal Statistical Society, Series B* **61**, 265–285.

Boscardin, W. J., and Weiss, R. E. (2004). Fitting unstructured covariance matrices to longitudinal data. Technical Report, Department of Biostatistics, University of California, Los Angeles.

Bosk, C. L. (1981). *Forgive and Remember: Managing Medical Failure*. University of Chicago Press.

Braitman, L. E., and Rosenbaum, P. R. (2002). Rare outcomes, common treatments: analytic strategies using propensity scores. *Annals of Internal Medicine* **137**, 693–695.

Brand, C. J., and Keith, L. B. (1979). Lynx demography during a snowshoe hare decline in Alberta. *Journal of Wildlife Management* **43**, 827–849.

Breslow, N. E., and Clayton, D. G. (1993). Approximate inference in generalized linear mixed models. *Journal of the American Statistical Association* **88**, 9–25.

Brewer, K. R. W., and Hanif, M. (1983). *Sampling with Unequal Probabilities*. New York: Springer-Verlag.

Brogan, D. R., and Kutner, M. H. (1980). Comparative analysis of pretest/posttest research designs. *American Statistician* **34**, 229–232.

Brooks-Gunn, J., Liaw, F. R., and Klebanove, P. K. (1992). Effects of early intervention on cognitive function of low birth weight preterm infants. *Journal of Pediatrics* **120**, 350–359.

Bruneau, J., Lamothe, F., and Franco, E. (1997). High rates of HIV infection among injection drug users participating in needle exchange programs in Montreal: results of a cohort study. *American Journal of Epidemiology* **146**, 994–1002.

Bulmer, M. G. (1974). A statistical analysis of the 10-year cycle in Canada. *Journal of Animal Ecology* **43**, 701–715.

Buyse, M., and Molenberghs, G. (1998). The validation of surrogate endpoints in randomized experiments. *Biometrics* **54**, 1014–1029.

Buyse, M., Molenberghs, G., Burzykowski, T., Renard, D., and Geys, H. (2000). The validation of surrogate endpoints in meta-analyses of randomized experiments. *Biostatistics* **1**, 49–68.

Campbell, D. T. (1969). Prospective: artifact and control. In *Artifact in Behavioral Research*, eds. R. Rosenthal and R. L. Rosnow. New York: Academic Press.

Campbell, M. J., and Walker, A. M. (1977). A survey of statistical work on the MacKenzie River series on annual Canadian lynx trappings for the years 1821–1934 and a new analysis. *Journal of the Royal Statistical Society A* **140**, 411–431.

Cappé, O., and Robert, C. P. (2000). Markov chain Monte Carlo: 10 years and still running. *Journal of the American Statistical Association* **95**, 1282–1286.

Candes, E. J., and Donoho, D. L. (1999). Ridgelets: a key to higher-dimensional intermittency? *Philosophical Transactions of the Royal Society of London A* **357**, 2495–2509.

Card, D. (1993). *Using Geographic Variation in College Proximity to Estimate the Return to Schooling*. National Bureau of Economic Research, Paper No. 4483.

Card, D. (1995). Using geographic variation in college proximity to estimate the return to schooling. In *Aspects of Labor Market Behaviour: Essays in Honour of John Vanderkamp*, eds. E. K. G. Christofides and R. Swidinsky. University of Toronto Press.

Card, D., and Krueger, A. (1994). Minimum wages and employment: a case study of the fast-food industry in New Jersey and Pennsylvania. *American Economic Review* **84** 772–793.

Carlin, B. P., and Hodges, J. S. (1999). Hierarchical proportional hazards regression models for highly stratified data. *Biometrics* **55**, 1162–1170.

Carlin, B. P., and Louis, T. A. (1996). *Bayes and Empirical Bayes Methods for Data Analysis*. London: Chapman & Hall.

Carnahan, B., Luther, H., and Wilks, J. O. (1969). *Applied Numerical Methods*. New York: Wiley.

Celeux, G., Hurn, M., and Robert, C. P. (2000). Computational and inferential difficulties with mixture posterior distributions. *Journal of the American Statistical Association* **95**, 957–970.

Chan, J. S. K., and Kuk, A. Y. C. (1997). Estimation for probit-linear mixed models with correlated random effects. *Biometrics* **53**, 86–97.

Chan, K.-S., and Ledolter, J. (1995). Monte Carlo EM estimation of time series models involving counts. *Journal of the American Statistical Association* **90**, 242–252.

Chang, H., and Stout, W. (1993). The asymptotic posterior normality of the latent trait in an IRT model. *Psychometrika* **58**, 37–52.

Chao, M. T. (1982). A general purpose unequal probability sampling plan. *Biometrika* **69**, 653–656.

Chung, H., Loken, E., and Schafer, J. L. (2004). Difficulties in drawing inferences with finite-mixture models: a simple example with a simple solution. *American Statistician* **58**, 152–158.

Chase, G. R. (1968). On the efficiency of matched pairs in Bernoulli trials. *Biometrika* **55**, 365–369.

Chen, C. F. (1979). Bayesian inference for a normal dispersion matrix and its application to stochastic multiple regression analysis. *Journal of the Royal Statistical Society B* **41**, 235–248.

Chen, G., and Keller-McNulty, S. (1998). Estimation of identification disclosure disk in microdata. *Journal of Official Statistics* **14**, 79–95.

Chen, M. H., Shao, Q. M., and Ibrahim, J. G. (2000). *Monte Carlo Methods in Bayesian Computation*. New York: Springer-Verlag.

Cheng, B., and Titterington, D. M. (1994). Neural networks: a review from a statistical perspective (with discussion). *Statistical Science* **9**, 2–54.

Chi, Z. (2001). Stationary self-similar random fields on the integer lattice. *Stochastic Processes and Applications* **91**, 99–113.

Chib, S., and Greenberg, E. (1998). Analysis of multivariate probit models. *Biometrika* **85**, 347–361.

Choldin, H. M. (1994). *Looking for the Last Percent: The Controversy Over Census Undercounts*. Piscataway, N.J.: Rutgers University Press.

Chung, H., Loken, E., and Schafer, J. L. (2004). Difficulties in drawing inferences with finite-mixture models: a simple example with a simple solution. *American Statistician* **58**, 152-158.

Citro, C. F., and Cohen, M. L., eds. (1985). *The Bicentennial Census: New Directions for Methodology in 1990*. Washington, D.C.: National Academies Press.

Citro, C. F., Cork, D. L., and Norwood, J. L., eds. (2004). *The 2000 Census: Counting Under Adversity*. Washington, D.C.: National Academies Press.

Clayton, D. G. (1996). Generalized linear mixed models. In *Markov Chain Monte Carlo in Practice*, W. R. Gilks, S. Richardson, and D. J. Spiegelhalter, eds., 275–301. London: Chapman & Hall.

Cleveland, W. S. (1979). Robust locally weighted regression and smoothing scatterplots. *Journal of the American Statistical Association* **74**, 829–836.

Clogg, C. C., Rubin, D. B., Schenker, N., Schultz, B., and Weidman, L. (1991). Multiple imputation of industry and occupation codes in Census public-use samples using Bayesian logistic regression. *Journal of the American Statistical Association* **86**, 68–78.

Cochran, W. G. (1968). The effectiveness of adjustment by subclassification in removing bias in observational studies. *Biometrics* **24**, 205–213.

Cochran, W. G. (1977). *Sampling Techniques*, third edition. New York: Wiley.

Cochran, W. G., and Rubin, D. B. (1973). Controlling bias in observational studies: a review. *Sankhya A* **35**, 417–446.

Cole, S. R., Chu, H., and Greenland, S. (2004). Using multiple-imputation for measurement error correction in pediatric chronic kidney disease. *American Journal of Epidemiology* **159**.

Collins, L. M., Schafer, J. L., and Kam, C.-M. (2001). A comparison of inclusive and restrictive strategies in modern missing-data procedures. *Psychological Methods* **6**, 330–351.

Conners, A. F., Speroff, S. A., Dawson, N. V., Thomas, C., Harrell, F. E., Wagner, D., et al. (1996). The effectiveness of right heart catheterization in the initial care of critically ill patients. *Journal of the American Medical Association* **276**, 889–997.

Contoyannis, P., and Rice, N. (2001). The impact of health on wages: evidence from the British Household Panel Survey. *Empirical Economics* **26** **4**, 599–622.

Copas, J. B., and Eguchi, S. (2001). Local sensitivity approximations for selectivity bias. *Journal of the Royal Statistical Society B* **63**, 871–895.

Copas, J. B., and Li, H. G. (1997). Inference for non-random samples (with discussion). *Journal of the Royal Statistical Society B* **59**, 55–95.

Corduneanu, A., and Bishop, C. M. (2001). Variational Bayesian model selection for mixture distributions. In *Proceedings of 8th International Conference. Artificial Intelligence and Statistics*, ed. T. Richardson and T. Jaakkola, 27–34. San Mateo, Calif.: Morgan Kaufmann.

Cox, D. R. (1958). *Planning of Experiments*. New York: Wiley.

Cox, D. R. (1972). Regression models and life tables (with discussion). *Journal of the Royal Statistical Society B* **34**, 187–220.

Cox, D. R. (1992). Causality: some statistical aspects. *Journal of the Royal Statistical Society A* **155**, part 2, 291–301.

Cox, D. R., and Hinkley, D. V. (1974). *Theoretical Statistics*. New York: Chapman & Hall.

Cox, D. R., and Oakes, D. (1984). *Analysis of Survival Data*. New York: Chapman & Hall.

Crager, M. R. (1987). Analysis of covariance in parallel-group clinical trials with pretreatment baseline. *Biometrics* **43**, 895–901.

Craiu, R. V., and Duchesne, T. (2004). Inference based on the EM algorithm for the competing risk model with masked causes of failure. *Biometrika*.

Cressie, N. (1993). *Statistics for Spatial Data*, revised edition. New York: Wiley.

Cruces, G., and Galiani, S. (2003). Generalizing the causal effect of fertility on female labor supply. Economics Working Paper Archive at Washington University, St. Louis, Labor and Demography Series, No. 310002.

Curley, C., McEachern, J. E., and Speroff, T. (1998). A firm trial of interdisciplinary rounds on the inpatient medical wards: an intervention designed using continuous quality improvement. *Medical Care* **36**, AS4–12.

D'Agostino Jr., R. B. (1998). Tutorial in biostatistics: propensity score methods for bias reduction in the comparison of a treatment to a non-randomized control group. *Statistics in Medicine* **17**, 2265–2281.

D'Agostino Jr., R. B., and Rubin D. B. (2000). Estimating and using propensity scores with partially missing data. *Journal of the American Statistical Association* **95**, 749–759.

Dalenius, T., and Reiss, S. P. (1982). Data-swapping: a technique for disclosure control. *Journal of Statistical Planning and Inference* **6**, 73–85.

Damien, P., Wakefield, J., and Walker, S. (1999). Gibbs sampling for Bayesian non-conjugate and hierarchical models using auxilliary variables. *Journal of the Royal Statistical Society B* **61**, 331–344.

Daniels, M. J., and Kass, R. E. (2001). Shrinkage estimators for covariance matrices. *Biometrics* **57**, 1173–1184.

Daniels, M. J., and Pourahmadi, M. (2002). Bayesian analysis of covariance matrices and dynamic models for longitudinal data. *Biometrika* **89**, 553–566.

Darroch, J. N., Fienberg, S. E., Glonek, G. F. V., and Junker, B. W. (1993). A three-sample multiple-recapture approach to census population estimation with heterogeneous catchability. *Journal of the American Statistical Association* **88**, 1137–1148.

Davis, M., Poisal, J., Chulis, G., Zarabozo, C., and Cooper, B. (1999). Prescription drug coverage, utilization, and spending among Medicare beneficiaries. *Health Affairs* **18**, 231–243.

Davison, A. C., Hinkley, D. V., and Schechtman, E. (1986). Efficient bootstrap simulation. *Biometrika* **73**, 555–566.

Dawid, A. P. (1979). Conditional independence in statistical theory. *Journal of the Royal Statistical Society B* **41**, 1–15.

Dawid, A. P. (2000). Causal inference without counterfactuals (with discussion). *Journal of the American Statistical Association* **95**, 407–448.

Defays, D., and Anwar, M. N. (1998). Masking microdata using micro-aggregation. *Journal of Official Statistics* **14**, 449–461.

Dehejia, R. H. (2004). Program evaluation as a decision problem. *Journal of Econometrics*.

Dehejia, R. H., and Wahba, S. (1999). Causal effects in nonexperimental studies: reevaluating the evaluation of training programs. *Journal of the American Statistical Association* **94**, 1053–1062.

Dehejia, R. H., and Wahba, S. (2002). Propensity score matching methods for non-experimental causal studies. *Review of Economics and Statistics* **84**, 151–161.

Dempster, A. P., Laird, N. M., and Rubin, D. B. (1977). Maximum likelihood from incomplete data via the EM algorithm (with discussion). *Journal of the Royal Statistical Society B* **39**, 1–38.

Des Jarlais, D. C., and Friedman, S. R. (1988). HIV infection among persons who inject illicit drugs: problems and prospects. *Journal of Acquired Immune Deficiency Syndromes* **1**, 267–273.

Derigs, U. (1988). Solving non-bipartite matching problems via shortest path techniques. *Annals of Operations Research*, **13**, 225–261.

Dewanji, A. (1992). A note on a test for competing risks with missing failure type. *Biometrika* **79**, 855–857.

Dickey, J. M., Lindley, D. V., and Press, S. J. (1985). Bayesian estimation of the dispersion matrix of a multivariate normal distribution. *Communications in Statistics A* **14**, 1019–1034.

Diffendal, G. (1988). The 1986 test of adjustment related operations in central Los Angeles County. *Survey Methodology* **14**, 71–86.

Diggle, P. J., Heagerty, P., Liang, K.-Y., and Zeger, S. L. (2002). *Analysis of Longitudinal Data*, second edition. Oxford University Press.

Diggle, P., and Kenward, M. G. (1994). Informative drop-out in longitudinal data analysis. *Journal of the Royal Statistical Society C* **43**, 49–73.

Dinse, G. E. (1986). Nonparametric prevalence and mortality estimators for animal experiments with incomplete cause-of-death data. *Journal of the American Statistical Association* **81**, 328–335.

Doretto, G., Chiuso, A., Wu, Y. N., and Soatto, S. (2003). Dynamic textures. *International Journal of Computer Vision* **51**, 91–109.

Doucet, A., de Freitas, N., and Gordon, N. (2001). *Sequential Monte Carlo Methods in Practice*. New York: Springer.

Drake, C. (1993). Effects of misspecification of the propensity score on estimators of treatment effect. *Biometrics* **49**, 1231–1236.

Drucker, E., Lurie, P., Wodak, A., and Alcabes, P. (1998). Measuring harm reduction: the effects of needle and syringe exchange programs and methadone maintenance on the ecology of HIV. *AIDS* **12**, S217–S230.

Drucker, P. F. (1985). *Innovation and Entrepreneurship*. New York: Harper & Row.

Duan, N., Kanouse, D. E., and Berry, S. H. (1992). Weighting a probability sample of street prostitutes. Technical report, RAND Corporation.

Dunn, G., Dowrick, C., Ayuso-Mateos, J. L., Dalgard, O. S., Page, H., Lehtinen, V., Casey, P., Wilkinson, C., Vazquez-Barquero, J. L., and Wilkinson, G. (2003). Estimating psychological treatment effects from a randomised controlled trial with both non-compliance and loss to follow-up. *British Journal of Psychiatry* **183**, 323–331.

Durbin, J. (1954). Errors in variables. *Review of the International Statistical Institute* **22**, 23–32.

Dykstra, R. L., and Laud, P. (1981). A Bayesian nonparametric approach to reliability. *Annals of Statistics* **9**, 356–367.

Eddy, D. M., Hasselblad, V., and Schachter, R. (1992). *Meta-Analysis by the Confidence Profile Method*. New York: Academic Press.

Efron, B. (2004). Statistics as a unified discipline. *Amstat News* **319** (January), 2–3.

Elliott, M. R., and Little, R. J. A. (2000). A Bayesian approach to combining information from a census, a coverage measurement survey, and demographic analysis. *Journal of the American Statistical Association* **95**, 351–362.

Elton, C., and Nicholson, M. (1942). The ten-year cycle in numbers of the lynx in Canada. *Journal of Animal Ecology* **11**, 215–244.

Esch, D. N. (2003). Extensions and applications of three statistical models. Ph.D. thesis, Department of Statistics, Harvard University.

Esch, D. N., Connors, A., Karovska, M., and van Dyk, D. A. (2004). An image reconstruction technique with error estimates. Manuscript in progress.

Estroff, S. E. (1985). *Making it Crazy: An Ethnography of Psychiatric Clients in an American Community*. Berkeley, Calif.: University of California Press.

Ettner, S. L. (1996). Adverse selection and the purchase of medigap insurance by the elderly. *Journal of Health Economics* **16**, 543–562.

Fay, R. E. (1991). A design-based perspective on missing-data variance. *Proceedings of the 1991 Annual Research Conference, U.S. Bureau of the Census*, 429–440.

Fay, R. E. (1992). When are inferences from multiple imputation valid? *Proceedings of the Survey Research Methods Section, American Statistical Association*, 227–232.

Fay, R. E. (1993). Valid inferences from imputed survey data. *Proceedings of the Survey Research Methods Section, American Statistical Association*, 41–48.

Fay, R. E. (1996). Alternative paradigms for the analysis of imputed survey data. *Journal of the American Statistical Association* **91**, 490–498.

Federman, A. D., Adams, A. S., Ross-Degnan, D., Soumerai, S. B., and Ayanian J. Z. (2001). Supplemental insurance and use of effective cardiovascular drugs among elderly Medicare beneficiaries with coronary heart disease. *Journal of the American Medical Association* **286**, 1732–1739.

Fein, D. J., and West, K. K. (1988). The sources of census undercount: findings from the 1986 Los Angeles test census. *Survey Methodology* **14**, 223–240.

Fellegi, I. P., and Sunter, A. B. (1969). A theory for record linkage. *Journal of the American Statistical Association* **64**, 1883–1210.

Fessler, J. A., and Hero, A. O. (1994). Space-alternating generalized expectation-maximization algorithm. *IEEE Transactions on Signal Processing* **42**, 2664–2677.

Finkbeiner, C. (1979). Estimation for the multiple factors model when data are missing. *Psychometrika* **44**, 409–420.

Finney, D. J. (1947). The estimation from individual records of the relationship between dose and quantal response. *Biometrika*, **34**, 320–334.

Fisher, R. A. (1918). The causes of human variability. *Eugenics Review* **10**, 213–220.

Fisher, R. A. (1925). *Statistical Methods for Research Workers*. Edinburgh: Oliver and Boyd.

Flehinger, B. J., Reiser, B., and Yashchin, E. (1998). Survival with competing risks and masked causes of failures. *Biometrika* **85**, 151–164.

Flehinger, B. J., Reiser, B., and Yashchin, E. (2002). Parametric modeling for survival with competing risks and masked failure causes. *Lifetime Data Analysis* **8**, 177–203.

Fokoué, E., and Titterington, D. M. (2003). Mixtures of factor analysers: Bayesian estimation and inference by stochastic simulation. *Machine Learning* **50**, 73–94.

Fortini, M., Liseo, B., Nuccitelli, A., and Scanu, M. (2001). On Bayesian record linkage. *Research in Official Statistics* **4**, 185–198.

Fortini, M., Nuccitelli, A., Liseo, B., and Scanu, M. (2002). Modelling issues in record linkage: a Bayesian perspective. *Proceedings of the American Statistical Association, Survey Research Methods Section*, 1008–1013.

Fraker, T., and Maynard, R. (1987). The adequacy of comparison group design for evaluations of employment-related programs. *Journal of Human Resources* **22**, 194–227.

Frangakis, C. E., and Baker, S. G. (2001). Compliance subsampling designs for comparative research: estimation and optimal planning. *Biometrics* **57**, 899–908.

Frangakis, C. E., Brookmeyer, R. S., Varadhan, R., Mahboobeh, S., Vlahov, D., and Strathdee, S. A. (2004). Methodology for evaluating a partially controlled longitudinal treatment using principal stratification, with application to a needle exchange program. *Journal of the American Statistical Association.*

Frangakis, C. E., and Rubin, D. B. (1999). Addressing complications of intention-to-treat analysis in the combined presence of all-or-none treatment-noncompliance and subsequent missing outcomes. *Biometrika* **86**, 365–379.

Frangakis, C. E., and Rubin, D. B. (2002). Principal stratification in causal inference. *Biometrics* **58**, 21–29.

Freedman, L. S., Graubard, B. I, and Schatzkin, A. (1992). Statistical validation of intermediate endpoints for chronic diseases. *Statistics in Medicine* **11**, 167–178.

Friedlander, D., and Robins, P. K. (1995). Evaluating program evaluations—new evidence on commonly used nonexperimental methods. *American Economic Review* **85**, 923–937.

Gail, M., Pfeiffer, R., Houwelingen, H., and Carroll, R. J. (2000). On meta-analytic assessment of surrogate outcomes. *Biostatistics* **1**, 231–246.

Galecki, A. T. (1994). General class of covariance structures for two or more repeated factors in longitudinal data analysis. *Communications in Statistics A* **23**, 3105–3119.

Galil, Z. (1986). Efficient algorithms for finding maximum matching in graphs. *Computing Surveys* **18**, 23–38. elib.zib.de/pub/Packages/mathprog/ matching/weighted/index.html.

Gasbarra, D., and Karia, S. R. (2000). Analysis of competing risks by using Bayesian smoothing. *Scandinavian Journal of Statistics* **27**, 605–617.

Gelfand, A. E. (2000). Gibbs sampling. *Journal of the American Statistical Association* **95**, 1300–1304.

Gelfand, A. E., and Ghosh, S. K. (1998). Model choice: a minimum posterior predictive loss approach. *Biometrika* **85**, 1–11.

Gelfand, A. E., and Smith, A. F. M. (1990). Sampling-based approaches to calculating marginal densities. *Journal of the American Statistical Association* **85**, 398–409.

Gelman, A. (1992). Iterative and non-iterative simulation algorithms. *Computing Science and Statistics* **24**, 433–438.

Gelman, A., Carlin, J. B., Stern, H. S., and Rubin, D. B. (2003). *Bayesian Data Analysis*, second edition. London: CRC Press.

Gelman, A., and Huang, Z. (2004). Estimating incumbency advantage and its variation, as an example of a before/after study. Under revision for *Journal of the American Statistical Association.*

Gelman, A., and King, G. (1990). Estimating incumbency advantage without bias. *American Journal of Political Science* **34**, 1142–1164.

Gelman, A., and King, G. (1994). Enhancing democracy through legislative redistricting. *American Political Science Review* **88**, 541–559.

Gelman, A., and Meng, X.-L. (1996). Model checking and model improvement. In *Practical Markov Chain Monte Carlo*, eds. W. R. Gilks, S. Richardson, and D. J. Spiegelhalter, 189–201. New York: Chapman & Hall.

Gelman, A., and Meng, X.-L. (1998). Simulating normalizing constants: from importance sampling to bridge sampling to path sampling. *Statistical Science* **13**, 163–185.

Gelman, A., Meng, X.-L., and Stern, H. S. (1996). Posterior predictive assessment of model fitness via realized discrepancies (with discussion). *Statistica Sinica* **6**, 733–807.

Gelman, A., and Rubin, D. B. (1992). Inference from iterative simulation using multiple sequences (with discussion). *Statistical Science* **7**, 457–511.

Geman, S., and Geman, D. (1984). Stochastic relaxation, Gibbs distributions, and the Bayesian restoration of images. *IEEE Transactions on Pattern Analysis and Machine Intelligence* **6**, 721–741.

Geman, S., Potter, D. F., and Chi, Z. (2002). Composition system. *Quarterly of Applied Mathematics* **60**, 707–736.

Genovese, C. R., Lazar, N. A., and Nichols, T. E. (2002). Thresholding of statistical maps in functional neuroimaging using the false discovery rate. *NeuroImage* **15**, 870–878.

George, E. I., and McCulloch, R. E. (1997). Approaches to Bayesian variable selection. *Statistica Sinica* **7**, 339–373.

Gerber, A. S., and Green, D. P. (1999). Does canvassing increase voter turnout? a field experiment. *Proceedings of the National Academy of Sciences U.S.A* **96**, 10939–10942.

Gerber, A. S., and Green, D. P. (2000a). The effects of canvassing, direct mail, and telephone contact on voter turnout: a field experiment. *American Political Science Review* **94**, 653–663.

Gerber, A. S., and Green, D. P. (2000b). The effect of a nonpartisan get-out-the-vote drive: an experimental study of leafletting. *Journal of Politics* **62**, 846–857.

Geweke, J. (1989). Bayesian inference in econometric models using Monte Carlo integration. *Econometrica* **57**, 1317–1339.

Geweke, J., and Zhou, G. (1996). Measuring the pricing error of the arbitrage pricing theory. *Review of Financial Studies* **9**, 557–587.

Ghahramani, Z. and Beal, M. (2000). Variational inference for Bayesian mixtures of factor analysers. In *Advances in Neural Information Processing*, volume 12, ed. S. A. Solla, T. K. Leen, and K. R. Muller, Cambridge, Mass.: MIT Press.

Gilbert, P. B., Bosch, R. J., and Hudgens, M. G. (2003). Sensitivity analysis for the assessment of causal vaccine effects on viral load in AIDS vaccine trials. *Biometrics* **59**, 531–541.

Gilks, W. R., Richardson, S., and Spiegelhalter, D., eds. (1996). *Practical Markov Chain Monte Carlo*. New York: Chapman & Hall.

Gilks, W. R., and Wild, P. (1992). Adaptive rejection sampling for Gibbs sampling. *Applied Statistics* **41**, 337–348.

Gill, R. D. (1985). Discussion of the paper by Clayton and Cuzick. *Journal of the Royal Statistical Society A* **148**, 108–109.

Glenn, T., Kelly, D., Boscardin, W. J., McArthur, D., Vespa, P., Oertel, M., Hovda, D., Bergsneider, M., Hillered, L., and Martin, N. (2003). Energy dysfunction as a predictor of outcome after moderate or severe head injury: indices of oxygen, glucose, and lactate metabolism. *Journal of Cerebral Blood Flow and Metabolism* **23**, 1239–1250.

Goetghebeur, E., and Ryan, L. (1990). A modified log rank test for competing risks with missing failure types. *Biometrika* **77**, 151–164.

Goetghebeur, E., and Ryan, L. (1995). Analysis of competing risks survival data when some failure types are missing. *Biometrika* **82**, 821–833.

Goetghebeur, E., and van Houwelingen, H. C., eds. (1998). Analyzing noncompliance in clinical trials. *Statistics in Medicine* **17**, 247–389.

Gomatam, S., and Larsen, M. D. (2004). Record linkage and counterterrorism. *Chance*.

Goodman, L. A. (1968). The analysis of cross-classified data: independence, quasi-independence, and interactions in contingency tables with or without missing entries. *Journal of the American Statistical Association* **63**, 1091–1131.

Goodman, L. A. (1981). Association models and the bivariate normal for contingency tables with ordered categories. *Biometrika*, **68**, 347–355.

Gordon, N. J., Salmond, D. J., and Smith, A. F. M. (1993). Novel approach to nonlinear/non-Gaussian Bayesian state estimation. *IEE Proceedings-F (Radar and Signal Processing)* **140**, 107–113.

Gouweleeuw, P. K., Willenborg, L. C. R. J., and de Wolf, P.-P. (1998). Post randomization for statistical disclosure control: theory and implementation. *Journal of Official Statistics* **14**, 463–478.

Graham, P. (2000). Bayesian inference for a generalized population attributable fraction. *Statistics in Medicine* **19**, 937–956.

Gray, R. (1994). A Bayesian analysis of institutional effects in a multicenter cancer clinical trial. *Biometrics* **50**, 244–253.

Gray, R. (1995). Tests for variation over groups in survival data. *Journal of the American Statistical Association* **90**, 198–203.

Green, D. A., and Warburton, W. P. (2001). Tightening a welfare system: the effects of benefit denial on future welfare receipt. Department of Economics, University of British Columbia, Discussion Paper No. 02-07.

Green, P. J., and Silverman, B. W. (1994). *Nonparametric Regression and Generalized Linear Models*. London: Chapman & Hall.

Greenberg, B. (1987). Rank swapping for masking ordinal microdata. U.S. Census Bureau. Unpublished manuscript.

Greenland, S. (1987). Interpretation and choice of effect measures in epidemiologic analysis. *American Journal of Epidemiology* **125**, 761–768.

Greenland, S. (1991). On the logical justification of conditional tests for two-by-two contingency tables. *American Statistician* **45**, 248–251.

Greenland, S. (1998). Induction versus Popper: substance versus semantics. *International Journal of Epidemiology* **27**, 543–548.

Greenland, S. (2000). Causal analysis in the health sciences. *Journal of the American Statistical Association* **95**, 286–289.

Greenland, S. (2001). Sensitivity analysis, Monte-Carlo risk analysis, and Bayesian uncertainty assessment. *Risk Analysis* **21**, 579–583.

Greenland, S. (2003). The impact of prior distributions for uncontrolled confounding and response bias. *Journal of the American Statistical Association* **98**, 47–54.

Greenland S. (2004). Multiple-bias modeling for observational studies (with discussion). *Journal of the Royal Statistical Society A*.

Greenland, S., and Brumback, B. A. (2002). An overview of relations among causal modelling methods. *International Journal of Epidemiology* **31**, 1030–1037.

Greenland, S., and Robins, J. M. (2000). Epidemiology, justice, and the probability of causation. *Jurimetrics* **40**, 321–340.

Greenland, S., Robins, J. M., and Pearl, J. (1999). Confounding and collapsibility in causal inference. *Statistical Science* **14**, 29–46.

Grenander, U. (1993). *General Pattern Theory*. Oxford University Press.

Grenander, U., and Miller, M. I. (1994). Representation of knowledge in complex systems. *Journal of the Royal Statistical Society B* **56**, 549–603.

Gu, X. S., and Rosenbaum, P. R. (1993). Comparison of multivariate matching methods: structures, distances and algorithms. *Journal of Computational and Graphical Statistics* **2**, 405–420.

Guo, C. E., Zhu, S. C., and Wu, Y. N. (2003). Towards a mathematical theory of primal sketch and sketchability. *Proceedings of 9th International Conference on Computer Vision*, 1228–1235.

Gupta, A. K., and Nagar, D. K. (2000). *Matrix Variate Distributions*. New York: CRC Press.

Gustafson, P. (2003). *Measurement Error and Misclassification in Statistics and Epidemiology.* New York: Chapman & Hall.

Haavelmo, T. (1943). The statistical implications of a system of simultaneous equations. *Econometrica* **11**, 1–12.

Haavelmo, T. (1944). The probability approach in econometrics. *Econometrica* **12**, 1–115.

Hahn, J. (1998). On the role of the propensity score in efficient semiparametric estimation of average treatment effects. *Econometrica* **66**, 315–331.

Hansen, M., and Yu, B. (2000). Wavelet thresholding via MDL for natural images. *IEEE Transactions on Information Theory* (Special Issue on Information Theoretic Imaging) **46**, 1778–1788.

Hansen, M. H., Madow, W. G., and Tepping, B. J. (1983). An evaluation of model-dependent and probability-sampling inferences in sample surveys (with discussion). *Journal of the American Statistical Association* **78**, 776–807.

Härdle, W., and Linton, O. (1994). Applied nonparametric regression. In *Handbook of Econometrics*, volume 4, eds. R. Engle and D. L. McFadden, 2295–2339. Amsterdam: Elsevier.

Hartley, H. O. (1966). Systematic sampling with unequal probability and without replacement. *Journal of the American Statistical Association* **61**, 739–748.

Hastie, T., and Tibshirani, R. (1990). *Generalized Additive Models.* London: Chapman & Hall.

Hastie, T., Tibshirani, R., and Friedman, J. (2001). *Elements of Statistical Learning: Data Mining, Inference and Prediction.* New York: Springer.

Hastings, W. K. (1970). Monte Carlo sampling methods using Markov chains and their applications. *Biometrika* **57**, 97–109.

Heckman, J. J., and Hotz, J. (1989). Choosing among alternative nonexperimental methods for estimating the impact of social programs: the case of manpower training. *Journal of the American Statistical Association* **84**, 862–874.

Heckman, J. J., Ichimura, H., Smith, J., and Todd, P. (1996). Sources of selection bias in evaluating social programs: an interpretation of conventional measures and evidence on the effectiveness of matching as a program evaluation method. *Proceedings of the National Academy of Sciences of the United States of America* **93**, 13416–13420.

Heckman, J. J., Ichimura, H., and Todd, P. (1997). Matching as an econometric evaluation estimator: evidence from a job training programme. *Review of Economic Studies* **64**, 605–654.

Heckman, J. J., Ichimura, H., and Todd, P. (1998). Matching as an econometric evaluations estimator. *Review of Economic Studies* **65**, 261–294.

Heckman, J. J., and Robb, R. (1985). Alternative methods for evaluating the impact of interventions. In *Longitudinal Analysis of Labor Market Data*, eds. J. Heckman and B. Singer. New York: Cambridge University Press.

Heeger, D. J., and Bergen, J. R. (1995). Pyramid based texture analysis/synthesis. *Computer Graphics Proceedings*, 229–238.

Heeringa, S. G., Little, R. J. A., and Raghunathan, T. E. (2002). Multivariate imputation of coarsened survey data on household wealth. In *Survey Nonresponse*, eds. R. M. Groves, D. A. Dillman, J. L. Eltinge, and R. J. A. Little, chapter 24. New York: Wiley.

Heitjan, D. F. (1997). Ignorability, sufficiency and ancillarity. *Journal of the Royal Statistical Society B* **59**, 375–381.

Heitjan, D. F., and Basu, S. (1996). Distinguishing "missing at random" and "missing completely at random." *American Statistician* **50**, 207–213.

Hernan, M. A., Brumback, B. A., and Robins, J. M. (2001). Marginal structural models to estimate the joint causal effect of nonrandomized treatments. *Journal of the American Statistical Association* **96**, 440–448.

Hill, A. B. (1965). The environment and disease: association or causation? *Proceedings of the Royal Society of Medicine* **58**, 295–300.

Hill, J. L. (2004). Reducing bias in treatment effect estimation in observational studies suffering from missing data. Columbia University Institute for Social and Economic Research and Policy (ISERP), Working Paper 04-01.

Hill, J. L., Brooks-Gunn, J., and Waldfogel, J. (2003). Sustained effects of high participation in an early intervention for low-birth-weight premature infants. *Developmental Psychology* **39**, 730–744.

Hill, J. L., Waldfogel, J., and Brooks-Gunn, J. (2002). Assessing the differential impacts of high-quality child care: a new approach for exploiting post-treatment variables. *Journal of Policy Analysis and Management* **21**, 601–627.

Hirano, K., Imbens, G. W., and Ridder, G. (2000). *Efficient Estimation of Average Treatment Effects Using the Estimated Propensity Score*. National Bureau of Economics, Research Technical Working Paper Series No. 251.

Hirano, K., Imbens, G. W., and Ridder, G. (2003). Efficient estimation of average treatment effects using the estimated propensity score. *Econometrica* **71**, 1161–1189.

Hjort, N. L. (1990). Nonparametric Bayes estimators based on beta processes for life history data. *Annals of Statistics* **18**, 1259–1294.

Hoeting, J., Madigan, D., Raftery, A. E., and Volinsky, C. (1999). Bayesian model averaging (with discussion). *Statistical Science* **14**, 382–417.

Hogan, H. (1993). The 1990 post-enumeration survey: operations and results. *Journal of the American Statistical Association*, **88**, 1047–1060.

Hogan, H. (2000). The accuracy and coverage evaluation: theory and application. *Proceedings of the American Statistical Association, Survey Research Methods Section*, 31–40.

Holland, P. (1986). Statistics and causal inference (with discussion). *Journal of the American Statistical Association* **81**, 945–960.

Hollister, R., Kemper, P., and Maynard, R. (1984). *The National Supported Work Demonstration*, Madison, University of Wisconsin Press.

Hume, D. (1748). *An Enquiry Concerning Human Understanding*. Reprinted by Open Court Press (1988).

Ibrahim, J. G., Chen, M. H., Sinha, D. (2001). *Bayesian Survival Analysis*, New York: Springer-Verlag.

Ichino, A., and Winter-Ebmer, R. (1998). Lower and upper bounds of returns to schooling: an exercise in IV estimation with different instruments. CEPR Discussion Paper No. 2007.

Imai, K., and van Dyk, D. (2004). Causal inference with general treatment regimes: generalizing the propensity score. *Journal of the American Statistical Association*.

Imbens, G. W. (2000). The role of the propensity score in estimating dose-response functions. *Biometrika* **83**, 706–710.

Imbens, G., and Angrist, J. (1992). Identification and estimation of local average treatment effects. *Econometrica* **62**, 467–476.

Imbens, G. W., and Angrist, J. (1994). Identification and estimation of local average treatment effects. *Econometrica* **62**, 467–475.

Imbens, G. W., and Rubin, D. B. (1994). Causal inference with instrumental variables. Discussion Paper No. 1676. Cambridge, Mass.: Harvard Institute of Economic Research.

Imbens, G. W., and Rubin, D. B. (1997a). Bayesian inference for causal effects in randomized experiments with noncompliance. *Annals of Statistics* **25**, 305–327.

Imbens, G. W., and Rubin, D. B. (1997b). Estimating outcome distributions for compliers in instrumental variables models. *Review of Economic Studies* **64**, 555–574.

Imbens, G. W., Rubin, D. B., and Sacerdote, B. (2001). Estimating the effect of unearned income on labor supply, earnings, savings and consumption: evidence from a survey of lottery players. *American Economic Review* **91**, 778–794.

Ingram, D. D., Parker, J. D., Schenker, N., Weed, J. A., Hamilton, B., Arias, E., and Madans, J. H. (2003). U.S. Census 2000 population with bridged race categories. *Vital and Health Statistics*, volume 2 (135). Hyattsville, Md.: National Center for Health Statistics.

Infant Health and Development Program (IHDP). (1990). Enhancing the outcomes of low-birth-weight, premature infants. *Journal of the American Medical Association* **22**, 3035–3042.

Ishwaran, H., Berry, S., Duan, N., and Kanouse, D. (1991). Replicate interviews in the Los Angeles women's health risk study: searching for the three-faced Eve. Technical Report, RAND Corporation.

James, W., and Stein, C. (1961). Estimation with quadratic loss. *Proceedings of the Fourth Berkeley Symposium on Mathematical Statistics and Probability*, volume 1, 361–379; Berkeley, Calif.: University of California Press.

Jaro, M. A. (1989). Advances in record-linkage methodology as applied to matching in the 1985 census of Tampa, Florida. *Journal of the American Statistical Association* **84**, 414–420.

Jennrich, R. I. (1978). Rotational equivalence of factor loading matrices with specified values. *Psychometrika* **43**, 421–426.

Jeon, Y. S. (1998). Inference in structural equation models with missing data. Ph.D. thesis, Department of Statistics, Iowa State University.

Jo, B. (1999). Power to detect intervention effects in randomized trials with noncompliance. Technical Report, Graduate School of Education and Information Studies, University of California, Los Angeles.

Joffe, M., and Rosenbaum, P. R. (1999a). Invited commentary: propensity scores. *American Journal of Epidemiology* **150**, 1–7.

Joffe, M., and Rosenbaum, P. R. (1999b). Propensity scores. *American Journal of Epidemiology* **150**, 327–333.

Johansen, S. (1993). An extension of Cox's regression model. *International Statistical Review* **51**, 258–262.

Johnson, J. (1984). *Econometric Methods.* New York: McGraw-Hill.

Jordan, M. I., Ghahramani, Z., Jaakkola, T. S., and Saul, L. K. (1999). An introduction to variational methods for graphical models. In *Learning in Graphical Models*, ed. M. I. Jordan, 105–161. Cambridge, Mass.: MIT Press.

Jordan, M. I., and Jacobs, R. A. (1994). Hierarchical mixtures of experts and the EM algorithm. *Neural Computation* **6**, 181–214.

Joreskog, K. G. (1969). A general approach to confirmatory maximum likelihood analysis. *Psychometrika* **34**, 183–202.

Joreskog, K. G., and Sorbom, D. (1979). *Advances in Factor Analysis and Structural Equation Models.* Cambridge, Mass.: Abt Books.

Julesz, B. (1981). Textons, the elements of texture perception and their interactions. *Nature* **290**, 91–97.

Kagan, J. (1994). *Galen's Prophecy.* New York: Basic Books.

Kaiser Family Foundation. (2004). *Prescription Drug Coverage for Medicare Beneficiaries: An Overview of the Medicare Prescription Drug, Improvement, and Modernization Act of 2003 (Public Law 108-173).* A report prepared by Health Policy Alternatives, Inc. for the Kaiser Family Foundation, Menlo Park, Calif.

Kalbfleisch, J. D. (1978). Non-parametric Bayesian analysis of survival time data. *Journal of the Royal Statistical Society B* **40**, 214–221.

Kanouse, D. E., Berry, S. H., Duan, N., Lever, J., Carson, S., Perlman, J. F., and Levitan, B. (1999). Drawing a probability sample of female street prostitutes in Los Angeles County. *Journal of Sex Research* **36**, 45–51.

Kanouse, D. E., Berry, S. H., Duan, N., Richwald, G., and Yano, E. M. (1992). Markers for HIV-1, hepatitis B, and syphilis in a probability sample of street prostitutes in Los Angeles County, California. *VIII International Conference on AIDS/III STD World Congress, Poster Abstracts*, volume 2, Amsterdam: PoC 4192.

Kaplan, E. H. (1994). A method for evaluating needle exchange programmes. *Statistics in Medicine* **13**, 2179–2187.

Karim, M. R., and Zeger, S. L. (1992). Generalized linear models with random effects; salamander mating revisited. *Biometrics* **48**, 631–644.

Kass, R. E., Carlin, B. P., Gelman, A., and Neal, R. (1998). Markov chain Monte Carlo in practice: a roundtable discussion. *American Statistician* **52**, 93–100.

Kass, R. E., and Raftery, A. E. (1995). Bayes factors and model uncertainty. *Journal of the American Statistical Association* **90**, 773–795.

Katz, J. (1999). *How Emotions Work*. University of Chicago Press.

Keende, J. M., Stimson, G. V., Jones, S., and Parry-Langdon, N. (1993). Evaluation of syringe-exchange for HIV prevention among injecting drug users in rural and urban areas of Wales. *Addiction* **88**, 1063–1070.

Keith, L. B. (1990). Dynamics of snowshoe hare populations. In *Current Mammalogy*, volume 2, ed. H. H. Genoways, 119–195. New York: Plenum Press.

Kennickell, A. B. (1991). Imputation of the 1989 survey of consumer finances: stochastic relaxation and multiple imputation. *American Statistical Association Proceedings of the Survey Research Methods Section*.

Kitagawa, G. (1996). Monte Carlo filter and smoother for non-Gaussian nonlinear state space models. *Journal of Computational and Graphical Statistics* **5**, 1–25.

Klebanove, P. K., Brooks-Gunn, J., and McCormick, M. C. (1994a). Classroom behavior of very low birth weight elementary school children. *Pediatrics* **94**, 700–708.

Klebanove, P. K., Brooks-Gunn, J., and McCormick, M. C. (1994b). School achievement and failure in very low birth weight children. *Journal of Developmental and Behavioral Pediatrics* **15**, 248–256.

Klein, J. P. (1992). Semiparametric estimation of random effects using the Cox model based on the EM algorithm. *Biometrics* **48**, 795–806.

Kodell, R. L., and Chen, J. J. (1987). Handling cause of death in equivocal cases using the EM algorithm (with discussion). *Communications in Statistics: Theory and Methods* **16**, 2565–2585.

Koffka, K. (1935). *Principles of Gestalt Psychology*. Harcourt.

Koop, G., and Poirier, D. J. (2001). Testing for optimality in job search models, *Econometrics Journal* **4**, 257–272.

Kott, P. S. (1992). A note on a counter-example to variance estimation using multiple imputation. Technical Report, U.S. National Agricultural Statistical Service.

Krzanowski, W. J. (1980). Mixtures of continuous and categorical variables in discriminant analysis. *Biometrics* **36**, 493–499.

Krzanowski, W. J. (1982). Mixtures of continuous and categorical variables in discriminant analysis: a hypothesis testing approach. *Biometrics* **38**, 991–1002.

Kuo, L., and Yang, T. Y. (2000). Bayesian reliability modeling for masked system lifetime data. *Statistics and Probability Letters* **47**, 229–241.

Lagakos, S. W. (1982). An evaluation of some two-sample tests used to analyze animal carcinogenicity experiments. *Utilitas Mathematicae* **21**, B239–B260.

Lagakos, S. W., and Louis, T. A. (1988). Use of tumour lethality to interpret tumorigenicity experiments lacking cause-of-death data. *Applied Statistics* **37**, 169–179.

Lahiri, P., and Larsen, M. D. (2000). Model-based analysis of records linked using mixture models. *Proceedings of the American Statistical Association, Survey Research Methods Section*, 11–19.

Lahiri, P., and Larsen, M. D. (2004). Regression analysis with linked data. Under revision for *Journal of the American Statistical Association*.

Lai, D. (1996). Comparison study of AR models on the Canadian lynx data: a close look at the BDS statistic. *Computational Statistics and Data Analysis* **22**, 409–423.

Laird, N. M. (1983). Further comparative analysis of pretest-posttest research designs. *American Statistician* **37**, 329–330.

Laird, N. M., and Ware, J. H. (1982). Random-effects models for longitudinal data. *Biometrics* **38**, 963–974.

Lalonde, R. (1986). Evaluating the econometric evaluations of training programs. *American Economic Review* **76**, 604–620.

Lalonde, R., and Maynard, R. (1987). How precise are evaluations of employment and training programs: evidence from a field experiment. *Evaluation Review* **11**, 428–451.

Lancaster, T. (1997). Exact structural inference in optimal job-search models. *Journal of Business and Economic Statistics* **15**, 165–179.

Lange, K. (2004). Computational statistics and optimization theory at UCLA. *American Statistician* **58**, 9–11.

Lange, K., and Carson, R. (1984). EM reconstruction algorithms for emission and transmission tomography. *Journal of Computer Assisted Tomography* **8**, 306–316.

Lange, K. L., Little, R. J. A., and Taylor, J. M. G. (1989). Robust statistical modeling using the t distribution. *Journal of the American Statistical Association* **84**, 881–896.

Larsen, M. D. (1994). Data augmentation with Bayesian iterative proportional fitting applied to a Census Bureau latent-class problem. *Proceedings of the American Statistical Association, Government Statistics Section*, 116–121.

Larsen, M. D. (1996). Bayesian approaches to finite mixture models. Ph.D. thesis, Harvard University.

Larsen, M. D. (1999). Multiple imputation analysis of records linked using mixture models. *Proceedings of the Statistical Society of Canada, Survey Methods Section*, 65–71.

Larsen, M. D. (2002). Comments on hierarchical Bayesian record linkage. *Proceedings of the American Statistical Association, Survey Research Methods Section*, 1995–2000.

Larsen, M. D., and Rubin, D. B. (2001). Iterative automated record linkage using mixture models. *Journal of the American Statistical Association* **96**, 32–41.

Lash, T. L., and Fink, A. K. (2003). Semi-automated sensitivity analysis to assess systematic errors in observational epidemiologic data. *Epidemiology* **14**, 451–458.

Lechner, M. (2001). Identification and estimation of causal effects of multiple treatments under the conditional independence assumption. In *Econometric Evaluation of Active Labor Market Policies in Europe*, ed. M. Lechner and F. Pfeiffer, 43–58. Heidelberg: Physica.

Lee, D. (1997). Selecting sample sizes for the sampling/importance resampling filter. *ASA Proceedings of the Section on Bayesian Statistical Science*, 72–77.

Lee, Y., and Nelder, J. A. (1996). Hierarchical generalized linear models. *Journal of the Royal Statistical Society B* **58**, 619–678.

Lewicki, M. S., and Olshausen, B. A. (1999). Probabilistic framework for the adaptation and comparison of image codes. *Journal of the Optical Society of America* **16**, 1587–1601.

Lewis, D. K. (1973). Causation. *Journal of Philosophy* **70**, 556–567.

Li, K. H. (1994). A computer implementation of the Yates-Grundy draw by draw procedure. *Journal of Statistical Computation and Simulation* **50**, 147–151.

Li, F., Frangakis, C. E., and Varadhan, R. (2004). Polydesigns for partially controlled studies: motivation and definition. *American Statistical Association Proceedings of the Biopharmaceutical Section*.

Li, Y. P., Propert, K. J., and Rosenbaum, P. R. (2001). Balanced risk set matching. *Journal of the American Statistical Association* **96**, 870–882.

Lieberman, E., Cohen, A., Lang, J. M., D'Agostino, R. B., Datta, S., and Frigoletto, F. D. (1996). The association of epidural anesthesia with cesarean delivery in nulliparas. *Obstetrics and Gynecology* **88**, 993–1000.

Lim, K. S. (1987). A comparative study of various univariate time series models for Canadian lynx data. *Journal of Time Series Analysis* **8**, 161–176.

Lin, D. Y., Fleming, T. R., and De Gruttola, V. (1997). Estimating the proportion of treatment effect explained by a surrogate marker. *Statistics in Medicine* **16**, 1515–1527.

Lindeberg, T. (1994). *Scale-Space Theory in Computer Vision*. Dordrecht, Netherlands: Kluwer.

Little, R. J. A. (1993). Statistical analysis of masked data. *Journal of Official Statistics* **9**, 407–426.

Little, R. J. A., and Liu, F. (2003a). *Selective Multiple Imputation for Statistical Disclosure Control in Microdata*. Research Report, Institute for Survey Research, University of Michigan.

Little, R. J. A., and Liu, F. (2003b). Selective multiple imputation of keys for statistical disclosure control. *Department of Biostatistics Working Paper, University of Michigan*.

Little, R. J. A., and Rubin, D. B. (2002). *Statistical Analysis with Missing Data*, second edition. New York: Wiley.

Little R. J. A., and Schluchter, M. D. (1985). Maximum likelihood estimation for mixed continuous and categorical data with missing values. *Biometrika* **72**, 497–512.

Liu, C. (1995). Missing data imputation using the multivariate t distribution. *Journal of Multivariate Analysis* **48**, 198–206.

Liu, C. (2000). Comment on "The Art of Data Augmentation" by Meng and van Dyk. *Journal of Computational and Graphical Statistics* **10**, 75–81.

Liu, C. (2001). Bayesian analysis of multivariate probit models. Discussion of "The art of data augmentation" by van Dyk and Meng. *Journal of Computational and Graphical Statistics* **10**, 1–50.

Liu, C., and Rubin, D. B. (1994). The ECME algorithm: a simple extension of EM and ECM with faster monotone convergence. *Biometrika* **81**, 633–648.

Liu, C., and Rubin, D. B. (1995). ML estimation of the t distribution using EM and its extensions, ECM and ECME. *Statistic Sinica* **5**, 19-39

Liu, C., and Rubin, D. B. (1998). Ellipsoidally symmetric extensions of the general location model for mixed categorical and continuous data. *Biometrika* **85**, 673–688.

Liu, C., Rubin, D. B., and Wu, Y. N. (1998). Parameter expansion to accelerate EM: The PX-EM algorithm. *Biometrika* **85**, 755–770.

Liu, F. (2003). Bayesian methods for statistical disclosure control in microdata. Ph.D. dissertation, Department of Biostatistics, University of Michigan.

Liu, F., and Little, R. J. A. (2002a). Selective multiple imputation for statistical disclosure control in microdata. *Proceedings of American Statistical Association, Section on Survey Research Methodology*.

Liu, F., and Little, R. J. A. (2002b). Multiple imputation and statistical disclosure control in microdata. *American Statistical Association Proceedings*, 2133–2138.

Liu, I. (2003). Genetic and environmental contributions to the development of alcohol dependence. Doctoral thesis, Department of Epidemiology, Harvard School of Public Health.

Liu, J. S. (1994). The collapsed Gibbs sampler in Bayesian computations with applications to a gene regulation problem. *Journal of the American Statistical Association* **89**, 958–966.

Liu, J. S., and Hodges, J. S. (2003). Posterior bimodality in the balanced one-way random effects model. *Journal of the Royal Statistical Society B* **65**, 247–255.

Lo, S. H. (1991). Estimating a survival function with incomplete cause-of-death data. *Journal of Multivariate Analysis* **39**, 217–235.

Lopes, H. F., Moreira, A. R. B., and Schmidt, A. M. (1999). Hyperparameter estimation in forecast models. *Computational Statistics and Data Analysis* **29**, 387–410.

Louis, T. A. (1982). Finding the observed information matrix when using the EM algorithm. *Journal of the Royal Statistical Society B* **44**, 190–200.

Lu, B., and Rosenbaum, P. R. (2004). Optimal pair matching with two control groups. *Journal of Computational and Graphical Statistics* **13**.

Lu, B., Zanutto, E., Hornik, R., and Rosenbaum, P. R. (2001). Matching with doses in an observational study of a media campaign against drug abuse. *Journal of the American Statistical Association* **96**, 1245–1253.

Lucy, L. B. (1974). An iterative technique for the rectification of observed distributions. *Astronomical Journal* **79**, 745–754.

Lynch, M., and Walsh, B. (1998). *Genetics and Analysis of Quantitative Traits*, chapter 26. Sunderland, Mass.: Sinauer Associates, Inc.

Lytle, B. W., Blackstone, E. H., Loop, F. D., Houghtaling, P. L., Arnold, J. H., McCarthy, P. M., and Cosgrove D. M. (1999). Two internal thoracic artery grafts are better than one. *Journal of Thoracic and Cardiovascular Surgery* **117**, 855–872.

MacLaren, M. D., and Marsaglia, G. (1965). Uniform random number generators. *Journal of the ACM* **12**, 83–89.

MacMahon, B., and Pugh, T. F. (1967). Causes and entities of disease. In *Preventive Medicine*, ed. D. W. Clark and B. MacMahon, 11–18. Boston, Mass.: Little, Brown.

Malik, J., and Perona, R. (1990). Preattentive texture discrimination with early vision mechanisms. *Journal of the Optical Society of America* **7**, 923–932.

Mallat, S. (1989). A theory of multiresolution signal decomposition: the wavelet representation. *IEEE Transactions on Pattern Analysis and Machine Intelligence* **11**, 674–693.

Mallat, S., and Zhang, Z. (1993). Matching pursuit in a time-frequency dictionary. *IEEE Signal Processing* **41**, 3397–3415.

Mandelbrot, B. B. (1982). *The Fractal Geometry of Nature*. San Francisco, Calif.: Freeman.

Manpower Demonstration Research Corporation. (1983). *Summary and Findings of the National Supported Work Demonstration*, Cambridge: Ballinger.

Marr, D. (1982). *Vision*. San Francisco, Calif.: Freeman.

Mazzeo, J., Johnson, E., Bowker, D., and Fong, Y. (1992). The use of collateral information in proficiency estimation for the trial state assessment. Paper presented at the American Educational Research Association Annual Meeting.

McAllister, M. K., Pikitch, E. K., Punt, A. E., and Hilborn, R. (1994). A Bayesian approach to stock assessment and harvest decisions using the sampling/

importance resampling algorithm. *Canadian Journal of Fisheries and Aquatic Sciences* **51**, 2673–2687.

McClellan, M., McNeil, B. J, and Newhouse, J. P. (1994). Does more intensive treatment of acute myocardial infarction in the elderly reduce mortality? analysis using instrumental variables. *Journal of the American Medical Association* **272**, 859–866.

McCord, C. M. and Cardoza, J. E. (1982). Bobcat and lynx (Felis rufus and F. lynx). In *Wild Mammals of North America: Biology, Management, Economics*, ed. J. A. Chapman and G. A. Feldhamer. Baltimore, M.d.: Johns Hopkins University Press.

McCullagh, P. (1980). Regression models for ordinal data. *Journal of the Royal Statistical Society B* **42**, 109–142.

McCullagh, P., and Nelder, J. A. (1989). *Generalized Linear Models*, second edition. London: Chapman and Hall.

McCulloch, C. E. (1997). Maximum likelihood algorithms for generalized linear mixed models. *Journal of the American Statistical Association* **92**, 162–170.

McCulloch, C. E., and Searle, S. R. (2001). *Generalized, Linear, and Mixed Models*. New York.

McLachlan, G., and Peel, D. (2000). *Finite Mixture Models*. New York: Wiley.

Meng, X.-L. (1994a). Multiple-imputation inferences with uncongenial sources of input (with discussion). *Statistical Science* **9**, 538–573.

Meng, X.-L. (1994b). Posterior predictive p-values. *Annals of Statistics* **22**, 1142–1160.

Meng, X., and Gregory, R. G. (1999). *Impact of Interrupted Education on Earnings: The Educational Cost of the Chinese Cultural Revolution*. Canadian International Labour Network Working Papers No. 40.

Meng, X.-L., and Rubin, D. B. (1991). Using EM to obtain asymptotic variance-covariance matrices: the SEM algorithm. *Journal of the American Statistical Association* **86**, 899–909.

Meng, X.-L., and Rubin, D. B. (1993). Maximum likelihood estimation via the ECM algorithm: a general framework. *Biometrika* **80**, 267–278.

Meng, X.-L., and Schilling, S. (1996). Fitting full-information item factor models and an empirical investigation of bridge sampling. *Journal of the American Statistical Association* **91**, 1254–1267.

Meng, X.-L., and van Dyk, D. A. (1997). The EM algorithm—an old folk song sung to a fast new tune (with discussion). *Journal of the Royal Statistical Society B* **59**, 511–567.

Meng, X.-L., and van Dyk, D. (1998). Fast EM-type implementations for mixed-effects models. *Journal of the Royal Statistical Society B* **60**, 559–578.

Meng, X.-L., and Wong, W. H. (1996). Simulating ratios of normalizing constants via a simple identity: a theoretical exploration. *Statistica Sinica* **6**, 831–860.

Meng, X.-L., and Zaslavsky, A. M. (2002). Single observation unbiased priors. *Annals of Statistics*, **30**, 1345–1375.

Mill, J. S. (1843). *A System of Logic, Ratiocinative and Inductive.* Reprinted by Longman and Greens, London (1956).

Ming, K., and Rosenbaum, P. R. (2000). Substantial gains in bias reduction from matching with a variable number of controls. *Biometrics* **56**, 118–124.

Ming, K., and Rosenbaum, P. R. (2001). A note on optimal matching with variable controls using the assignment algorithm. *Journal of Computational and Graphical Statistics* **10**, 455–463.

Mira, A., and Tierney, L. (2002). On the use of auxiliary variables in Markov chain Monte Carlo sampling. *Scandinavian Journal of Statistics* **29**, 1–12.

Mislevy, R. (1992). Scaling procedures. In *NAEP 1990 Technical Report.* Washington, D.C.: National Center for Educational Statistics.

Mislevy, R., and Bock, D. (1982). *BILOG: Item analysis and test scoring with binary logistic models.* Mooresville, Ind.: Scientific Software.

Mislevy, R., Johnson, E., and Muraki, E. (1992). Estimating population characteristics from sparse matrix samples of item responses. *Journal of Educational Statistics.* **17**, 131–154.

Morrell, R. W., Park, D. C., Kidder D. P., and Martin, M. (1997). Adherence to antihypertensive medications across the life span. *Gerontologist* **37**, 609–619.

Morris, C. N. (1983). Parametric empirical Bayes inference: theory and applications (with discussion). *Journal of the American Statistical Association* **78**, 47–65.

Morrison, A. S. (1985). *Screening in Chronic Disease.* Oxford University Press.

Morrison, R., and McCammmon, D. (1983). Interstellar photoelectric absorption cross sections, 0.03-10 kev. *Astrophysical Journal* **270**, 119–122.

Mudholkar, G. S., and George, E. O. (1978). A remark on the shape of the logistic distribution. *Biometrika* **65**, 667–668.

Mumford, D. (1994). Pattern theory: a unifying perspective. *Proceedings of 1st European Congress of Mathematics*, Boston, Mass.: Birkhauser.

Mumford, D., and Gidas, B. (2001). Stochastic models for generic images. *Quarterly of Applied Mathematics* **59**, 85–111.

Muraki, E. (1992). A generalized partial credit model: application of an EM algorithm. *Applied Psychological Measurements* **16**, 159–176.

Murphy, S. A. (1994). Consistency in a proportional hazards model incorporating a random effect. *Annals of Statistics* **22**, 712–731.

Murphy, S. A. (1995). Asymptotic theory for the frailty model. *Annals of Statistics* **23**, 182–198.

Nakamura, Y., Moss, A. J., Brown, M. W., Kinoshita, M., and Kawai, C. (1999). Long term nitrate use may be deleterious in ischemic heart disease: a study using the databases from two large scale postinfarction studies. *American Heart Journal* **138**, 577–85.

Nance, R. E., and Overstreet, C. (1978). Some experimental observations on the behavior of composite random number generators. *Operations Research* **26**, 915–935.

Neal, R. M. (2003). Slice sampling. *Annals of Statistics* **31**, 705–767.

Newcombe, H. B., Kennedy, J. M., Axford, S. J., and James, A. P. (1959). Automatic linkage of vital records. *Science* **130**, 954–959.

Newton, M. A., and Raftery, A. E. (1994). Approximate Bayesian inference with the weighted likelihood bootstrap (with discussion). *Journal of the Royal Statistical Society B* **56**, 3–48.

Neyman, J. (1923). On the application of probability theory to agricultural experiments. Essay on principles. Section 9, Translated and edited by D. M. Dabrowska and T. P. Speed. *Statistical Science* **5**, 463–480 (1990).

Nielsen, G., Gill, R. D., Andersen, P. K., and Sorensen, T. I. A. (1992). A counting process approach to maximum likelihood estimation in frailty models. *Scandinavian Journal of Statistics* **19**, 25–44.

Nieto-Barajas, L. E., and Walker, S. G. (2002). Markov beta and gamma processes for modelling hazard rates. *Scandinavian Journal of Statistics* **29**, 413–424.

Normand, S. T., Landrum, M. B., Guadagnoli, E., et al. (2001). Validating recommendations for coronary angiography following acute myocardial infarction in the elderly: a matched analysis using propensity scores. *Journal of Clinical Epidemiology* **54**, 387–398.

Office of Management and Budget (1977). Race and ethnic standards for Federal statistics and administrative reporting. Statistical Policy Directive No. 15.

Office of Management and Budget (1997). Revisions to the standards for the classification of Federal data on race and ethnicity. *Federal Register* **62**, 58781–58790.

Office of Management and Budget (2000). Provisional guidance on the implementation of the 1997 standards for the collection of Federal data on race and ethnicity. December 15. www.whitehouse.gov/omb/inforeg/re_guidance2000update.pdf

Oh, M.-S., and Berger, J. O. (1992). Adaptive importance sampling in Monte Carlo integration. *Journal of Statistical Computation and Simulation* **41**, 143–168.

Olkin, I., and Tate, R. F. (1961). Multivariate correlation models with mixed discrete and continuous variables. *Annals of Mathematical Statistics* **32**, 448–465.

Olshausen, B. A., and Field, D. J. (1996). Emergence of simple-cell receptive field properties by learning a sparse code for natural images. *Nature* **381**, 607–609.

Olshausen, B. A., and Millman, K. J. (2000). Learning sparse codes with a mixture-of-Gaussians prior. In *Advances in Neural Information Processing Systems* volume 12, ed. S. A. Solla, T. K. Leen, and K. R. Muller, 841–847. Cambridge, Mass.: MIT Press.

Orchard, T., and Woodbury, M. A. (1972). A missing information principle, theory and application. *Proceedings of the 6th Berkeley Symposium on Mathematical Statistics and Probability* **1**, 697–715.

O'Quigley, J., and Stare, J. (2002). Proportional hazards models with frailties and random effects. *Statistics in Medicine* **21**, 3219–3233.

Park, T., Siemiginowska, A., and van Dyk, D. A. (2004). Fitting narrow emission lines in x-ray spectra: computation and methods. Manuscript in progress.

Park, T., and van Dyk, D. A. (2004). Spectral analysis with delta functions emission lines. Technical Report, Department of Statistics, Harvard University.

Parker, J. D., Schenker, N., Ingram, D. D., Weed, J. A., Heck, K. E., and Madans, J. H. (2004). Bridging between two standards for collecting information on race and ethnicity: an application to Census 2000 and vital rates. *Public Health Reports* **119**, 192–205.

Parner, E. (1998). Asymptotic theory for the correlated Gamma-frailty model. *Annals of Statistics* **26**, 183–214.

Patil, G. P., and Rao, C. R. (1977). The weighted distributions: a survey of their applications. In *Applications of Statistics*, ed. P. R. Krishnaiah, 383-405. Amsterdam: North-Holland.

Pearl, J. (1995). Causal diagrams for empirical research. *Biometrika* **82**, 669–710.

Pearl, J. (2000). *Causality*. Cambridge University Press.

Phillips, C. V. (2003). Quantifying and reporting uncertainty from systematic errors. *Epidemiology* **14**, 459–466.

Phillips, C. V., and Goodman, K. (2003). The messed lessons of Sir Austin Bradford Hill. www.epiphi.com/papers/ phillips-goodman_abhill_nov03.pdf.

Pinheiro, J. C., and Bates, D. M. (2000). *Mixed Effects Models in S and S-Plus*. New York: Springer-Verlag.

Poole, K. G. (1994). Characteristics of an unharvested lynx population during a snowshoe hare decline. *Journal of Wildlife Management* **58**, 608–618.

Portilla, J., and Simoncelli, E. P. (2000). A parametric texture model based on joint statistics of complex wavelet coefficients. *International Journal of Computer Vision* **40**, 49–71.

Pourahmadi, M., and Daniels, M. J. (2002). Dynamic conditionally linear mixed models for longitudinal data. *Biometrics* **58**, 225–231.

Pregibon, D. (1982). Resistant fits for some commonly used logistic models with medical applications. *Biometrics* **38**, 485–498.

Prentice, R. L. (1989). Surrogate endpoints in clinical trials: definition and operational criteria. *Statistics in Medicine* **8**, 431–440.

Press, W.H., Flannery, P. P., Teukolsky, S. A., and Vetterling, W. T. (1992). *Numerical Recipes in C: The Art of Scientific Computing*. Cambridge University Press.

Racine-Poon, A. H., and Hoel, D. G. (1984). Nonparametric estimation of the survival function when cause of death is uncertain. *Biometrics* **40**, 1151–1158.

Raghunathan, T. E., Lepkowski, J. M., Van Hoewyk, J., and Solenberger, P. (2001). A multivariate technique for multiply imputing missing values using a sequence of regression models, *Survey Methodology* **27**, 85–95.

Raghunathan, T. E., and Rubin, D. B. (1990). An application of Bayesian statistics using sampling/importance resampling for a deceptively simple problem in quality control. In *Data Quality Control: Theory and Pragmatics*, ed. G. Liepins and V. R. R. Uppuluri, 229–244. New York: Marcel Dekker.

Raghunathan, T. E., and Rubin, D. B. (2000). Bayesian multiple imputation to preserve confidentiality in public-use data sets. *Presented at the Sixth World Meeting of the International Society for Bayesian Analysis*.

Raghunathan, T. E., Reiter, J. P., and Rubin, D. B. (2003). Multiple imputation for statistical disclosure limitation. *Journal of Official Statistics* **19**, 1–16.

Reiser, B., Guttman, I., Lin, D. K. J., Guess, F. M., and Usher, J. S. (1995). Bayesian inference for masked system lifetime data. *Applied Statistics* **44**, 79–90.

Reiter, J. P. (2002). Satisfying disclosure restrictions with synthetic data sets. *Journal of Official Statistics* **18**, 531–543.

Reiter, J. P. (2004). Inferences for partially synthetic, public use microdata sets. *Survey Methodology*.

Rich, S. S. (1998). Analytic options for asthma genetics, *Clinical and Experimental Allergy* **28**, 108–110.

Richardson, W. H. (1972). Bayesian-based iterative method of image restoration. *Journal of the Optical Society of America* **62**, 55–59.

Richardson, S., and Green, P. J. (1997). On Bayesian analysis of mixtures with an unknown number of components. *Journal of the Royal Statistical Society Series B* **59**, 731–792.

Ripatti, S., and Palmgren, J. (2000). Estimation of multivariate frailty models using penalized likelihood. *Biometrics* **56**, 1016–1022.

Robins, J. M. (1986). A new approach to causal inference in mortality studies with sustained exposure periods—application to control of the healthy worker survivor effect. *Mathematical Modelling* **7**, 1393–1512.

Robins, J. M. (1987). A graphical approach to the identification and estimation of causal parameters in mortality studies with sustained exposure periods. *Journal of Chronic Diseases* **40**(Suppl. 2), 139s–161s.

Robins, J. M. (1988). Confidence intervals for causal parameters. *Statistics in Medicine* **7**, 773–785.

Robins, J. M. (1997). Causal inference from complex longitudinal data. In *Latent Variable Modeling and Applications to Causality*, Lecture Notes in Statistics 120, ed. M. Berkane, 69–117. New York: Springer-Verlag.

Robins, J. M. (1998). Structural nested failure time models. In *Encyclopedia of Biostatistics*, ed. P. Armitage and T. Colton, 4372–4389. New York: Wiley.

Robins, J. M. (1999). Marginal structural models versus structural nested models as tools for causal inference. In *Statistical Models in Epidemiology: The Environment and Clinical Trials*, IMA Volume 116, ed. M. E. Halloran and D. Berry, 95–134. New York: Springer-Verlag.

Robins, J. M., and Greenland, S. (1992). Identifiability and exchangeability for direct and indirect effects. *Epidemiology* **3**, 143–155.

Robins, J. M., and Greenland, S. (1994). Adjusting for differential rates of prophylaxis therapy for PCP in high-versus low-dose AZT treatment arms in an AIDS randomized trial. *Journal of the American Statistical Association* **89**, 737–479.

Robins, J. M., Blevins, D., Ritter, G., and Wulfson, M. (1992). G-estimation of the effect of prophylaxis therapy for Pneumocystis carinii pneumonia on the survival of AIDS patients. *Epidemiology* **3**, 319-336; Errata: *Epidemiology* **4**, 189.

Robins, J. M., Greenland, S., and Hu, F. C. (1999). Estimation of the causal effect of a time-varying exposure on the marginal mean of a repeated binary outcome (with discussion). *Journal of the American Statistical Association* **94**, 687–712.

Robins, J. M., Hernan, M. A., and Brumback, B. (2000). Marginal structural models and causal inference in epidemiology. *Epidemiology* **11**, 550–560.

Robins, J. M., Mark, S. D., and Newey, W. K. (1992). Estimating exposure effects by modeling the expectation of exposure conditional on confounders. *Biometrics* **48**, 479–495.

Robins, J. M., and Ritov, Y. (1997). Toward a curse of dimensionality appropriate asymptotic theory for semi-parametric models. *Statistics in Medicine* **16**, 285–319.

Robins, J. M., Rotnitzky, A., and Scharfstein, D. O. (1999). Sensitivity analysis for selection bias and unmeasured confounding in missing data and causal inference models. In *Statistical Models in Epidemiology*, ed. M. E. Halloran and D. A. Berry, 1–92. New York: Springer-Verlag.

Robins, J. M., Rotnitzky, A., and Zhao, L. P. (1994). Estimation of regression coefficients when some regressors are not always observed. *Journal of the American Statistical Association* **89**, 846–866.

Robins, J. M., Rotnitzky, A., and Zhao, L. P. (1995). Analysis of semiparametric regression models for repeated outcomes in the presence of missing data. *Journal of the American Statistical Association* **90**, 106–121.

Rogers, E. M. (1995). *Diffusion of Innovations*, fourth Edition. New York: Free Press.

Rosenbaum, P. R. (1984a). Conditional permutation tests and the propensity score in observational studies. *Journal of the American Statistical Association* **79**, 565–574.

Rosenbaum, P. R. (1984b). The consequences of adjustment for a concomitant variable that has been affected by the treatment. *Journal of the Royal Statistical Society A* **147**, 656–666.

Rosenbaum, P. R. (1986). Dropping out of high school in the United States: an observational study. *Journal of Educational Statistics* **11**, 207–224.

Rosenbaum, P. R. (1987a). Model-based direct adjustment. *Journal of the American Statistical Association* **82**, 387–394.

Rosenbaum, P. R. (1987b). The role of a second control group in an observational study (with discussion). *Statistical Science* **2**, 292–316.

Rosenbaum, P. R. (1989). Optimal matching in observational studies. *Journal of the American Statistical Association* **84**, 1024–1032.

Rosenbaum, P. R. (1991). A characterization of optimal designs for observational studies. *Journal of the Royal Statistical Society B*, **53** 597–610.

Rosenbaum, P. R. (2002a). Covariance adjustment in randomized experiments and observational studies. *Statistical Science* **17**, 286–327.

Rosenbaum, P. R. (2002b). *Observational Studies*, second edition. New York: Springer-Verlag.

Rosenbaum, P. R. (2003). Does a dose-response relationship reduce sensitivity to hidden bias? *Biostatistics* **4**, 1–10.

Rosenbaum, P. R. (2004). Design sensitivity in observational studies. *Biometrika* **91**.

Rosenbaum, P. R., and Rubin, D. B. (1983a). The central role of the propensity score in observational studies for causal effects. *Biometrika* **70**, 41–55.

Rosenbaum, P. R., and Rubin, D. B. (1983b). Assessing sensitivity to an unobserved binary covariate in an observational study with binary outcome. *Journal of the Royal Statistical Society B* **45**, 212–218.

Rosenbaum, P. R., and Rubin, D. B. (1984). Reducing bias in observational studies using subclassification on the propensity score. *Journal of the American Statistical Association* **79**, 516–524.

Rosenbaum, P. R., and Rubin, D. B. (1985a). Constructing a control group using multivariate matched sampling methods that incorporate the propensity score. *American Statistician* **39**, 33–38.

Rosenbaum, P. R., and Rubin, D. B. (1985b). The bias due to incomplete matching. *Biometrics* **41**, 103–116.

Rosenbaum, P. R., and Silber, J. H. (2001). Matching and thick description in an observational study of mortality after surgery. *Biostatistics* **2**, 217–232.

Rothman, K. J., and Greenland, S. (1998). *Modern Epidemiology*, second edition. Philadelphia: Lippincott.

Rotnitzky, A., Robins, J. M., and Scharfstein, D. O. (1998). Semiparametric regression for repeated outcomes with nonignorable nonresponse. *Journal of the American Statistical Association* **93**, 1321–1339.

Rubin, D. B. (1973a). Matching to remove bias in observational studies. *Biometrics* **29**, 159–183. Correction: **30**, 728 (1974).

Rubin, D. B. (1973b). The use of matched sampling and regression adjustment to remove bias in observational studies. *Biometrics* **29**, 185–203.

Rubin, D. B. (1974). Estimating causal effects of treatments in randomized and nonrandomized studies. *Journal of Educational Psychology* **66**, 688–701.

Rubin, D. B. (1976a). Inference and missing data. *Biometrika* **3**, 581–592.

Rubin, D. B. (1976b). Multivariate matching methods that are equal percent bias reducing, I: some examples. *Biometrics* **32**, 109–120.

Rubin, D. B. (1976c). Multivariate matching methods that are equal percent bias reducing, II: maximums on bias reduction for fixed sample sizes. *Biometrics* **32**, 121–132.

Rubin, D. B. (1977). Assignment to treatment group on the basis of a covariate. *Journal of Educational Statistics* **2**, 1–26.

Rubin, D. B. (1978a). Bayesian inference for causal effects: the role of randomization. *Annals of Statistics* **6**, 34–58.

Rubin, D. B. (1978b). Multiple imputations in sample surveys: a phenomenological Bayesian approach to nonresponse (with discussion). *Proceedings of the Survey Research Methods Section, American Statistical Association*, 20–34.

Rubin, D. B. (1979). Using multivariate matched sampling and regression adjustment to control bias in observational studies. *Journal of the American Statistical Association* **74**, 318–328.

Rubin, D. B. (1980a). Bias reduction using Mahalanobis metric matching. *Biometrics* **36**, 293–298.

Rubin, D. B. (1980b). Discussion of "Randomization analysis of experimental data: the Fisher randomization test," by Basu. *Journal of the American Statistical Association* **75**, 591–593.

Rubin, D. B. (1981a). Estimation in parallel randomized experiments. *Journal of Educational Statistics* **6**, 377–401.

Rubin, D. B. (1981b). The Bayesian bootstrap. *Annals of Statistics* **9**, 130–134.

Rubin, D. B. (1983). Progress report on project for multiple imputation of 1980 codes. Manuscript delivered to the U.S. Bureau of the Census, the U.S. National Science Foundation, and the Social Science Research Foundation.

Rubin, D. B. (1984). Bayesianly justifiable and relevant frequency calculations for the applied statistician. *Annals of Statistics*, **12**, 1151–1172.

Rubin, D. B. (1986). Discussion of "Statistics and causal inference" by Holland. *Journal of the American Statistical Association* **81**, 961–962.

Rubin, D. B. (1987a). A noniterative sampling/importance resampling alternative to the data augmentation algorithm for creating a few imputations when fractions of missing information are modest: the SIR algorithm. Discussion of Tanner and Wong (1987). *Journal of the American Statistical Association* **82**, 543–546.

Rubin, D. B. (1987b). *Multiple Imputation for Nonresponse in Surveys*. New York: Wiley.

Rubin, D. B. (1988). Using the SIR algorithm to simulate posterior distributions. In *Bayesian Statistics 3*, ed. J. M. Bernardo, M. H. DeGroot, D. V. Lindley, and A. F. M. Smith, 395–402. Oxford University Press.

Rubin, D. B. (1990). Discussion of "On the application of probability theory to agricultural experiments. Essay on principles. Section 9," by J. Neyman. *Statistical Science* **5**, 472–480.

Rubin, D. B. (1991). Practical implications of modes of statistical inference for causal effects and the critical role of the assignment mechanism. *Biometrics*, **47**, 1213–1234.

Rubin, D. B. (1993). Satisfying confidentiality constraints through use of synthetic multiply-imputed microdata. *Journal of Official Statistics* **9**, 461–468.

Rubin, D. B. (1996). Multiple imputation after 18+ years (with discussion). *Journal of the American Statistical Association* **91**, 473–520.

Rubin, D. B. (1997). Estimating causal effects from large data sets using propensity scores. *Annals of Internal Medicine* **127**, 757–763.

Rubin, D. B. (2000). Discussion of "Causal inference without counterfactuals," by A. P. Dawid. *Journal of the American Statistical Association* **95**, 435–437.

Rubin, D. B., and Schenker, N. (1986). Multiple imputation for interval estimation from simple random samples with ignorable nonresponse. *Journal of the American Statistical Association* **81**, 366–374.

Rubin, D. B., and Schenker, N. (1987a). Interval estimation from multiply-imputed data: a case study using census agriculture industry codes. *Journal of Official Statistics* **3**, 375–387.

Rubin, D. B., and Schenker, N. (1987b). Logit-based interval estimation for binomial data using the Jeffreys prior. *Sociological Methodology* **17**, 131–144.

Rubin, D. B., and Stern, H. (1994). Testing in latent class models using a posterior predictive check distribution. In *Latent Variables Analysis*, eds. A. von Eye and C. C. Clogg, 420–438. Thousand Oaks, Calif.: Sage.

Rubin, D. B., and Thayer, D. T. (1982). EM algorithms for ML factor analysis. *Psychometrika* **47**, 69–76.

Rubin, D. B., and Thayer, D. T. (1983). More on EM for ML factor analysis. *Psychometrika* **48**, 253–257.

Rubin, D. B., and Thomas, N. (1992). Affinely invariant matching methods with ellipsoidal distributions. *Annals of Statistics* **20**, 1079–93.

Rubin, D. B., and Thomas, N. (1996). Matching using estimated propensity scores: relating theory to practice. *Biometrics* **52**, 249–264.

Rubin, D. B., and Thomas, N. (2000). Combining propensity score matching with additional adjustments for prognostic covariates. *Journal of the American Statistical Association* **95**, 573–585.

Rubin, D. B., and Zaslavsky, A. M. (1989). An overview of representing within-household and whole-household misenumerations in the census by multiple imputations. *Proceedings of the Bureau of the Census Annual Research Conference* **5**, 109–117.

Ruderman, D. L., and Bialek, W. (1994). Statistics of natural images: scaling in the woods. *Physical Review Letters* **73**, 814–817.

Ruppert, D., Stefanski, L. A., and Carroll, R. J. (1995). *Measurement Error in Nonlinear Models*. New York: Chapman and Hall.

Sacerdote, B. (1997). The lottery winner survey, crime and social interactions, and why is there more crime in cities. Ph.D. thesis, Department of Economics, Harvard University.

Salinas-Torres, V. H., Pereira, C. A. B., and Tiwari, R. C. (2002). Bayesian non-parametric estimation in a series system or a competing-risk model. *Journal of Nonparametric Statistics* **14**, 449–458.

Samuels, S. M. (1998). A Bayesian, species-sampling-inspired approach to the unique problem in microdata disclosure risk assessment. *Journal of Official Statistics* **14**, 373–383.

Sanson-Fisher, R. W., and Clover, K. (1995). Compliance in the treatment of hypertension: a need for action. *American Journal of Hypertension* **8**, 82S–88S.

Sargent, D. J. (1998). A general framework for random effects survival analysis in the Cox proportional hazards setting. *Biometrics* **54**, 1486–1497.

Sargent, D. J., and Hodges, J. S. (1997). Smoothed ANOVA with applications to subgroup analysis. Technical Report, Division of Biostatistics, University of Minnesota.

Schafer, J. L. (1995). Model-based imputation of census short form items. *Proceedings of the 1995 Census Annual Research Conference*, 267–299.

Schafer, J. L. (1997). *Analysis of Incomplete Multivariate Data*. London: Chapman and Hall.

Schafer, J. L., and Schenker, N. (2000). Inference with imputed conditional means. *Journal of the American Statistical Association* **95**, 144–154.

Scharfstein, D., Rotnitzky, A., and Robins, J. M. (1999). Adjusting for nonignorable dropout using semiparametric models (with discussion). *Journal of the American Statistical Association* **94**, 1096–1146.

Schechter, M. T., Strathdee, S. A., Cornelisse, P. G., Currie, S., Patrick, D. M., Rekart, M. L., and O'Shaughnessy, M. V. (1999). Do needle exchange programmes increase the spread of HIV among injection drug users: an investigation of the Vancouver outbreak. *AIDS* **13**, F45–F51.

Schenker, N. (2003). Assessing variability due to race bridging: application to census counts and vital rates for the year 2000. *Journal of the American Statistical Association* **98**, 818–828.

Schenker, N., and Parker, J. D. (2003). From single-race reporting to multiple-race reporting: using imputation methods to bridge the transition. *Statistics in Medicine* **22**, 1571–1587.

Schenker, N., Treiman, D. J., and Weidman, L. (1993). Analyses of public-use decennial census data with multiply imputed industry and occupation codes. *Applied Statistics* **42**, 545–556.

Scheuren, F., and Winkler, W. E. (1993). Regression analysis of data files that are computer matched. *Survey Methodology* **19**, 39–58.

Scheuren, F., and Winkler, W. E. (1997). Regression analysis of data files that are computer matched—part II. *Survey Methodology* **23**, 157–165.

Schluchter, M. D. (1992). Methods for the analysis of informatively censored longitudinal data. *Statistics in Medicine* **11**, 1861–1870.

Schmidt, A. M., Gamerman, D., and Moreira, A. R. B. (1999). An adaptive resampling scheme for cycle estimation. *Journal of Applied Statistics* **26**, 619–641.

Schneider, E. C., Cleary, P. D., Zaslavsky, A. M., and Epstein, A. M. (2001). Racial disparity in influenza vaccination: does managed care narrow the gap between African Americans and Whites? *Journal of the American Medical Association* **286**, 1455–1460.

Servidea, J. D. (2002). Bridge sampling with dependent random draws: techniques and strategy. Ph.D. thesis, Department of Statistics, University of Chicago.

Shedler, G. S. (1993). *Regenerative Stochastic Simulation*. San Diego, Calif.: Academic Press.

Shepp, L. A., and Vardi, Y. (1982). Maximum likelihood reconstruction for emission tomography. *IEEE Transactions on Image Processing* **2**, 113–122.

Shi, M., Weiss, R. E., and Taylor, J. M. G. (1996). An analysis of pediatric AIDS CD4 counts using flexible random curves. *Applied Statistics* **45**, 151–163.

Silber, J. H., Rosenbaum, P. R., Trudeau, M. E., Even-Shoshan, O., Chen, W., Zhang, X., and Mosher, R. E. (2001). Multivariate matching and bias reduction in the surgical outcomes study. *Medical Care* **39**, 1048–1064.

Simon, H. A., and Rescher, N. (1966). Cause and counterfactual. *Philosophy of Science* **33**, 323–340.

Simoncelli, E. P., and Olshausen, B. A. (2001). Natural image statistics and neural representation. *Annual Review of Neuroscience* **24**, 1193–1216.

Singer, J. M., and Andrade, D. F. (1997). Regression models for the analysis of pretest/posttest data. *Biometrics* **53**, 729–735.

Skinner, C. J., Holt, D., and Smith, T. M. F., eds. (1989). *Analysis of Complex Surveys*. New York: Wiley.

Skinner, C. J., Marsh, C., Openshaw, S., and Wymer, C. (1994). Disclosure control for census microdata. *Journal of Official Statistics* **10**, 31–51.

Slough, B. G., and Mowat, G. (1996). Lynx population dynamics in an untrapped refugium. *Journal of Wildlife Management* **60**, 946–961.

Smith, A. F. M. (1995). A conversation with Dennis Lindley. *Statistical Science* **10**, 305–319.

Smith, A. F. M., and Gelfand, A. E. (1992). Bayesian statistics without tears: a sampling-resampling perspective. *American Statistician* **46**, 84–88.

Smith, A. F. M., and Roberts, G. O. (1993). Bayesian computation via the Gibbs sampler and related Markov chain Monte Carlo methods. *Journal of the Royal Statistical Society B* **55**, 3–23.

Smith, H. (1997). Matching with multiple controls to estimate treatment effects in observational studies. *Sociological Methodology* **27**, 325–353.

Smith, N. L., Reiber, G. E., Psaty, B. M., Heckbert, S. R., Siscovick, D. S., Ritchie, J. L., Every, N. R., and Koepsell, T. D. (1998). Health outcomes associated with beta-blocker and diltiazem treatment of unstable angina. *Journal of the American College of Cardiology* **32**, 1305–1311.

Snedecor, G. W., and Cochran, W. G. (1989). *Statistical Methods*, eighth edition. Ames: Iowa State University Press.

Snyder, T. L., and Steele, J. M. (1990). Worst case greedy matching in the unit d-cube. *Networks* **20**, 779–800.

Sommer, A., and Zeger, S. (1991). On estimating efficacy from clinical trials. *Statistics in Medicine* **10**, 45–52.

Srinivasan, R. (2002). *Importance Sampling*. New York: Springer.

Srivastava, A., Grenander, U., and Liu, X. (2002). Universal analytical forms for modeling image probabilities. *IEEE Transactions on Pattern Analysis and Machine Intelligence* **24**, 1200–1214.

Srivastava, A., Lee, A., Simoncelli, E., and Zhu, S. C. (2003). On advances in statistical modeling of natural images. *Journal of Mathematical Imaging and Vision* **18**, 17–33.

Stanek, E. J. (1988). Choosing a pretest-posttest analysis. *American Statistician* **42**, 178–183.

Steenland, K., and Greenland, S. (2004). Monte-Carlo sensitivity analysis and Bayesian analysis of smoking as an unmeasured confounder in a study of silica and lung cancer. *American Journal of Epidemiology* **160**.

Stein, R. A. (1989). Adjusting treatment effects for baseline and other predictor variables. *Proceedings of the American Statistical Association Biopharmaceutical Section*, 274–280.

Stephens, M. (2000). Dealing with label switching in mixture models. *Journal of the Royal Statistical Society Series B* **62**, 795–809.

Stern, H. S., Arcus, D., Kagan, J., Rubin, D. B., and Snidman, N. (1994). Statistical choices in temperament research. *Behaviormetrika* **21**, 1–17.

Stern, H. S., Arcus, D., Kagan, J., Rubin, D. B., and Snidman, N. (1995). Using mixture models in temperament research. *International Journal of Behavioral Development* **18**, 407–423.

Storey, J. D., and Tibshirani, R. (2003). Statistical significance for genome-wide studies. *Proceedings of the National Academy of Sciences* **100**, 9440–9445.

Stoyan, D., Kendall, W., and Mecke, J. (1987). *Stochastic Geometry and Its Applications*. New York: Wiley.

Stuart, B., and Zacker C. (1999). Who bears the burden of Medicaid drug copayment policies? *Health Affairs* **18**, 201-302.

Stuart, B., Shea D., and Briesacher B. (2000). Prescription drug costs for Medicare beneficiaries: coverage and health status matter. Issue Brief, Commonwealth Fund, New York.

Susser, M. (1988). Falsification, verification and causal inference in epidemiology: reconsideration in light of Sir Karl Popper's philosophy. In *Causal Inference*, ed. K. J. Rothman, 33–57. Boston, Mass.: Epidemiology Resources.

Susser, M. (1991). What is a cause and how do we know one? A grammar for pragmatic epidemiology. *American Journal of Epidemiology* **133**, 635–648.

Takizawa T., Haga, M., Yagi, N., Terashima, M., Uehara, H., Yokoyama, A., and Kurita Y. (1999). Pulmonary function after segmentectomy for small peripheral carcinoma of the lung. *Journal of Thoracic and Cardiovascular Surgery* **118**, 536–541.

Tan, M., Tian, G.-L., and Ng, K. W. (2003). A noniterative sampling method for computing posteriors in the structure of EM-type algorithms. *Statistica Sinica* **13**, 625–639.

Tanizaki, H. (2001). Nonlinear and non-Gaussian state space modeling using sampling techniques. *Annals of the Institute of Statistical Mathematics* **53**, 63–81.

Tanner, M. A., and Wong, W. H. (1987). The calculation of posterior distributions by data augmentation (with discussion). *Journal of the American Statistical Association* **82**, 528–550.

Taylor, J. M. G. (1995). Semi-parametric estimation in failure time mixture models. *Biometrics* **51**, 899–907.

Thall, P. F., Simon, R., and Estey, E. H. (1995). Bayesian sequential monitoring designs for single-arm clinical trials with multiple outcomes. *Statistics in Medicine* **14**, 357–379.

Thomas, N. (2002). The role of secondary covariates when estimating latent trait population distributions. *Psychometrika* **67**, 33–48.

Thomas, N., and Gan, N. (1997). Generating multiple imputations for matrix sampling data analyzed with item response models. *Journal of Educational and Behavioral Statistics* **22**, 425–445.

Tierney, L., and Kadane, J. B. (1986). Accurate approximations for posterior moments and marginal densities. *Journal of the American Statistical Association* **81**, 82–86.

Titterington, D. M. (2004). Bayesian methods for neural networks and related models. *Statistical Science* **19**.

Tong, H. (1990). *Non-linear Time Series: A Dynamical System Approach*. Oxford: Clarendon Press.

Tong, H., and Lim, K. S. (1980). Threshold autoregression, limit cycles, and cyclical data. *Journal of the Royal Statistical Society B* **42**, 245–292.

Treiman, D. J., Bielby, W. T., and Cheng, M. (1988). Evaluating a multiple-imputation method for recalibrating 1970 U.S. Census detailed industry codes to the 1980 standard. *Sociological Methodology* **18**, 309–345.

Treiman, D. J., and Rubin, D. B. (1983). *Multiple Imputation of Categorical Data to Achieve Calibrated Public-Use Samples*. Proposal to the National Science Foundation.

Troxel, A. B., Ma, G., and Heitjan, D. F. (2004). An index of sensitivity to non-ignorability. *Statistica Sinica*.

Tu, Z. W., and Zhu, S. C. (2002). Image segmentation by data driven Markov chain Monte Carlo. *IEEE Transactions on Pattern Analysis and Machine Intelligence* **24**, 657–673.

Ueda, N., and Ghahramani, Z. (2003). Bayesian model search for mixture models based on optimizing variational bounds. *Neural Networks* **16**, 1223–1241.

U.S. General Accounting Office (1994). Breast conservation versus mastectomy: patient survival in day-to-day medical practice and randomized studies. Report to the chairman, Subcommittee on Human Resources and intergovernmental Relations, Committee on Government Operations, House of Representatives. Washington, D.C.: U.S. General Accounting Office, Report GAO-PEMD-95-9.

United States Pharmacopeial Convention (1999). *United States Pharmacopeia Drug Information (USP DI)*. Micromedix.

Vach, W., and Blettner, M. (1995). Logistic regression with incompletely observed categorical covariates—investigating the sensitivity against violation of the missing at random assumption. *Statistics in Medicine* **14**, 1315–1329.

Vaida, F. (1998). At the confluence of the EM algorithm and Markov chain Monte Carlo: theory and applications. Ph.D. thesis, Department of Statistics, University of Chicago.

Vaida, F., and Xu, R. (2000). Proportional hazards model with random effects. *Statistics in Medicine* **19**, 3309–3324.

van Ameijden, E. J. C., van den Hoek, J. A. R., and Coutinho, R. A. (1994). Injecting risk behavior among drug users in Amsterdam, 1986 to 1992, and its relationship to AIDS prevention programs. *American Journal of Public Health* **84**, 275–281.

van Buuren, S., Boshuizen, H. C., and Knook, D. L. (1999). Multiple imputation of missing blood pressure covariates in survival analysis. *Statistics in Medicine* **18**, 681–694.

van Dyk, D. A. (2003). Hierarchical models, data augmentation, and Markov chain Monte Carlo (with discussion). In *Statistical Challenges in Modern Astronomy III*, ed. E. Feigelson and G. Babu, 41–56. New York: Springer-Verlag.

van Dyk, D. A., Connors, A., Kashyap, V., and Siemiginowska, A. (2001). Analysis of energy spectra with low photon counts via Bayesian posterior simulation. *Astrophysical Journal* **548**, 224–243.

van Dyk, D. A., Connors, A., Esch, D. N., Freeman, P., Kang, H., Karovska, M., Kashyap, V., Siemiginowska, A., and Zezas, A. (2004). Deconvolution in high energy astrophysics: science, instrumentation, and methods. *Bayesian Analysis*, submitted.

van Dyk, D. A., and Hans, C. M. (2002). Accounting for absorption lines in images obtained with the Chandra X-ray Observatory. In *Spatial Cluster Modelling*, ed. D. Denison and A. Lawson, 175–198. London: CRC Press.

van Dyk, D. A., and Kang, H. (2003). Highly structured models for spectral analysis in high energy astrophysics. *Statistical Science*.

Vaupel, J. W., Manton, K. G., and Stallard E. (1979). The impact of heterogeneity in individual frailty on the dynamics of mortality. *Demography* **16**, 439–454.

Verbeke, G., Molenberghs, G., Thijs, H., Lesaffre, E., and Kenward, M. G. (2001). Sensitivity analysis for nonrandom dropout: a local influence approach. *Biometrics* **57**, 7–14.

Vlahov, D., Anthony, J. C., Munoz, A., Margolick, J., Nelson, K. E., Celentano, D. D., Solomon, L., and Polk, B. F. (1991). The ALIVE study: a longitudinal study of HIV infection among injection drug users: description of methods. NIDA Research Monograph **107**, 75–100.

Vlahov, D., Junge, B., Brookmeyer, R., Cohn, S., Riley, E., Armenian, H., and Beilenson, P. (1997). Reduction in high-risk drug use behaviors among participants in the Baltimore Needle Exchange Program. *Journal of Acquired Immune Deficiency Syndromes and Human Retrovirology* **16**, 400–406.

Vose, D. (2000). *Risk Analysis*. New York: Wiley.

Wahba, G. (1978). Improper priors, spline smoothing and the problem of guarding against model errors in regression. *Journal of the Royal Statistical Society B* **40**, 364–372.

Walker, S. G., and Mallick, B. K. (1997). Hierarchical generalized linear models with Bayesian nonparametric mixing. *Journal of the Royal Statistical Society B* **59**, 845–860.

Wainer, H. (1989). Eelworms, bullet holes, and Geraldine Ferraro: some problems with statistical adjustment and some solutions. *Journal of Educational Statistics* **14**, 121–140.

Wasserman, L. (2000). Comment. *Journal of the American Statistical Association* **95**, 442–443.

Weed, D. L. (1986). On the logic of causal inference. *American Journal of Epidemiology* **123**, 965–979.

Wei, G., and Tanner, M. A. (1990). A Monte Carlo implementation of the EM algorithm and the poor man's data augmentation algorithm. *Journal of the American Statistical Association* **85**, 699–704.

Weidman, L. (1989). Final report: industry and occupation imputation. Report Census/SRD/89/03, Statistical Research Division, U.S. Bureau of the Census.

Weiss, N. S. (1981). Inferring causal relationships: elaboration of the criterion of "dose-response." *American Journal of Epidemiology* **113**, 487–490.

Weiss, N. S. (2002). Can "specificity" of an association be rehabilitated as a basis for supporting a causal hypothesis? *Epidemiology* **13**, 6–8.

Willenborg, L., and De Waal, T. (1996). *Statistical Disclosure Control in Practice.* Lecture Notes in Statistics 111. New York: Springer-Verlag.

Willoughby, A., Graubard, B. I., Hocker, A., Storr, C., Vletza, P., Thackaberry, J. M., et al. (1990). Population-based study of the developmental outcome of children exposed to chloride-deficient infant formula. *Pediatrics* **85**, 485–490.

Winkler, W. E. (1994). Advanced methods of record linkage. *Proceedings of the American Statistical Association, Section on Survey Research Methods Statistical Association,* 467–472.

Winkler, W. E. (1995). Matching and record linkage. In *Business Survey Methods,* ed. B. G. Cox, D. A. Binder, B. N. Chinnappa, A. Christianson, M. J. Colledge, and P. S. Kott, 355–384. New York: Wiley.

Winkler, W. E. (2003). Methods for evaluating and creating data quality. *Journal of Information Sciences.*

Wolfinger, R., Stroup, W., Milliken, M., and Littell, L. (1990). Mixed procedure for SAS. Cary, N.C.: SAS Institute.

Wu, Y. N., Zhu, S. C., and Guo, C. E. (2002). Statistical modeling of texture sketch. *Proceedings of 7th European Conference of Computer Vision,* 240–254.

Wu, Y. N., Zhu, S. C., and Liu, X. (1999). Equivalence of Julesz and Gibbs ensembles. *Proceedings of 7th International Conference on Computer Vision,* 1025–1032.

Wu, Y. N., Zhu, S. C., and Liu, X. (2000). Equivalence of Julesz ensemble and FRAME model. *International Journal of Computer Vision* **38**, 245–261.

Xie, H. (2003). An index of sensitivity to nonignorability: extensions and applications. Doctoral dissertation, Department of Biostatistics, Columbia University.

Xue, X., and Brookmeyer, R. (1996). Bivariate frailty model for the analysis of multivariate survival time. *Lifetime Data Analysis* **2**, 277–289.

Yanagawa, T. (1984). Case-control studies: assessing the effect of a confounding factor. *Biometrika* **71**, 191–194.

Yang, L., and Tsiatis, A. A. (2001). Efficiency study of estimators for a treatment effect in a pretest-posttest trial. *American Statistician* **55**, 314–321.

Yates, F., and Grundy, P. M. (1953). Selection without replacement from within strata with probability proportional to size. *Journal of the Royal Statistical Society B* **15**, 253–261.

Yuan, K.-H., and Bentler, P. M. (1996). Mean and covariance structure analysis with missing data. In *Multidimensional Statistical Analysis and Theory of Random Matrices: Proceedings of the Sixth Eugene Lukacs Symposium,* eds. A. Gupta and V. Girko, 307–326. Utrecht, Netherlands: VSP.

Zaslavsky, A. M. (1988). Representing local area adjustments by reweighting of households. *Survey Methodology* **14**, 265–288.

Zaslavsky, A. M. (1989). Representing the census undercount: reweighting and imputation methods. Ph.D. dissertation, Department of Mathematics, Massachusetts Institute of Technology.

Zaslavsky, A. M., Schenker, N., and Belin, T. R. (2001). Downweighting influential clusters in surveys: application to the 1990 post enumeration survey. *Journal of the American Statistical Association* **96**, 858–869.

Zeger, S. L., and Karim, M. R. (1991). Generalized linear models with random effects: a Gibbs sampling approach. *Journal of the American Statistical Association* **86**, 79–86.

Zeger, S. L., Liang, K.-Y., and Albert, P. S. (1988). Models for longitudinal data: a generalized estimating equation approach. *Biometrics* **44**, 1049–1060.

Zeger, L., and Thomas, N. (1997). Efficient matrix sampling instruments for correlated latent traits: examples from the National Assessment of Educational Progress. *Journal of the American Statistical Association* **92**, 416–425.

Zhang, J. (1992). The mean field theory in EM procedures for Markov random fields. *IEEE Transactions on Signal Processing* **40**, 2570–2583.

Zhang, J. (2002). Causal inference with principal stratification: some theory and application. Ph.D. thesis, Department of Statistics, Harvard University.

Zhang, X., Boscardin, W. J., and Belin, T. R. (2004). Sampling correlation matrices in Bayesian models with correlated latent variables. Technical Report, Department of Biostatistics, University of California, Los Angeles.

Zhu, S. C., Wu, Y. N., and Mumford, D. (1997). Minimax entropy principle and its applications in texture modeling. *Neural Computation* **9**, 1627–1660.

Zhuang, D., Schenker, N., Taylor, J. M. G., Mosseri, V., and Dubray, B. (2000). Analysing the effects of anaemia on local recurrence of head and neck cancer when covariate values are missing. *Statistics in Medicine* **19**, 1237–1249.

Index

Adaptive rejection sampling, 253–264
Adjustment, 129–140
AECM algorithm, 291–294
Agriculture, 117–128
AIDS, 319–329
Ancillary, 175–186
Applied Bayesian inference, 279–342
Approximate Bayesian bootstrap, 319–329
Approximation, 117–128
AR, 253–264
Artificial neural networks, 189–194
Ashenfelter dip, 25–35
Astronomy, 285–296
Asymptotic standard errors, 319–329
Average treatment effect, 25–35

Background contamination, 285–296
Balanced designs, 153–162
Balanced multiple sampling, 274–275
Bayes factor, 239–251
Bayesian computing, 279–284
Bayesian education, 284
Bayesian estimation, 117–128, 195–202, 279–342
Bayesian model choice, 282
Bayesian p-values, 282
Before–after data, 195–202
Beta distribution, 309–318

Bias, 49–60
Bias modeling, 3–13
Binary response, 253–264
Binomial, 117–128
Binomial regression, 288
Blocking, 309–318
Boltzmann machines, 189–194
Box–Cox transformations, 319–329
Bridge sampling, 239–251, 253–264
Bridging, 117–128
Bureau of the Census, 117–128
Burn-in, 253–264

Calibration, 319–329
Canadian lynx series, 297–308
Categorical variable, 117–128
Causal effects, 25–35, 61–71
Causal inference, 3–13, 15–24, 97–108
Causal models, 3–13
Cause-specific hazard, 239–251
Censored data, 136, 253–264
Census, 117–140, 309–329
Census 2000 Modified Race Data Summary File, 117–128
Centers for Disease Control and Prevention, 117–128
Chandra X-ray Observatory, 285–296
Change, 117–128
Children's educational expenses and attainment, 61–71
Classification system, 117–128

Applied Bayesian Modeling and Causal Inference from Incomplete-Data Perspectives.
Edited by A. Gelman and X.-L. Meng © 2004 John Wiley & Sons, Ltd ISBN: 0-470-09043-X

Clinical trial, 175–186, 253–264, 283
Clustered data, 253–264
Coding, 117–128
Combined model, 117–128
Comparability, 117–128
Comparison group, 25–35
Competing-risk model, 239, 251
Complier average causal effect (CACE), 85–96
Computational methods, 285–296
Computational statistics, 117–128
Confidentiality, 319–329
Confirmatory factor model
    Bayesian inference, 334–339
    Gibbs sampling, 337–338
    model checking, 341
    specification, 332–334
Confounding, 3–13
Congeniality of imputation models, 141–152
Conservative, 117–128
Constructed observational study, 49–60
Contextual predictor, 117–128
Contingency table, 117–128
Continuous treatments, 73–84
Control group, 61–71
Convergence, 253–264, 291–294
Correlation in before–after data, 195–202
Correlation matrix, 215–226
Counterfactual, 3–13
Counterterrorism, 309–318
County, 117–128
Covariance matrix, 215–226
Covariate, 117–128
Coverage rate, 117–128
Credibility interval, 253–264
Crossed design, 253–264
Cumulative hazard, 253–264

Data augmentation, 140, 239–251, 253–264
Decennial census, 117–128

Default priors, 282
Default statistical models, 195–202
Diagnostic probability, 239–251
Dichotomy, 117–128
Differences-in-differences, 61–71
Dirichlet distribution, 135
Disclosure avoidance, 319–329
Disclosure protection, 141–152
Discrete gamma process, 239–251
Distance metric, 319–329
Divorce, 61–71
DNA sequencing, 189–194
Double-coded, 117–128
Drug expenditures, 37–47
Duplicates, 319–329

ECME algorithm, 291–294
Education, 153–162
Effect, 3–13
Efficient computation, 285–296
EM algorithm, 189–194, 227–251, 253–264, 319–329
EM-type algorithms, 285–296
Empirical Bayes, 117–128
Employment, 117–128
Epidemiology, 117–128
Estimated propensity scores, 61–71
Exclusion restriction, 85–96
Experiment, 49–60
Exposure, 239–251

Factor analysis, 189–194, 203–213
False match, 319–329
Federal statistical system, 117–128
Fellegi–Sunter, 309–329
Finite mixture model, 285–296
First-order approximation, 117–128
Fisher information, 253–264
Fisher's equation, 253–264
Fraction of missing information, 117–128
Frailty, 253–264
Frequency evaluation, 140
Frequentist inference, 175–186
Functional imaging, 189–194

g-estimation, 3–13
Gamma distribution, 253–264
Generalized estimating equations,
        189–194
Generalized linear mixed-effects
        model, 253–264
Generalized linear model, 189–194,
        253–264
Generalized propensity score, 73–84
Genetics, 253–264
Gibbs sampler, 134, 189–194,
        253–264, 280, 285–296
Government, 111–115, 117–128
Graphical models, 189–194

Hazard function, 253–264
Health care, 37–47
Hidden Markov models, 189–194
Hierarchical mixtures of experts
        models, 189–194
Hierarchical model, 131–137, 140,
        309–318
Hierarchical models, 195–202
Hill considerations, 3–13
HIV, 319–329
Hot-deck imputation, 319–329
Household composition, 129–140
Hybrid algorithms, 291–294

Identifiability, 239–251, 297
Ignorability, 175–186
Image reconstruction, 289
Imbalances, 61–71
Importance ratio, 266
Importance sampling, 189–194, 266,
        269–271
Importance weighted regenerative
        process, 275–276
Imputation, 117–140
Independent multiple sampling,
        274–275
Industry, 117–128
Infant Health Development
        Program, 49–60
Inference, 117–128

Information loss, 141–152
Information loss in missing data
        imputation, 141–152
Informative prior distribution,
        319–329
Instrumental variables, 85–108
Intention to treat, 49–60
Interactions, 195–202
Inverse-Wishart distribution,
        215–226
Iowa Youth and Families Project,
        339–342
IRT, 153–162
Item response model, 153–162

Jackpot prize, 61–71
James–Stein estimation, 280
Jeffreys prior, 117–128
Job, 117–128
Jumbo shrimp, 279

Labor force, 117–128
Labor supply, 61–71
Labor training programs, 25–35
Laplace approximations, 189–194
Latent class model, 309–318
Latent class models, 203–213
Latent variable model, 153–162,
        332–334
Latent-variable model, 189–194
Least squares, 25–35
Likelihood ratio test, 294–296
Linear approximation, 117–128
Linear mixed-effects model,
        253–264
Linear regression model, 309–318
Link function, 253–264
Local average treatment effect
        (LATE), 85–96
Log-concave, 253–264
Loglinear model, 288
Logistic regression, 117–128,
        227–238
Logit, 253–264
Loglinear model, 134, 288

Longitudinal data, 189–194, 215–226
Los Angeles Women's Health Risk Study, 319–329
Lotka–Volterra equations, 297–308
Lottery, 61–71
Low birth weight infants, 49–60
Lung cancer, 253–264

Machine learning, 189–194
Marginal likelihood, 253–264
Marginal structural model, 3–13
Markov chain Monte Carlo, 117–128, 189–194, 279–296
Masked failure causes, 239–251
Masking group, 239–251
Masking probability, 239–251
Matched sample, 49–60
Matched sampling, 15–24
Matching, 25–35, 49–60
Matching error, 309–318
Maximum likelihood estimation, 117–128, 189–194, 253–264, 285–296, 309–318
MCMC convergence, 281
Mean-field approximations, 189–194
Measurement error, 3–13, 309–318
Methodologic modeling, 3–13
Metropolis algorithm, 280, 297–308
Metropolis–Hastings algorithm, 135, 189–194, 215–226, 253–264
Microarray analysis, 189–194
Misclassified observations, 131
Missing at random, 335
Missing completely at random, 175–186
Missing data, 3–13, 49–60, 117–128, 136–137, 140, 175–186, 253–264
Missing-data mechanism, 285–296, 335
Mixed effects model, 253–264

Mixing sets for disclosure protection, 141–152
Mixture model, 189–194, 203–213, 309–329
Mixtures of factor analyzers, 189–194
Model comparisons, 294–296
Model diagnostics, 294–296, 304–306
Moment-generating function, 175–186
Monotonicity, 85–96
Monte Carlo EM, 253–264
Multilevel models, 195–202, 285–296
Multimodality, 203–213
Multinomial logit, 117–128
Multinomial models, 288
Multiple imputation, 49–60, 117–128, 130, 134, 136, 141–162, 189–194, 319–329
    combining rules, 141–152
    for disclosure protection, 141–152
Multiple imputation of keys (MIKe), 141–152
Multiple-race reporting, 117–128
Multivariate repeated measures, 215–226

National Assessment of Educational Progress (NAEP), 153, 162
National Center for Health Statistics, 117–128
National Health Interview Survey, 117–128
National Longitudinal Survey of Youth, 49–60
National Science Foundation, 117–128
Natural experiments, 61–71
Needle exchange, 97–108
Noncomparability, 117–128

Nonexperimental studies, 25–35
Nonignorable missing data,
      175–186, 285–296
Nonlinear regression, 189–194
Nonparametric analysis, 25–35
Nonparametric Bayes, 133
Nonparametric regression, 189–194
Nonresponse, 111–115, 117–128
Normal approximation, 117–128
Normal distribution, 253–264

Observational studies, 3–13, 15–35,
      49–60
Occupation, 117–128
Odds ratio, 253–264
Office of Management and Budget,
      117–128
Offset, 253–264
One-pass algorithm, 272, 274
Ordinal data, 215–226

Parameter expansion, 215–226
Partial likelihood, 253–264
Partially controlled studies, 97–108
Particle filtering, 189–194
Penalized likelihood criteria, 282
Penalized quasi-likelihood, 253–264
Person-level predictor, 117–128
Piecewise constant hazard, 239–251
Planned missing data, 153–162
Plateau, 253–264
Poisson image analysis, 285–296
Polytomous logistic regression,
      117–128, 134, 135
Population count, 117–128
Post-Enumeration Survey (PES),
      129, 132, 134–138
Post-enumeration survey, 309–318
Posterior bimodality, 282
Posterior mode, 117–128, 319–329
Posterior predictive checks, 282,
      297–308
Posterior-predictive p-values,
      294–296
Potential outcomes, 3–13, 97–108

Power, 175–186
Power law, 287
PQL, 253–264
Prediction, 117–128
Predictive distribution, 117–128
Pretreatment covariates, 25–35
Principal stratification, 85–108
Prior distribution, 135
Probit model, 215–226
Probit regression, 227–238
Profile likelihood, 253–264
Program evaluation, 25–35
Propensity score, 15–35, 37–47,
      49–71, 73–84
Proper, 117–128
Proportional hazards assumption,
      239–251
Proportional hazards model,
      253–264
Public-use data files, 117–128,
      319–329
PX-EM algorithm, 227–238

Quasar PG1634+706, 291–296

Raking ratio estimation, 130
Random effects, 189–194, 253–264
Randomized experiment, 25–35,
      49–60
Record linkage, 309–329
Regression, 49–60, 134
Rejection sampling, 253–264
Relative risk, 253–264
Relative standard error, 117–128
Repeated measures, 215–226
Resampling, 319–329
Residual plots, 294–296
Reweighting, 130, 139
Robit regression, 227–238
Robust regression, 227–238

Salamander mating data, 253–264
Sample size, 117–128
Sampling/importance resampling
      (SIR) algorithm, 117–128,
      265–276

SAS Proc Mixed, 281
Score function, 253–264
Selection, 61–71
Selection bias, 3–13, 37–47
Selection of predictors, 117–128
Selective multiple imputation of
        keys (SMIKe), 141–152
SEM algorithm, 239–251, 319–
        329
Semiparametric model, 133,
        189–194
Sensitivity analysis, 175–186
Separate model, 117–128
Simulated annealing, 297–308
Simulation, 130, 135, 138–139
Single imputation, 117–128
Single observation unbiased prior,
        136–137
Single-race reporting, 117–128
Slice sampling, 253–264
Small area, 132, 140
Social sciences, 331–342
SOUP, 136–137
Sparse data, 117–128
Spatial epidemiology, 283
Spectral analysis, 285–296
Standard error, 117–128
Standards, 117–128
Starting values, 291–294
State space models, 297–308
Statistical disclosure control,
        141–152
Statistical education, 284
Statistical image analysis, 189–194
Stratification, 25–35
Structural equation model, 3–13,
        331–342
Structural nested failure-time model,
        3–13
Surrogate, 97–108
Surveys, 111–115
Survival analysis, 253–264
Symmetry assumption, 239–
        251

Time series, 297–308
Time-varying masking probability,
        239–251
Training data, 319–329
Training programs, 25–35
Transformations, 319–329
Transition, 117–128, 253–264
Treatment and control groups,
        195–202
Treatment effect, 25–35, 49–71,
        195–202
Treatment group, 61–71
Triplicates, 319–329
True match, 319–329
Truncated data, 133, 136
Truncated normal distribution,
        253–264
Two-step procedure, 117–128

Unbiased estimator, 136–137
Uncertainty, 117–128
Undercount, 129–140
Uniform distribution, 253–264
Unimodal, 117–128

Variability, 117–128
Variance standardization, 253–
        264
Variance/covariance matrix,
        117–128
Variational approximations, 189–
        194
Vital event, 117–128

Wavelets, 189–194
Weak unconfoundedness, 73–84
Weighted bootstrap, 266
Weighted sampling, 272
        Chao's algorithm, 274
        multinomial sampling, 273
        ordered systematic procedure,
                273–274
        simple weighted sampling with
                replacement, 273

tight sampling algorithm,
        268–269, 272–274
    Yates–Grundy draw by draw
        procedure, 273
Weighting, 130, 139
Weighting adjustments, 319–329

Weighting by the propensity score,
        61–71
WinBUGS software, 280
Wishart distribution, 215–226

X-ray astronomy, 285–296

WILEY SERIES IN PROBABILITY AND STATISTICS

ABRAHAM and LEDOLTER · Statistical Methods for Forecasting
AGRESTI · Analysis of Ordinal Categorical Data
AGRESTI · An Introduction to Categorical Data Analysis
AGRESTI · Categorical Data Analysis, *Second Edition*
ALTMAN, GILL, and McDONALD · Numerical Issues in Statistical Computing
    for the Social Scientist
AMARATUNGA and CABRERA · Exploration and Analysis of DNA Microarray and Protein Array
    Data
ANDĚL · Mathematics of Chance
ANDERSON · An Introduction to Multivariate Statistical Analysis, *Third Edition*
*ANDERSON · The Statistical Analysis of Time Series
ANDERSON, AUQUIER, HAUCK, OAKES, VANDAELE, and WEISBERG ·
    Statistical Methods for Comparative Studies
ANDERSON and LOYNES · The Teaching of Practical Statistics
ARMITAGE and DAVID (editors) · Advances in Biometry
ARNOLD, BALAKRISHNAN, and NAGARAJA · Records
*ARTHANARI and DODGE · Mathematical Programming in Statistics
*BAILEY · The Elements of Stochastic Processes with Applications to the Natural
    Sciences
BALAKRISHNAN and KOUTRAS · Runs and Scans with Applications
BARNETT · Comparative Statistical Inference, *Third Edition*
BARNETT · Environmental Statistics: Methods & Applications
BARNETT and LEWIS · Outliers in Statistical Data, *Third Edition*
BARTOSZYNSKI and NIEWIADOMSKA-BUGAJ · Probability and Statistical Inference
BASILEVSKY · Statistical Factor Analysis and Related Methods: Theory and
    Applications
BASU and RIGDON · Statistical Methods for the Reliability of Repairable Systems
BATES and WATTS · Nonlinear Regression Analysis and Its Applications
BECHHOFER, SANTNER, and GOLDSMAN · Design and Analysis of Experiments for
    Statistical Selection, Screening, and Multiple Comparisons
BELSLEY · Conditioning Diagnostics: Collinearity and Weak Data in Regression
BELSLEY, KUH, and WELSCH · Regression Diagnostics: Identifying Influential
    Data and Sources of Collinearity

*Now available in a lower priced paperback edition in the Wiley Classics Library.

BENDAT and PIERSOL · Random Data: Analysis and Measurement Procedures, *Third Edition*

BERNARDO and SMITH · Bayesian Theory

BERRY, CHALONER, and GEWEKE · Bayesian Analysis in Statistics and Econometrics: Essays in Honor of Arnold Zellner

BHAT and MILLER · Elements of Applied Stochastic Processes, *Third Edition*

BHATTACHARYA and JOHNSON · Statistical Concepts and Methods

BHATTACHARYA and WAYMIRE · Stochastic Processes with Applications

BILLINGSLEY · Convergence of Probability Measures, *Second Edition*

BILLINGSLEY · Probability and Measure, *Third Edition*

BIRKES and DODGE · Alternative Methods of Regression

BLISCHKE AND MURTHY (editors) · Case Studies in Reliability and Maintenance

BLISCHKE AND MURTHY · Reliability: Modeling, Prediction, and Optimization

BLOOMFIELD · Fourier Analysis of Time Series: An Introduction, *Second Edition*

BOLLEN · Structural Equations with Latent Variables

BOROVKOV · Ergodicity and Stability of Stochastic Processes

BOULEAU · Numerical Methods for Stochastic Processes

BOX · Bayesian Inference in Statistical Analysis

BOX · R. A. Fisher, the Life of a Scientist

BOX and DRAPER · Empirical Model-Building and Response Surfaces

*BOX and DRAPER · Evolutionary Operation: A Statistical Method for Process Improvement

BOX, HUNTER, and HUNTER · Statistics for Experimenters: An Introduction to Design, Data Analysis, and Model Building

BOX and LUCEÑO · Statistical Control by Monitoring and Feedback Adjustment

BRANDIMARTE · Numerical Methods in Finance: A MATLAB-Based Introduction

BROWN and HOLLANDER · Statistics: A Biomedical Introduction

BRUNNER, DOMHOF, and LANGER · Nonparametric Analysis of Longitudinal Data in Factorial Experiments

BUCKLEW · Large Deviation Techniques in Decision, Simulation, and Estimation

CAIROLI and DALANG · Sequential Stochastic Optimization

CHAN · Time Series: Applications to Finance

CHATTERJEE and HADI · Sensitivity Analysis in Linear Regression

CHATTERJEE and PRICE · Regression Analysis by Example, *Third Edition*

CHERNICK · Bootstrap Methods: A Practitioner's Guide

CHERNICK and FRIIS · Introductory Biostatistics for the Health Sciences

CHILÈS and DELFINER · Geostatistics: Modeling Spatial Uncertainty

CHOW and LIU · Design and Analysis of Clinical Trials: Concepts and Methodologies, *Second Edition*

CLARKE and DISNEY · Probability and Random Processes: A First Course with Applications, *Second Edition*

*COCHRAN and COX · Experimental Designs, *Second Edition*

CONGDON · Applied Bayesian Modelling

CONGDON · Bayesian Statistical Modelling

CONOVER · Practical Nonparametric Statistics, *Second Edition*

COOK · Regression Graphics

COOK and WEISBERG · Applied Regression Including Computing and Graphics

COOK and WEISBERG · An Introduction to Regression Graphics

CORNELL · Experiments with Mixtures, Designs, Models, and the Analysis of Mixture Data, *Third Edition*

COVER and THOMAS · Elements of Information Theory

COX · A Handbook of Introductory Statistical Methods

*COX · Planning of Experiments

CRESSIE · Statistics for Spatial Data, *Revised Edition*

CSÖRGÖ and HORVÁTH · Limit Theorems in Change Point Analysis

DANIEL · Applications of Statistics to Industrial Experimentation

DANIEL · Biostatistics: A Foundation for Analysis in the Health Sciences, *Sixth Edition*

---

*Now available in a lower priced paperback edition in the Wiley Classics Library.

*DANIEL · Fitting Equations to Data: Computer Analysis of Multifactor Data, *Second Edition*

DASU and JOHNSON · Exploratory Data Mining and Data Cleaning

DAVID and NAGARAJA · Order Statistics, *Third Edition*

*DEGROOT, FIENBERG, and KADANE · Statistics and the Law

DEL CASTILLO · Statistical Process Adjustment for Quality Control

DENISON, HOLMES, MALLICK and SMITH · Bayesian Methods for Nonlinear Classification and Regression

DETTE and STUDDEN · The Theory of Canonical Moments with Applications in Statistics, Probability, and Analysis

DEY and MUKERJEE · Fractional Factorial Plans

DILLON and GOLDSTEIN · Multivariate Analysis: Methods and Applications

DODGE · Alternative Methods of Regression

*DODGE and ROMIG · Sampling Inspection Tables, *Second Edition*

*DOOB · Stochastic Processes

DOWDY and WEARDEN, and CHILKO · Statistics for Research, *Third Edition*

DRAPER and SMITH · Applied Regression Analysis, *Third Edition*

DRYDEN and MARDIA · Statistical Shape Analysis

DUDEWICZ and MISHRA · Modern Mathematical Statistics

DUNN and CLARK · Applied Statistics: Analysis of Variance and Regression, *Second Edition*

DUNN and CLARK · Basic Statistics: A Primer for the Biomedical Sciences, *Third Edition*

DUPUIS and ELLIS · A Weak Convergence Approach to the Theory of Large Deviations

*ELANDT-JOHNSON and JOHNSON · Survival Models and Data Analysis

ENDERS · Applied Econometric Time Series

ETHIER and KURTZ · Markov Processes: Characterization and Convergence

EVANS, HASTINGS, and PEACOCK · Statistical Distributions, *Third Edition*

FELLER · An Introduction to Probability Theory and Its Applications, Volume I, *Third Edition*, Revised; Volume II, *Second Edition*

FISHER and VAN BELLE · Biostatistics: A Methodology for the Health Sciences

*FLEISS · The Design and Analysis of Clinical Experiments

FLEISS · Statistical Methods for Rates and Proportions, *Second Edition*

FLEMING and HARRINGTON · Counting Processes and Survival Analysis

FULLER · Introduction to Statistical Time Series, *Second Edition*

FULLER · Measurement Error Models

GALLANT · Nonlinear Statistical Models

GELMAN and MENG (editors) · Applied Bayesian Modeling and Casual Inference from Incomplete-Data Perspectives

GHOSH, MUKHOPADHYAY, and SEN · Sequential Estimation

GIESBRECHT and GUMPERTZ · Planning, Construction, and Statistical Analysis of Comparative Experiments

GIFI · Nonlinear Multivariate Analysis

GLASSERMAN and YAO · Monotone Structure in Discrete-Event Systems

GNANADESIKAN · Methods for Statistical Data Analysis of Multivariate Observations, *Second Edition*

GOLDSTEIN and LEWIS · Assessment: Problems, Development, and Statistical Issues

GREENWOOD and NIKULIN · A Guide to Chi-Squared Testing

GROSS and HARRIS · Fundamentals of Queueing Theory, *Third Edition*

*HAHN and SHAPIRO · Statistical Models in Engineering

HAHN and MEEKER · Statistical Intervals: A Guide for Practitioners

HALD · A History of Probability and Statistics and their Applications Before 1750

HALD · A History of Mathematical Statistics from 1750 to 1930

HAMPEL · Robust Statistics: The Approach Based on Influence Functions

HANNAN and DEISTLER · The Statistical Theory of Linear Systems

HEIBERGER · Computation for the Analysis of Designed Experiments

HEDAYAT and SINHA · Design and Inference in Finite Population Sampling

HELLER · MACSYMA for Statisticians

*Now available in a lower priced paperback edition in the Wiley Classics Library.

HINKELMAN and KEMPTHORNE: · Design and Analysis of Experiments, Volume 1:
  Introduction to Experimental Design

HOAGLIN, MOSTELLER, and TUKEY · Exploratory Approach to Analysis
  of Variance

HOAGLIN, MOSTELLER, and TUKEY · Exploring Data Tables, Trends and Shapes

*HOAGLIN, MOSTELLER, and TUKEY · Understanding Robust and Exploratory Data Analysis

HOCHBERG and TAMHANE · Multiple Comparison Procedures

HOCKING · Methods and Applications of Linear Models: Regression and the Analysis
  of Variance, *Second Edition*

HOEL · Introduction to Mathematical Statistics, *Fifth Edition*

HOGG and KLUGMAN · Loss Distributions

HOLLANDER and WOLFE · Nonparametric Statistical Methods, *Second Edition*

HOSMER and LEMESHOW · Applied Logistic Regression, *Second Edition*

HOSMER and LEMESHOW · Applied Survival Analysis: Regression Modeling of
  Time to Event Data

HUBER · Robust Statistics

HUBERTY · Applied Discriminant Analysis

HUNT and KENNEDY · Financial Derivatives in Theory and Practice, *Revised Edition*

HUSKOVA, BERAN, and DUPAC · Collected Works of Jaroslav Hajek—
  with Commentary

IMAN and CONOVER · A Modern Approach to Statistics

JACKSON · A User's Guide to Principle Components

JOHN · Statistical Methods in Engineering and Quality Assurance

JOHNSON · Multivariate Statistical Simulation

JOHNSON and BALAKRISHNAN · Advances in the Theory and Practice of Statistics: A
  Volume in Honor of Samuel Kotz

JUDGE, GRIFFITHS, HILL, LÜTKEPOHL, and LEE · The Theory and Practice of
  Econometrics, *Second Edition*

JOHNSON and KOTZ · Distributions in Statistics

JOHNSON and KOTZ (editors) · Leading Personalities in Statistical Sciences: From the
  Seventeenth Century to the Present

JOHNSON, KOTZ, and BALAKRISHNAN · Continuous Univariate Distributions,
  Volume 1, *Second Edition*

JOHNSON, KOTZ, and BALAKRISHNAN · Continuous Univariate Distributions,
  Volume 2, *Second Edition*

JOHNSON, KOTZ, and BALAKRISHNAN · Discrete Multivariate Distributions

JOHNSON, KOTZ, and KEMP · Univariate Discrete Distributions, *Second Edition*

JUREČKOVÁ and SEN · Robust Statistical Procedures: Asymptotics and Interrelations

JUREK and MASON · Operator-Limit Distributions in Probability Theory

KADANE · Bayesian Methods and Ethics in a Clinical Trial Design

KADANE AND SCHUM · A Probabilistic Analysis of the Sacco and Vanzetti Evidence

KALBFLEISCH and PRENTICE · The Statistical Analysis of Failure Time Data *Second
  Edition*

KARIYA and KURATA · Generalized Least Squares

KASS and VOS · Geometrical Foundations of Asymptotic Inference

KAUFMAN and ROUSSEEUW · Finding Groups in Data: An Introduction to Cluster
  Analysis

KEDEM and FOKIANOS · Regression Models for Time Series Analysis

KENDALL, BARDEN, CARNE, and LE · Shape and Shape Theory

KHURI · Advanced Calculus with Applications in Statistics, *Second Edition*

KHURI, MATHEW, and SINHA · Statistical Tests for Mixed Linear Models

KLEIBER and KOTZ · Statistical Size Distributions in Economics and Actuarial Sciences

KLUGMAN, PANJER, and WILLMOT · Loss Models: From Data to Decisions

KLUGMAN, PANJER, and WILLMOT · Solutions Manual to Accompany Loss Models:
  From Data to Decisions

*Now available in a lower priced paperback edition in the Wiley Classics Library.

KOTZ, BALAKRISHNAN, and JOHNSON · Continuous Multivariate Distributions, Volume 1, *Second Edition*

KOTZ and JOHNSON (editors) · Encyclopedia of Statistical Sciences: Volumes 1 to 9 with Index

KOTZ and JOHNSON (editors) · Encyclopedia of Statistical Sciences: Supplement Volume

KOTZ, READ, and BANKS (editors) · Encyclopedia of Statistical Sciences: Update Volume 1

KOTZ, READ, and BANKS (editors) · Encyclopedia of Statistical Sciences: Update Volume 2

KOVALENKO, KUZNETZOV, and PEGG · Mathematical Theory of Reliability of Time-Dependent Systems with Practical Applications

LACHIN · Biostatistical Methods: The Assessment of Relative Risks

LAD · Operational Subjective Statistical Methods: A Mathematical, Philosophical, and Historical Introduction

LAMPERTI · Probability: A Survey of the Mathematical Theory, *Second Edition*

LANGE, RYAN, BILLARD, BRILLINGER, CONQUEST, and GREENHOUSE · Case Studies in Biometry

LARSON · Introduction to Probability Theory and Statistical Inference, *Third Edition*

LAWLESS · Statistical Models and Methods for Lifetime Data, *Second Edition*

LAWSON · Statistical Methods in Spatial Epidemiology

LE · Applied Categorical Data Analysis

LE · Applied Survival Analysis

LEE and WANG · Statistical Methods for Survival Data Analysis, *Third Edition*

LePAGE and BILLARD · Exploring the Limits of Bootstrap

LEYLAND and GOLDSTEIN (editors) · Multilevel Modelling of Health Statistics

LIAO · Statistical Group Comparison

LINDVALL · Lectures on the Coupling Method

LINHART and ZUCCHINI · Model Selection

LITTLE and RUBIN · Statistical Analysis with Missing Data, *Second Edition*

LLOYD · The Statistical Analysis of Categorical Data

MAGNUS and NEUDECKER · Matrix Differential Calculus with Applications in Statistics and Econometrics, *Revised Edition*

MALLER and ZHOU · Survival Analysis with Long Term Survivors

MALLOWS · Design, Data, and Analysis by Some Friends of Cuthbert Daniel

MANN, SCHAFER, and SINGPURWALLA · Methods for Statistical Analysis of Reliability and Life Data

MANTON, WOODBURY, and TOLLEY · Statistical Applications Using Fuzzy Sets

MARCHETTE · Random Graphs for Statistical Pattern Recognition

MARDIA and JUPP · Directional Statistics

MASON, GUNST, and HESS · Statistical Design and Analysis of Experiments with Applications to Engineering and Science, *Second Edition*

McCULLOCH and SEARLE · Generalized, Linear, and Mixed Models

McFADDEN · Management of Data in Clinical Trials

McLACHLAN · Discriminant Analysis and Statistical Pattern Recognition

McLACHLAN and KRISHNAN · The EM Algorithm and Extensions

McLACHLAN and PEEL · Finite Mixture Models

McNEIL · Epidemiological Research Methods

MEEKER and ESCOBAR · Statistical Methods for Reliability Data

MEERSCHAERT and SCHEFFLER · Limit Distributions for Sums of Independent Random Vectors: Heavy Tails in Theory and Practice

*MILLER · Survival Analysis, *Second Edition*

MONTGOMERY, PECK, and VINING · Introduction to Linear Regression Analysis, *Third Edition*

MORGENTHALER and TUKEY · Configural Polysampling: A Route to Practical Robustness

*Now available in a lower priced paperback edition in the Wiley Classics Library.

MUIRHEAD · Aspects of Multivariate Statistical Theory

MURRAY · X-STAT 2.0 Statistical Experimentation, Design Data Analysis, and Nonlinear Optimization

MURTHY, XIE, and JIANG · Weibull Models

MYERS and MONTGOMERY · Response Surface Methodology: Process and Product Optimization Using Designed Experiments, *Second Edition*

MYERS, MONTGOMERY, and VINING · Generalized Linear Models. With Applications in Engineering and the Sciences

NELSON · Accelerated Testing, Statistical Models, Test Plans, and Data Analyses

NELSON · Applied Life Data Analysis

NEWMAN · Biostatistical Methods in Epidemiology

OCHI · Applied Probability and Stochastic Processes in Engineering and Physical Sciences

OKABE, BOOTS, SUGIHARA, and CHIU · Spatial Tesselations: Concepts and Applications of Voronoi Diagrams, *Second Edition*

OLIVER and SMITH · Influence Diagrams, Belief Nets and Decision Analysis

PALTA · Quantitative Methods in Population Health: Extensions of Ordinary Regressions

PANKRATZ · Forecasting with Dynamic Regression Models

PANKRATZ · Forecasting with Univariate Box-Jenkins Models: Concepts and Cases

*PARZEN · Modern Probability Theory and It's Applications

PEÑA, TIAO, and TSAY · A Course in Time Series Analysis

PIANTADOSI · Clinical Trials: A Methodologic Perspective

PORT · Theoretical Probability for Applications

POURAHMADI · Foundations of Time Series Analysis and Prediction Theory

PRESS · Bayesian Statistics: Principles, Models, and Applications

PRESS · Subjective and Objective Bayesian Statistics, *Second Edition*

PRESS and TANUR · The Subjectivity of Scientists and the Bayesian Approach

PUKELSHEIM · Optimal Experimental Design

PURI, VILAPLANA, and WERTZ · New Perspectives in Theoretical and Applied Statistics

PUTERMAN · Markov Decision Processes: Discrete Stochastic Dynamic Programming

*RAO · Linear Statistical Inference and Its Applications, *Second Edition*

RAUSAND and HØYLAND · System Reliability Theory: Models, Statistical Methods and Applications, *Second Edition*

RENCHER · Linear Models in Statistics

RENCHER · Methods of Multivariate Analysis, *Second Edition*

RENCHER · Multivariate Statistical Inference with Applications

RIPLEY · Spatial Statistics

RIPLEY · Stochastic Simulation

ROBINSON · Practical Strategies for Experimenting

ROHATGI and SALEH · An Introduction to Probability and Statistics, *Second Edition*

ROLSKI, SCHMIDLI, SCHMIDT, and TEUGELS · Stochastic Processes for Insurance and Finance

ROSENBERGER and LACHIN · Randomization in Clinical Trials: Theory and Practice

ROSS · Introduction to Probability and Statistics for Engineers and Scientists

ROUSSEEUW and LEROY · Robust Regression and Outlier Detection

RUBIN · Multiple Imputation for Nonresponse in Surveys

RUBINSTEIN · Simulation and the Monte Carlo Method

RUBINSTEIN and MELAMED · Modern Simulation and Modeling

RYAN · Modern Regression Methods

RYAN · Statistical Methods for Quality Improvement, *Second Edition*

SALTELLI, CHAN, and SCOTT (editors) · Sensitivity Analysis

*SCHEFFE · The Analysis of Variance

SCHIMEK · Smoothing and Regression: Approaches, Computation, and Application

---

*Now available in a lower priced paperback edition in the Wiley Classics Library.

SCHOTT · Matrix Analysis for Statistics

SCHOUTENS · Levy Processes in Finance: Pricing Financial Derivatives

SCHUSS · Theory and Applications of Stochastic Differential Equations

SCOTT · Multivariate Density Estimation: Theory, Practice, and Visualization

*SEARLE · Linear Models

SEARLE · Linear Models for Unbalanced Data

SEARLE · Matrix Algebra Useful for Statistics

SEARLE, CASELLA, and McCULLOCH · Variance Components

SEARLE and WILLETT · Matrix Algebra for Applied Economics

SEBER · Multivariate Observations

SEBER and LEE · Linear Regression Analysis, *Second Edition*

SEBER and WILD · Nonlinear Regression

SENNOTT · Stochastic Dynamic Programming and the Control of Queueing Systems

*SERFLING · Approximation Theorems of Mathematical Statistics

SHAFER and VOVK · Probability and Finance: Its Only a Game!

SMALL and McLEISH · Hilbert Space Methods in Probability and Statistical Inference

SRIVASTAVA · Methods of Multivariate Statistics

STAPLETON · Linear Statistical Models

STAUDTE and SHEATHER · Robust Estimation and Testing

STOYAN, KENDALL, and MECKE · Stochastic Geometry and Its Applications, *Second Edition*

STOYAN and STOYAN · Fractals, Random Shapes and Point Fields: Methods of Geometrical Statistics

STYAN · The Collected Papers of T. W. Anderson: 1943–1985

SUTTON, ABRAMS, JONES, SHELDON, and SONG · Methods for Meta-Analysis in Medical Research

TANAKA · Time Series Analysis: Nonstationary and Noninvertible Distribution Theory

THOMPSON · Empirical Model Building

THOMPSON · Sampling, *Second Edition*

THOMPSON · Simulation: A Modeler's Approach

THOMPSON and SEBER · Adaptive Sampling

THOMPSON, WILLIAMS, and FINDLAY · Models for Investors in Real World Markets

TIAO, BISGAARD, HILL, PEÑA, and STIGLER (editors) · Box on Quality and Discovery: with Design, Control, and Robustness

TIERNEY · LISP-STAT: An Object-Oriented Environment for Statistical Computing and Dynamic Graphics

TSAY · Analysis of Financial Time Series

UPTON and FINGLETON · Spatial Data Analysis by Example, Volume II: Categorical and Directional Data

VAN BELLE · Statistical Rules of Thumb

VESTRUP · The Theory of Measures and Integration

VIDAKOVIC · Statistical Modeling by Wavelets

WEISBERG · Applied Linear Regression, *Second Edition*

WELSH · Aspects of Statistical Inference

WESTFALL and YOUNG · Resampling-Based Multiple Testing: Examples and Methods for p-Value Adjustment

WHITTAKER · Graphical Models in Applied Multivariate Statistics

WINKER · Optimization Heuristics in Economics: Applications of Threshold Accepting

WONNACOTT and WONNACOTT · Econometrics, *Second Edition*

WOODING · Planning Pharmaceutical Clinical Trials: Basic Statistical Principles

WOOLSON and CLARKE · Statistical Methods for the Analysis of Biomedical Data, *Second Edition*

WU and HAMADA · Experiments: Planning, Analysis, and Parameter Design Optimization

YANG · The Construction Theory of Denumerable Markov Processes

*Now available in a lower priced paperback edition in the Wiley Classics Library.

*ZELLNER · An Introduction to Bayesian Inference in Econometrics

ZELTERMAN · Discrete Distributions: Applications in the Health Sciences

ZHOU, OBUCHOWSKI, and McCLISH · Statistical Methods in Diagnostic Medicine

*Now available in a lower priced paperback edition in the Wiley Classics Library.